HANDBOOK OF ERGONOMIC AND HUMAN FACTORS TABLES

Jon Weimer, Ph.D.

General Motors
Advanced Product Engineering

P T R Prentice Hall, Englewood Cliffs, New Jersey 07632

Library of Congress Cataloging-in-Publication Data

Weimer, Jon.
Handbook of ergonomic and human factors tables / Jon Weimer.
p. cm.
Includes index.
ISBN 0-13-374174-5
1. Human engineering--Tables. I. Title.
TA166.W45 1993
620.8'2--dc20

93-12554
CIP

Editorial production: *bookworks*
Acquisitions Editor: *Michael Hays*
Buyer: *Mary Elizabeth McCartney*

Footnotes to insert
[1] Courtesy of Photo Research.
[2] Reprinted from Thorell/Smith, *Using Computer Color Effectively*, 1990, Prentice Hall. Courtesy of Prentice Hall. Pages 25, 116, 126, 133, 213, 218, 219, 223, 226, 228, 230 and 233.

The publisher offers discounts on this book when ordered in bulk quantities. For more information contact:

Corporate Sales Department
P T R Prentice Hall
113 Sylvan Avenue
Englewood Cliffs, New Jersey 07632

Phone: 201-592-2863
FAX: 201-592-2249

Printed in the United States of America

10 9 8 7 6 5 4 3 2 1

ISBN 0-13-374174-5

ISBN 0-13-374174-5

Prentice-Hall International (UK) Limited, *London*
Prentice-Hall of Australia Pty. Limited, *Sydney*
Prentice-Hall Canada Inc., *Toronto*
Prentice-Hall Hispanoamericana, S.A., *Mexico*
Prentice-Hall of India Private Limited, *New Delhi*
Prentice-Hall of Japan, Inc., *Tokyo*
Simon & Schuster Asia Pte. Ltd., *Singapore*
Editora Prentice-Hall do Brasil, Ltda., *Rio de Janeiro*

For William Banks, Jr.: Mentor, Friend, Colleague, and Cohort.

You taught me that education begins when you leave the classroom. Among other things.

For Rebecca Fleischman: Friend and Colleague.

You gave me the idea for this, and you inspired me to write it. I hope that it's worthy of you.

For Mike Hays: Editor and Friend.

You continue to believe in my ideas and pay me for them. Thanks.

For Anna Weimer: Wife, Best Friend, and Master Shearer.

Thanks for your love and support. Most of all, though, thanks for helping to put together the rough draft. I couldn't have done this without you!

TABLE OF CONTENTS

Chapter 1: Anthropometry

Chapter 2: Workplace Tables

Chapter 3: Control Tables

Chapter 4: Sensing Tables

Chapter 5: Display Tables

Chapter 6: Work Physiology Tables

Chapter 7: Information Processing Tables

Chapter 8: Methods & Equations

PREFACE

This book was born out of frustration. As a practitioner, I would find myself at meetings, my two best reference books close at hand, and I would inevitably get asked a question such as, "If a display has yellow lights on it, what does a color blind person see? I don't know about you, but my mind is not a database of human factors information. Reaching for my trusty reference books, I would say something like, "It depends on the type of color deficiency." The engineers would bob their heads as if I had said something pithy. My hand placed squarely on my reference books, I prepared to further astound them with my knowledge of human limitations. Then I would realize that the reference book that I needed was sitting on a bookshelf back in my office.

This has happened to me on more occasions than I would comfortably care to admit. Since there is so much overlap in ergonomic references, it is easy to confuse what is one with what is in another. I suppose you could carry around CSERIAC'S ***Human Engineering Data Compendium***, but those 3 volumes are darn heavy, and you would soon look like Arnold Schwarzenegger.

Since I could not find a book of ergonomic and human factors tables, I photocopied a group of tables and made my own reference. One of my colleagues suggested that I should publish it. So I added some tables and formulas, and you hold the result in your hands. This book does not contain every table ever written in the fields of human factors and ergonomics. That was not my goal. I have tried my best to provide you with the most commonly needed tables, and a few uncommon ones (See Color Plate 12). I am sure you will occasionally need a table and be unable to find here. If that happens, I am sorry. Space demands forced me to selectively limit the number of tables, and there were a number of commonly used tables that I wanted to include, but their publisher would not grant me permission to reprint them. If you have table that you would like to see included in future editions, you can write me at the address below.

This book is meant to be your best friend. A book that you carry to meetings and cannot afford to do without. I hope you find it as useful as I do. I would love to hear from you.

Jon Weimer, Ph.D.

You can write me c/o:

PTR Editorial
Prentice-Hall
113 Sylvan Avenue, Route 9W
Englewood Cliffs, NJ 07632

HANDBOOK OF ERGONOMIC AND HUMAN FACTORS TABLES

CHAPTER 1: ANTHROPOMETRY

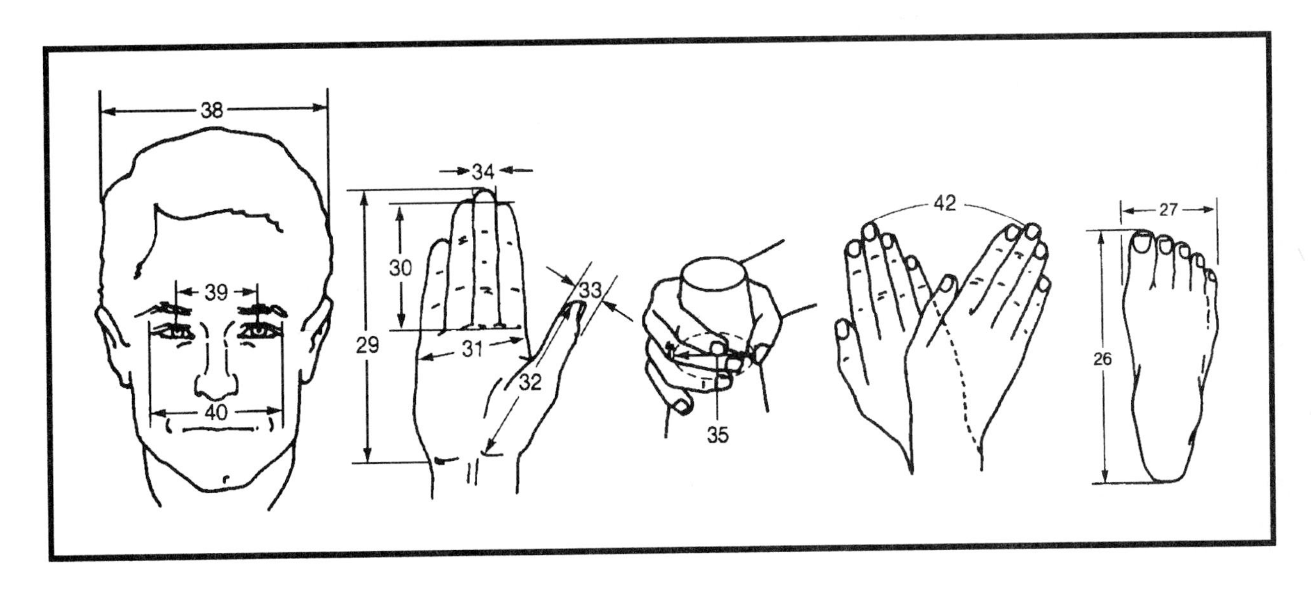

Section A: Military & Industrial

The data presented in the following section are for U.S. men and women, minimally clothed and standing or sitting erect. Because of a lack of data on industrial populations, most of the data are from Air Force or Army studies based upon thousands of subjects. For purposes of comparison, some measurements from industrial studies, performed at Eastman Kodak Company, have been included. However, since the industrial studies included small numbers of subjects, the data should be used cautiously. The main difference between the industrial data and the military data is that the population of military personnel is more restrictive due to selection; the industrial data is more likely to have a wider range of values, often reflecting population extremes (e.g., weight) that would not be seen in a military sample. For an alternate collection of anthropometric data, the reader is referred to the ***Anthropometric Source Book: Volumes 1, 2, & 3***.

The ***Anthropometric Source Book*** contains around a thousand pages of anthropometric tables and information. The ***Source Book*** volumes are U.S. Department of Commerce documents (Document #s: N79-11734, N79-13711, and N79-13712) and are available from the National Technical Information Service (NTIS) [Springfield, VA] for about $200. The ***Source Book*** is also referred to as NASA Reference Document 1024.

Figure ANTH-A1: Anthropometric Dimensions for Standing and Sitting

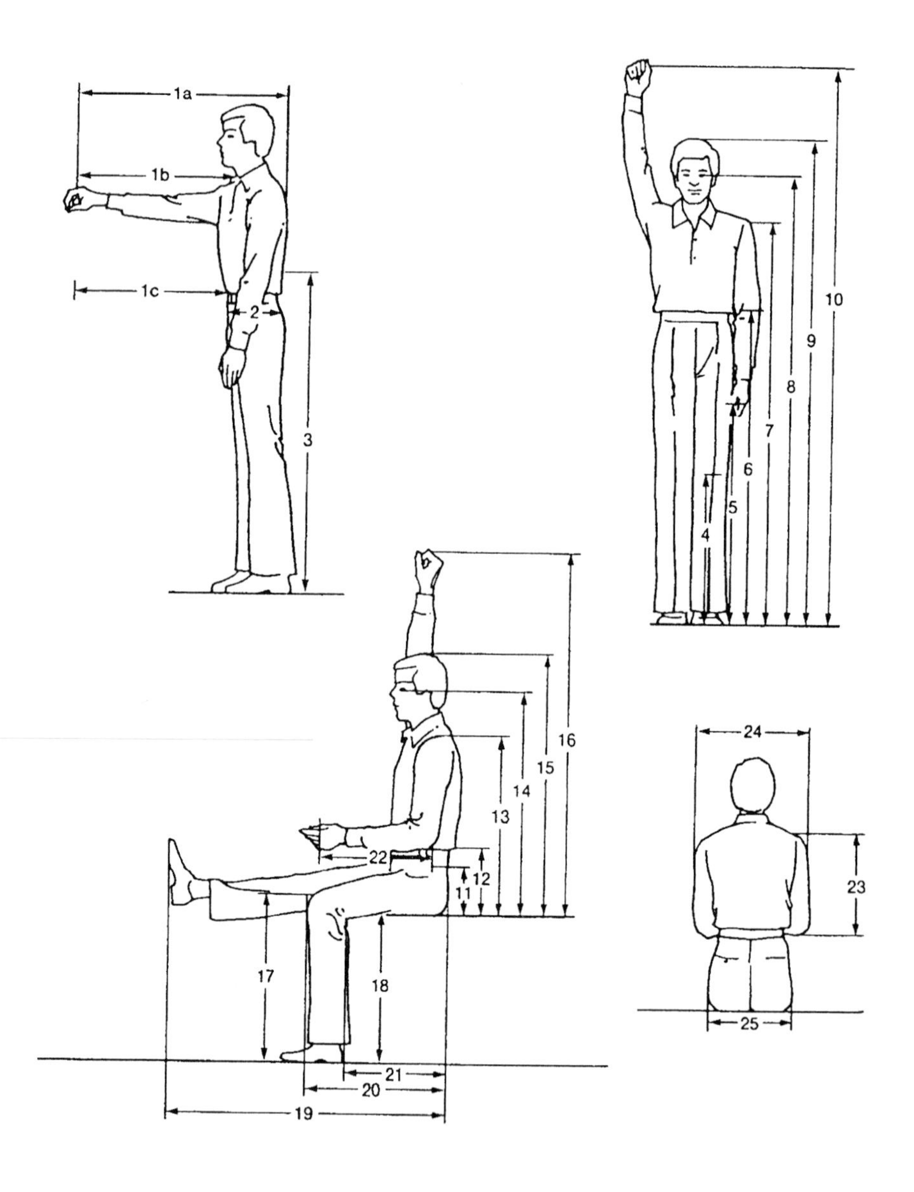

These illustrations correspond to the anthropometric measurements presented in Tables ANTH-A1 & ANTH-A2. Each number on this figure corresponds to a table number. This figure shows the dimensions that are important to the design of sitting and standing workplaces.

Figure ANTH-A2: Anthropometric Dimensions of the Hand, Face, and Foot.[1]

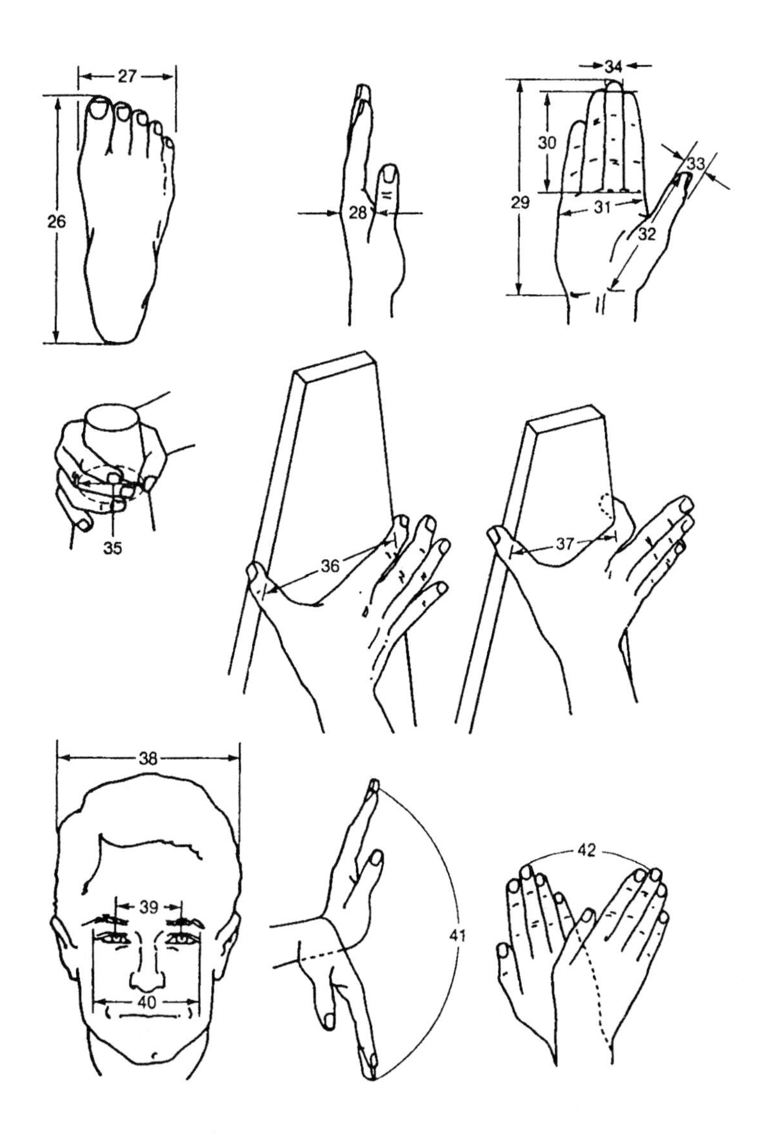

These illustrations show the dimensions of the hand, face, and foot. They also show the range of motion of the wrist and the functional grasp dimensions, depicting the measurements presented in Tables ANTH-A1 & ANTH-A2. Each number on this figure corresponds to a table number.

Table ANTH-A1: Anthropometric Data (cm).[1]

The 50th percentiles, plus or minus one standard deviation (S.D.), are shown for 43 anthropometric variables (Column 1). Variables 1-10 are standing heights, clearances, or reaches. Variables 11-25 are measurements for seated persons. Data on American men (Columns 2 & 3) and women (Columns 4 & 5) were statistically combined to derive the 5th, 50th, and 95th percentile variables for a 50/50 mix of these populations (Columns 6-8). These data were taken primarily from military studies of thousands of subjects. Entries shown in parentheses are from industrial studies, where 50-100 women and 100-150 men were measured. The data in the footnote are from a study of 50 women and 100 men in an industrial setting.

	Males		Females		Population Percentiles, 50/50 Males/Females		
Measurement	50th percentile	±1S.D.	50th percentile	±1S.D.	5th	50th	95th
STANDING							
1. Forward Functional Reach							
a. Includes body depth at shoulder	82.6 (79.3)	4.8 (5.6)	74.1 (71.3)	3.9 (4.4)	69.1 (65.5)	77.9 (74.8)	88.8 (86.5)
b. Acromial Process to Functional Pinch	63.8	4.3	62.5	3.4	57.5	65.0	74.5
c. Abdominal Extension to Functional Pinch†	(62.1)	(8.9)	(60.4)	(6.7)	(48.5)	(61.1)	(74.5)
2. Abdominal Extension Depth	23.1	2.0	20.9	2.1	18.1	22.0	25.8

* These values should be adjusted for clothing and posture.
† Add the following for bending forward from hips or waist: Male: waist, 25 ± 7; hips, 42 ± 8. Female: waist, 20 ± 5; hips, 36 ± 9.

Table ANTH-A1: (continued)

Measurement	Males 50th percentile	Males ±1S.D.	Females 50th percentile	Females ±1S.D.	Population Percentiles, 50/50 Males/Females 5th	50th	95th
3. Waist Height	106.3	5.4	101.7	5.0	94.9	103.9	113.5
	(104.8)	(6.3)	(98.5)	(5.5)	(91.0)	(101.4)	(113.0)
4. Tibial Height	45.6	2.8	42.0	2.4	38.8	43.6	49.2
5. Knuckle Height	75.5	4.1	71.0	4.0	65.7	73.2	80.9
6. Elbow Height	110.5	4.5	102.6	4.8	96.4	106.7	116.3
	(114.6)	(6.3)	(107.1)	(6.8)	(98.8)	(110.7)	(123.5)
7. Shoulder Height	143.7	6.2	132.9	5.5	124.8	137.4	151.7
	(146.4)	(7.8)	(135.3)	(6.6)	(126.6)	(140.4)	(156.4)
8. Eye Height	164.4	6.1	151.4	5.6	144.2	157.7	172.3
9. Stature	174.5	6.6	162.1	6.0	154.4	168.0	183.0
	(177.5)	(6.7)	(164.5)	(7.2)	(155.1)	(170.4)	(188.7)
10. Functional Overhead Reach	209.6	8.5	199.2	8.6	188.0	204.5	220.8
SEATED							
11. Thigh Clearance Height	14.7	1.4	12.4	1.2	10.8	13.5	16.5
12. Elbow Rest Height	24.1	3.2	23.1	3.0	18.4	23.6	28.9
13. Midshoulder Height	62.4	3.2	58.0	2.7	54.5	60.0	66.5
14. Eye Height	78.7	3.6	73.7	3.1	69.7	76.0	83.3
15. Sitting Height Normal	86.6	3.8	81.8	4.0	76.6	84.2	91.6
16. Functional Overhead Reach	128.4	8.5	119.8	6.6	110.6	123.6	139.3

Table ANTH-A1: (continued)

Measurement	Males		Females		Population Percentiles, 50/50 Males/Females		
	50th percentile	±1S.D.	50th percentile	±1S.D.	5th	50th	95th
17. Knee Height	54.0	2.7	51.0	2.6	47.5	52.5	57.7
18. Popliteal Height	44.6	2.5	41.0	1.9	38.6	42.6	47.8
19. Leg Length	105.1	4.8	100.7	4.3	94.7	102.8	111.4
20. Upper-Leg Length	59.4	2.8	57.4	2.6	53.7	58.4	63.3
21. Buttocks-to-Popliteal Length	49.8	2.5	48.0	3.2	43.8	49.0	53.6
22. Elbow-to-Fist Length	38.5	2.1	34.8	2.3	31.9	36.7	41.1
	(37.1)	(3.0)	(32.9)	(3.1)	(28.9)	(35.0)	(41.0)
23. Upper-Arm Length	36.9	1.9	34.1	2.5	31.0	35.7	39.4
	(37.0)	(2.5)	(33.8)	(2.1)	(28.9)	(35.0)	(41.0)
24. Shoulder Breadth	45.4	1.9	39.0	2.1	36.3	42.3	47.8
25. Hip Breadth	35.6	2.3	38.0	2.6	32.4	36.8	41.5
FOOT							
26. Foot Length	26.8	1.3	24.1	1.1	22.6	25.3	28.4
27. Foot Breadth	10.0	0.6	8.9	0.5	8.2	9.4	10.8
HAND							
28. Hand Thickness, Metacarpal III	3.3	0.2	2.8	0.2	2.7	3.0	3.6
29. Hand Length	19.0	1.0	18.4	1.0	17.0	18.7	20.4
30. Digit Two Length	7.5	0.7	6.9	0.8	5.8	7.2	8.5
31. Hand Breadth	8.7	0.5	7.7	0.5	7.0	8.2	9.3

Table ANTH-A1: (continued)

Measurement	Males		Females		Population Percentiles, 50/50 Males/Females		
	50th percentile	±1S.D.	50th percentile	±1S.D.	5th	50th	95th
32. Digit One Length	12.7	1.1	11.0	1.0	9.7	11.8	14.2
33. Breadth of Digit One Interphalangeal Joint	2.3	0.1	1.9	0.1	1.8	2.1	2.5
34. Breadth of Digit Three Interphalangeal Joint	1.8	0.1	1.5	0.1	1.4	1.7	2.0
35. Grip Breadth, Inside Diameter	4.9	0.6	4.3	0.3	3.8	4.5	5.7
36. Hand Spread, Digit One to Digit Two, 1st Phalangeal Joint	12.4	2.4	9.9	1.7	7.5	10.9	15.5
37. Hand Spread, Digit One to Digit Two, 2nd Phalangeal Joint	10.5	1.7	8.1	1.7	5.9	9.3	12.7
HEAD							
38. Head Breadth	15.3	0.6	14.5	0.6	13.8	14.9	16.0
39. Interpupillary Breadth	6.1	0.4	5.8	0.4	5.2	6.0	6.7
40. Biocular Breadth	9.2	0.5	9.0	0.5	8.3	9.1	10.0
OTHER MEASUREMENTS							
41. Flexion-Extension, Range of Motion of Wrist Radians (57 degrees/radian)	2.33	0.33	2.46	0.26	1.92	2.4	2.8
42. Ulnar-Radial Range of Motion of Wrist Radians (57 degrees/radian)	1.05	0.23	1.17	0.24	0.81	1.15	1.49
43. Weight, in kilograms	83.2	15.1	66.4	13.9	47.7	74.4	102.9

Table ANTH-A2: Anthropometric Data (in).[1]

This data is the same as that in Table ANTH-A1, except this data is expressed in inches rather than centimeters.

Measurement	Males		Females		Population Percentiles, 50/50 Males/Females		
	50th percentile	±1S.D.	50th percentile	±1S.D.	5th	50th	95th
STANDING							
1. Forward Functional Reach							
a. includes body depth at shoulder	32.5 (31.2)	1.9 (2.2)	29.2 (28.1)	1.5 (1.7)	27.2 (25.7)	30.7 (29.5)	35.0 (34.1)
b. Acromial Process to Functional Pinch	26.9	1.7	24.6	1.3	22.6	25.6	29.3
c. Abdominal Extension to Functional Pinch†	(24.4)	(3.5)	(23.8)	(2.6)	(19.1)	(24.1)	(29.3)
2. Abdominal Extension Depth	9.1	0.8	8.2	0.8	7.1	8.7	10.2
3. Waist Height	41.9 (41.3)	2.1 (2.1)	40.0 (38.8)	2.0 (2.2)	37.4 (35.8)	40.9 (39.9)	44.7 (44.5)
4. Tibial Height	17.9	1.1	16.5	0.9	15.3	17.2	19.4
5. Knuckle Height	29.7	1.6	28.0	1.6	25.9	28.8	31.9
6. Elbow Height	43.5 (45.1)	1.8 (2.5)	40.4 (42.2)	1.4 (2.7)	38.0 (38.5)	42.0 (43.6)	45.8 (48.6)
7. Shoulder Height	56.6 (57.6)	2.4 (3.1)	51.9 (56.3)	2.7 (2.6)	48.4 (49.8)	54.4 (55.3)	59.7 (61.6)

* These values should be adjusted for clothing and posture.

† Add the following for bending forward from hips or waist: Male: waist, 10 ± 3; hips, 16 ± 3. Female: waist, 8 ± 2; hips, 14 ± 4.

Table ANTH-A2: (continued)

Measurement	Males		Females		Population Percentiles, 50/50 Males/Females		
	50th percentile	±1S.D.	50th percentile	±1S.D.	5th	50th	95th
8. Eye Height	64.7	2.4	59.6	2.2	56.8	62.1	67.8
9. Stature	68.7	2.6	63.8	2.4	60.8	66.2	72.0
	(69.9)	(2.6)	(64.8)	(2.8)	(61.1)	(67.1)	(74.3)
10. Functional Overhead Reach	82.5	3.3	78.4	3.4	74.0	80.5	86.9
SEATED							
11. Thigh Clearance Height	5.8	0.6	4.9	0.5	4.3	5.3	6.5
12. Elbow Rest Height	9.5	1.3	9.1	1.2	7.3	9.3	11.4
13. Midshoulder Height	24.5	1.2	22.8	1.0	21.4	23.6	26.1
14. Eye Height	31.0	1.4	29.0	1.2	27.4	29.9	32.8
15. Sitting Height, Normal	34.1	1.5	32.2	1.6	32.0	34.6	37.4
16. Functional Overhead Reach	50.6	3.3	47.2	2.6	43.6	48.7	54.8
17. Knee Height	21.3	1.1	20.1	1.0	18.7	20.7	22.7
18. Popliteal Height	17.2	1.0	16.2	0.7	15.1	16.6	18.4
19. Leg Length	41.4	1.9	39.6	1.7	37.3	40.5	43.9
20. Upper-Leg Length	23.4	1.1	22.6	1.0	21.1	23.0	24.9
21. Buttocks-to-Popliteal Length	19.2	1.0	18.9	1.2	17.2	19.1	20.9
22. Elbow-to-Fist Length	14.2	0.9	12.7	1.1	12.6	14.5	16.2
	(14.6)	(1.2)	(13.0)	(1.2)	(11.4)	(13.8)	(16.2)

Table ANTH-A2: (continued)

Measurement	Males		Females		Population Percentiles, 50/50 Males/Females		
	50th percentile	±1S.D.	50th percentile	±1S.D.	5th	50th	95th
23. Upper-Arm Length	14.5 (14.6)	0.7 (1.0)	13.4 (13.3)	0.4 (0.8)	12.9 (12.1)	13.8 (13.8)	15.5 (16.0)
24. Shoulder Breadth	17.9	0.8	15.4	0.8	14.3	16.7	18.8
25. Hip Breadth	14.0	0.9	15.0	1.0	12.8	14.5	16.3
FOOT							
26. Foot Length	10.5	0.5	9.5	0.4	8.9	10.0	11.2
27. Foot Breadth	3.9	0.2	3.5	0.2	3.2	3.7	4.2
HAND							
28. Hand Thickness, Metacarpal III	1.3	0.1	1.1	0.1	1.0	1.2	1.4
29. Hand Length	7.5	0.4	7.2	0.4	6.7	7.4	8.0
30. Digit Two Length	3.0	0.3	2.7	0.3	2.3	2.8	3.3
31. Hand Breadth	3.4	0.2	3.0	0.2	2.8	3.2	3.6
32. Digit One Length	5.0	0.4	4.4	0.4	3.8	4.7	5.6
33. Breadth of Digit One Interphalangeal Joint	0.9	0.05	0.8	0.05	0.7	0.8	1.0
34. Breadth of Digit Three Interphalangeal Joint	0.7	0.05	0.6	0.04	0.6	0.7	0.8
35. Grip Breadth, Inside Diameter	1.9	0.2	1.7	0.1	1.5	1.8	2.2
36. Hand Spread, Digit One to Two, 1st Phalangeal Joint	4.9	0.9	3.9	0.7	3.0	4.3	6.1

Table ANTH-A2: (continued)

Measurement	Males		Females		Population Percentiles, 50/50 Males/Females		
	50th percentile	±1S.D.	50th percentile	±1S.D.	5th	50th	95th
37. Hand Spread, Digit One to Two, 2nd Phalangeal Joint	4.1	0.7	3.2	0.7	2.3	3.6	5.0
HEAD							
38. Head Breadth	6.0	0.2	5.7	0.2	5.4	5.9	6.3
39. Interpupillary Breadth	2.4	0.2	2.3	0.2	2.1	2.4	2.6
40. Biocular Breadth	3.6	0.2	3.6	0.2	3.3	3.6	3.9
OTHER MEASUREMENTS							
41. Flexion-Extension, Range of Motion of Wrist, Degrees	134	19	141	15	108	138	166
42. Ulnar-Radial Range of Motion of Wrist, Degrees	60	13	67	14	41	63	87
43. Weight, in pounds	183.4	33.2	146.3	30.7	105.3	164.1	226.8

Table ANTH-A3: Functional Hand Grasp Dimensions[1]

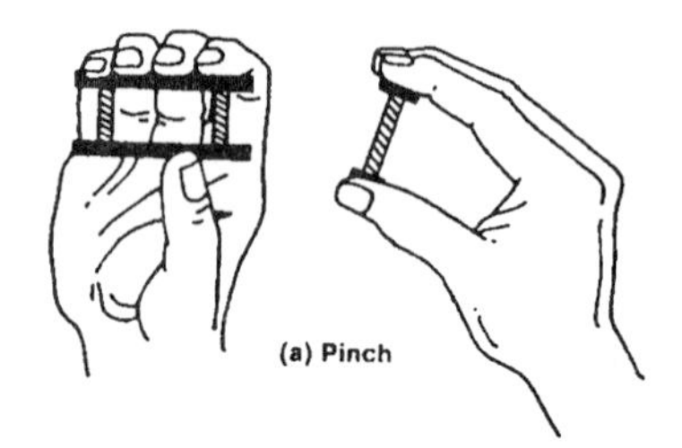

	Span, in centimeters (inches), 50/50 Male/Female Mix		
	5th Percentile	50th Percentile	95th Percentile
True	2.1 (0.8)	4.3 (1.7)	7.9 (3.1)
Maximum	10.8 (4.2)	12.5 (4.9)	15.0 (5.9)

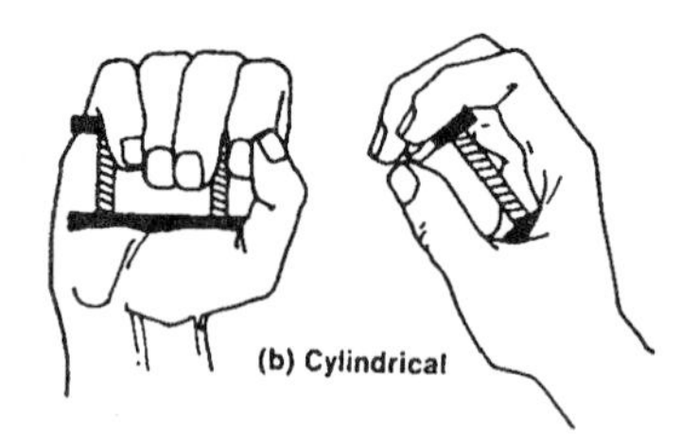

	Span, in centimeters (inches), 50/50 Male/Female Mix		
	5th Percentile	50th Percentile	95th Percentile
True	4.5 (1.8)	5.5 (2.2)	5.9 (2.3)
Maximum	9.5 (3.7)	11.0 (4.3)	13.0 (5.1)

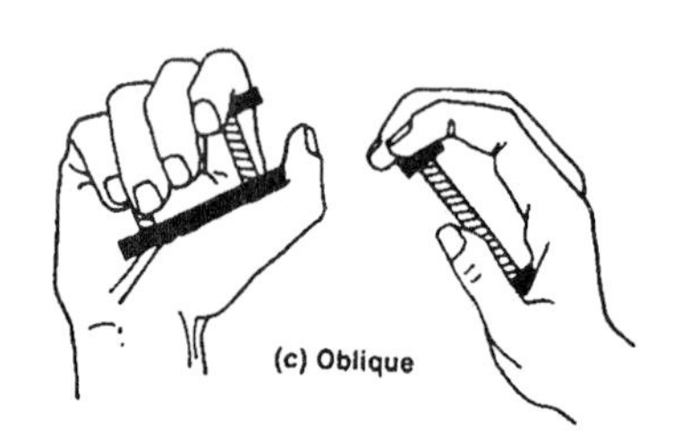

	Span, in centimeters (inches), 50/50 Male/Female Mix		
	5th Percentile	50th Percentile	95th Percentile
True	3.6 (1.4)	4.5 (1.8)	5.8 (2.3)
Maximum	9.5 (3.7)	11.0 (4.3)	13.0 (5.1)

The functional hand grasp spans for three types of grasp are shown: pinch (a), cylindrical grasp (b), and oblique grasp (c). The 5th, 50th, and 95th percentiles are based on a 50/50 ratio of men to women from industrial studies of 46 men and 38 women. Each value is given in centimeters, with its inches equivalent given in parentheses.

REFERENCES

1. Reprinted with permission from *Ergonomic Design for People at Work*, © Eastman Kodak Company, 1983, published by Van Nostrand Reinhold. Courtesy of Eastman Kodak Company.

CHAPTER 1: ANTHROPOMETRY

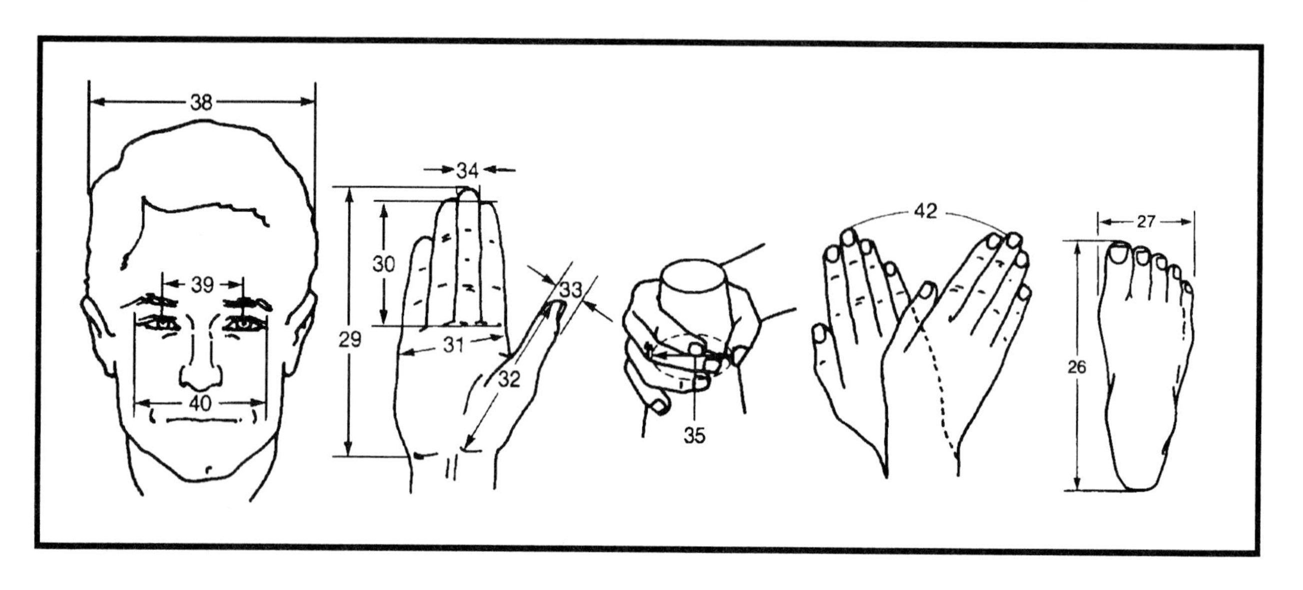

Section B: Office Workers

The data presented in the following section are for U.S. men and women, minimally clothed and standing or sitting erect. Caution should be exercised when using this data, since the original source did not list the size of the sample or any of the population parameters, other than to say that the subjects were male and female office workers. Nevertheless, this set of data is an interesting contrast with the military anthropometric data in Sections A & C. For alternative anthropometric data, the reader is referred to the ***Anthropometric Source Book: Volumes 1, 2, & 3***.

The ***Anthropometric Source Book*** contains around a thousand pages of anthropometric tables and information. The ***Source Book*** volumes are U.S. Department of Commerce documents (Document #s: N79-11734, N79-13711, and N79-13712) and are available from the National Technical Information Service (NTIS) [Springfield, VA] for about $200. The ***Source Book*** is also referred to as NASA Reference Document 1024.

Figure ANTH-B1: Anthropometric Dimensions for Standing and Sitting[1]

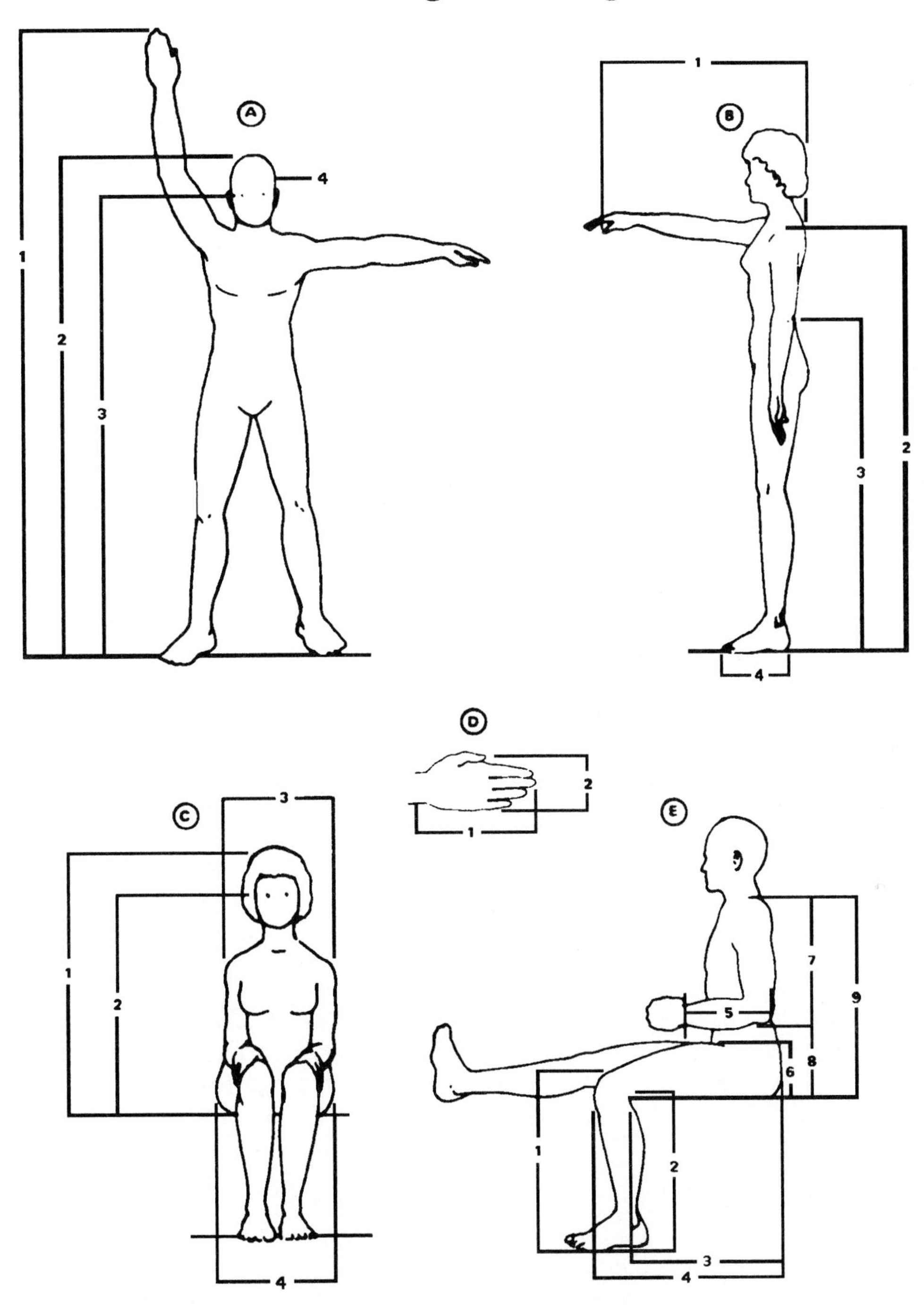

These illustrations correspond to the anthropometric measurements presented in Table ANTH-B1. Each number on this figure corresponds to a table number. This figure shows the dimensions that are important to the design of sitting and standing workplaces.

Table ANTH-B1: Anthropometric Dimensions of Office Workers in inches (cm).[1]

Legend	Dimensions	Estimated percentiles				
		2.5th	5th	50th	95th	97.5th
A-1	Vertical reach	73.5 (187)	75.0 (190)	80.5 (204)	86.0 (219)	87.5 (222)
A-2	Stature	61.5 (156)	62.0 (158)	66.5 (168)	70.5 (179)	71.5 (181)
A-3	Eye height, standing	56.5 (144)	57.5 (146)	62.0 (157)	66.0 (167)	67.0 (169)
A-4	Head circumference	21.5 (54)	22.0 (55)	22.5 (57)	23.0 (58)	23.5 (60)
B-1	Thumb tip reach	27.5 (69)	28.0 (71)	31.0 (78)	33.5 (86)	34.5 (87)
B-2	Shoulder height	50.0 (128)	51.0 (130)	55.0 (139)	59.0 (149)	59.5 (151)
B-3	Elbow to floor	39.5 (100)	40.0 (101)	42.5 (108)	45.0 (115)	45.5 (116)
B-4	Foot length	9.0 (23)	9.5 (24)	10.0 (25)	10.5 (27)	11.0 (28)
B-5	Foot width	3.0 (8)	3.0 (8)	3.5 (9)	4.0 (10)	4.0 (10)
C-1	Head to seat height	32.5 (83)	33.0 (84)	35.5 (90)	37.5 (96)	38.0 (97)
C-2	Eye height, sitting	28.0 (71)	28.5 (72)	30.5 (78)	32.5 (83)	33.0 (84)
C-3	Shoulder breadth	16.0 (40)	16.0 (41)	18.0 (46)	19.5 (50)	20 (51)
C-4	Hip breadth	13.0 (33)	13.5 (34)	15.0 (38)	17.0 (42)	17.0 (43)
D-1	Hand length	6.5 (17)	6.5 (17)	7.5 (19)	8.0 (21)	8.0 (21)
D-2	Hand width	3.0 (7)	3.0 (7)	3.5 (8)	3.5 (9)	4.0 (9)
E-1	Knee height	19.0 (48)	19.0 (49)	21.0 (53)	23.0 (58)	23.0 (59)
E-2	Popliteal height	15.0 (38)	15.0 (39)	17.0 (42)	18.0 (46)	18.5 (47)
E-3	Buttock to popliteal	17.5 (45)	18.0 (45)	19.5 (49)	21.0 (54)	21.5 (55)
E-4	Buttock to knee	21.5 (55)	22.0 (56)	23.5 (60)	25.0 (64)	25.5 (65)
E-5	Elbow to wrist	9.0 (23)	9.0 (23)	10.5 (26)	11.5 (29)	12.0 (30)
E-6	Thigh clearance	5.0 (12)	5.0 (13)	6.0 (15)	6.5 (17)	7.0 (18)
E-7	Shoulder to elbow	12.5 (31)	12.5 (32)	14.0 (35)	15.0 (38)	15.0 (39)
E-8	Elbow rest height	8.0 (19)	8.0 (20)	9.5 (24)	11.0 (28)	11.5 (29)
E-9	Shoulder to seat height	22.0 (56)	22.5 (57)	24.5 (62)	26.5 (67)	27.0 (68)

REFERENCES

1. Bailey, R.W. 1983. *Human Performance Engineering.* Englewood Cliffs, NJ: Prentice-Hall. Pages 80-85. Reprinted with the permission of Prentice-Hall.

CHAPTER 1: ANTHROPOMETRY

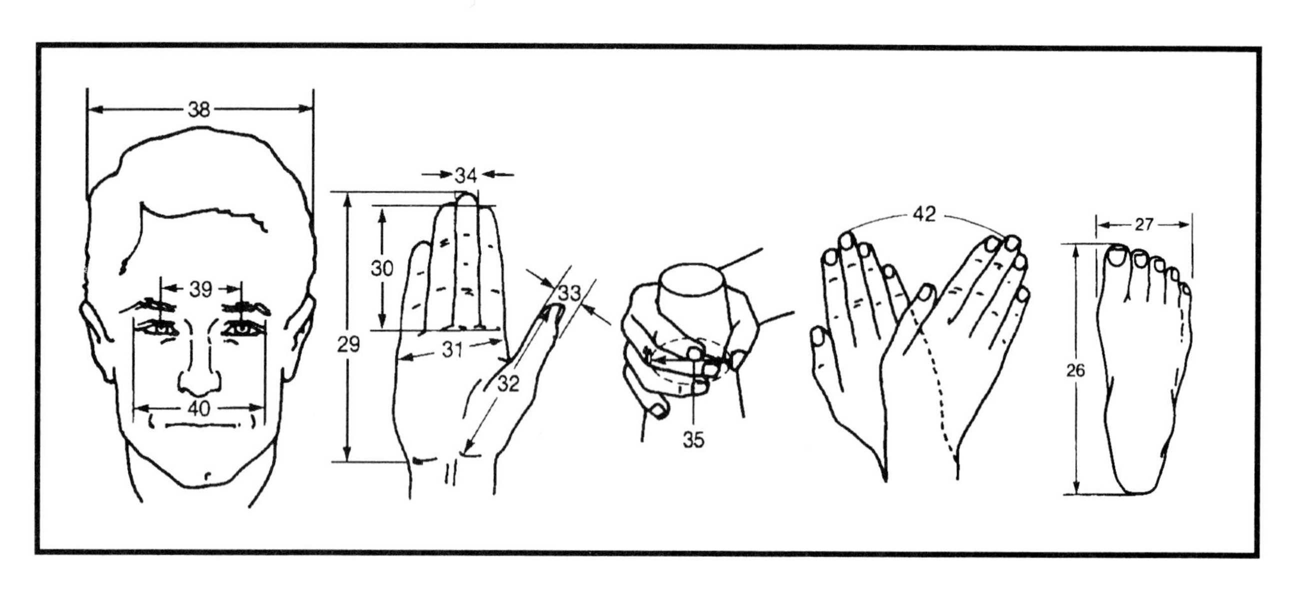

Section C: Military & Industrial

The data presented in the following section are for U.S. men and women, minimally clothed and standing or sitting erect. Caution should be exercised when using this data, since the data is at least 20 years old. It is presented here since this data is still referenced in many places, and is the most thorough set of anthropometric data that would fit within the scope of this book. For a more recent collection of anthropometric data, the reader is referred to the ***Anthropometric Source Book: Volumes 1, 2, & 3***.

The ***Anthropometric Source Book*** contains around a thousand pages of anthropometric tables and information. The ***Source Book*** volumes are U.S. Department of Commerce documents (Document #s: N79-11734, N79-13711, and N79-13712) and are available from the National Technical Information Service (NTIS) [Springfield, VA] for about $200. The ***Source Book*** is also referred to as NASA Reference Document 1024.

Table ANTH-C1: Comparision of Weight, Stature and Sitting Height of USAF Flying Personnel Measured in 1967 and 1950[1]†

Dimension	Year	N	Mean	SD	Mean	SD
Weight	1967	2420	173.60 lb	21.44	78.74 kg	9.72
	1950	4063	163.66 lb	20.86	74.23 kg	4.42
Increase	----	----	9.94 lb	----	4.51 kg	----
Stature	1967	2420	69.82 in	2.44	177.34 cm	6.19
	1950	4063	69.11 in	2.44	175.54 cm	6.19
Increase	----	----	0.71 in	----	1.80 cm	----
Sitting height	1967	2420	36.69 in	1.25	93.18 cm	3.18
	1950	4063	35.94 in	1.29	91.28 cm	3.27
Increase	----	----	0.75 in	----	1.90 cm	----

Table ANTH-C2: Percentiles of Weight, Stature and Sitting Height of USAF Flying Personnel. 1967 versus 1950[1]†

	Year	N	Percentiles				
			1st	5th	50th	95th	99th
Weight (lbs.)	1967	2420	127.6	140.2	172.4	210.8	227.7
	1950	4063	123.1	132.5	161.9	200.8	215.9
Increase	----	----	4.5	7.7	10.3	10.0	11.6
Stature (in.)	1967	----	64.3	65.9	69.8	73.9	75.6
	1950	----	63.5	65.2	69.1	73.1	74.9
Increase	----	----	.8	.7	.7	.8	.7
Sitting height (in.)	1967	----	33.9	34.7	36.7	38.8	39.6
	1950	----	32.9	33.8	36.0	38.0	38.9
Increase	----	----	1.0	.9	.7	.8	.7

† Note the change in stature and weight. The anthropometric dimensions of a given population tend to change over a number of years due to changes in nutritional habits, prevalence of preventative health care, and heterosis (change in size due to interbreeding of different physical types).

Table ANTH-C3: Height and Weight of White Male and Female Americans at Different Ages[1†]

	Male				Female			
	Height (in.)		Weight (lb)		Height (in.)		Weight (lb)	
Age (yr)	Mean	S.D.	Mean	S.D.	Mean	S.D.	Mean	S.D.
1	29.7	1.1	23	3	29.3	1.0	21	3
2	34.5	1.2	28	3	34.1	1.2	27	3
3	37.8	1.3	32	3	37.5	1.4	31	4
4	40.8	1.9	37	5	40.6	1.6	36	5
5	43.7	2.0	42	5	43.8	1.7	41	5
6	46.1	2.1	47	6	45.7	1.9	45	5
7	48.2	2.2	54	7	47.9	2.0	50	7
8	50.4	2.3	60	8	50.3	2.2	58	11
9	52.8	2.4	66	8	52.1	2.3	64	11
10	54.5	2.5	73	10	54.6	2.5	72	14
11	56.8	2.6	82	11	57.1	2.6	82	18
12	58.3	2.9	87	12	59.6	2.7	93	18
13	60.7	3.2	99	13	61.4	2.6	102	18
14	63.6	3.2	113	15	62.8	2.5	112	19
15	66.3	3.1	128	16	63.4	2.4	117	20
16	67.7	2.8	137	16	63.9	2.2	120	21
17	68.3	2.6	143	19	64.1	2.2	122	19
18	68.5	2.6	149	20	64.1	2.3	123	17
19	68.6	2.6	153	21	64.1	2.3	124	17
20–24	68.7	2.6	158	23	64.0	2.4	125	19
25–29	68.7	2.6	163	24	63.7	2.5	127	21
30–34	68.5	2.6	165	25	63.6	2.4	130	24
35–39	68.4	2.6	166	25	63.4	2.4	136	25
40–49	68.0	2.6	167	25	63.2	2.4	142	27
50–59	67.3	2.6	165	25	62.8	2.4	148	28
60–69	66.8	2.4	162	24	62.2	2.4	146	28
70–79	66.5	2.2	157	24	61.8	2.2	144	27
80–89	66.1	2.2	151	24				

BODY DIMENSIONS

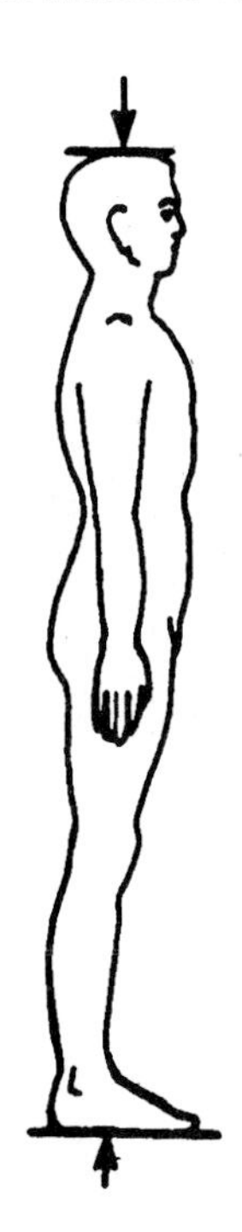

Figure ANTH-C1: Stature

Table ANTH-C4: Nude Stature of US Military Personnel[1]

Population	Percentiles (in.)					S.D.
	1st	5th	50th	95th	99th	
Air Force personnel	63.5	65.2	69.1	73.1	74.9	2.44
Pilots, multi-engine	64.4	65.9	69.4	73.3	74.9	2.31
Pilots, fighter	63.8	65.2	68.8	72.6	74.2	2.24
Cadets	63.6	65.2	69.2	73.1	74.7	2.45
Bombardiers	63.5	65.2	69.1	73.0	74.5	2.32
Navigators	63.5	65.2	69.2	73.3	75.0	2.46
Observers	63.8	65.4	69.1	72.8	74.2	2.44
Flight engineers	63.1	64.8	69.0	73.2	75.0	2.51
Gunners	62.4	64.2	68.3	72.2	73.7	2.43
Radio operators	63.0	64.6	68.3	71.8	73.2	2.37
Basic trainees	62.5	64.2	68.6	72.7	74.7	2.61
Army personnel:						
Inductees less than 20 years old	62.4	64.3	68.7	73.1	74.9	2.66
Inductees more than 20 years old	62.7	64.6	69.0	73.4	75.2	2.65
Separatees, white	62.7	64.3	68.5	72.6	74.5	2.52
Separatees, Negro	62.3	64.0	68.0	72.2	74.0	2.58
Marine Corps personnel	64.4	66.1	69.7	73.5	74.5	2.18
Recruits	63.0	64.6	68.6	72.5	74.1	2.40
Navy personnel	64.1	65.7	69.7	73.5	75.1	2.34
Recruits, 18 years old	62.8	64.5	68.5	72.6	74.2	2.50
Recruits, 17–25 years old	62.9	64.6	68.6	72.7	74.4	2.48
Enlisted men, general	63.2	64.8	69.5	73.5	75.5	2.48
Pilots, aircraft	65.1	66.2	69.9	73.9	75.3	2.33
Cadets, aviation	65.1	66.6	70.1	73.8	75.2	----
Army aviators	64.4	65.8	69.4	73.3	74.8	----

Table ANTH-C5: Nude Stature of Male Civilians (US, Canada)[1]

Population	Percentiles (in.)					
	1st	5th	50th	95th	99th	S.D.
Railroad travelers	*62.5	*64.5	*69.0	*73.8	*75.6	----
Truck and bus drivers	63.0	64.6	68.4	72.5	74.1	----
Airline pilots	64.4	66.0	70.0	73.9	75.6	2.40
Industrial workers	*64.4	*66.1	*70.3	*74.4	*76.2	2.46
College students:	62.5	64.4	68.7	73.1	74.9	2.68
Eastern 18 years old	64.5	66.1	69.9	73.8	75.4	2.38
Eastern 19 years old	65.0	66.5	70.2	74.0	75.5	2.30
Midwest, 18 years old	63.2	65.0	69.1	73.3	75.0	2.60
Midwest, 18–22 years old	64.2	65.9	70.0	74.1	75.8	2.49
Draft registrants						
18–19 years old	62.0	63.8	68.0	72.3	74.1	2.61
20–24 years old	62.1	63.9	68.2	72.4	74.2	2.60
25–29 years old	61.9	63.7	68.1	72.4	74.2	2.63
30–34 years old	61.7	63.5	67.8	72.1	73.9	2.66
35–37 years old	61.3	63.2	67.6	72.0	73.8	2.64
Civilian men	61.7	63.6	68.3	72.8	74.6	----
Canadians						
18–19 years old	62.4	64.1	68.2	72.1	73.7	----
20–24 years old	62.0	63.8	68.3	72.5	74.3	----
25–29 years old	60.6	62.9	68.3	74.0	76.2	----
30–34 years old	61.5	63.4	68.1	72.8	74.8	----
35–44 years old	60.5	62.7	67.6	72.6	74.7	----
45–54 years old	59.7	61.8	66.8	72.0	74.1	----
55–64 years old	58.4	60.6	66.0	71.3	73.6	----
More than 64 years old	58.6	60.6	65.1	69.8	71.8	----

* Includes shoes. Subtract 1 inch for nude height.

Table ANTH-C6: Nude Stature of US Female Military Personnel[1]

Population	Percentiles (in.)					
	1st	5th	50th	95th	99th	S.D.
Air Force personnel:						
WAF basic trainees	59.3	60.3	64.0	68.2	69.9	2.34
Pilots	60.8	61.7	64.9	68.3	70.0	----
Flight nurses	59.0	60.2	63.5	67.7	69.3	----
Army personnel	58.4	59.9	63.9	68.0	69.7	2.42
WAC enlisted women	58.3	60.0	63.9	68.0	69.6	2.40
WAC officers	59.2	61.0	64.5	68.9	70.6	2.40
Nurses	58.7	60.4	64.1	68.3	70.0	2.40

Table ANTH-C7: Nude Stature of Female Civilians (US, Canada)[1]

Population	Percentiles (in.)					
	1st	5th	50th	95th	99th	S.D.
Railroad travelers	59.1	60.8	64.9	69.1	70.8	----
Working women	58.1	59.7	63.6	67.5	69.2	2.43
City women:						
Eastern, white	56.5	58.4	62.8	67.3	69.2	2.69
Eastern, Negro	57.3	59.0	63.1	67.3	69.0	2.53
College students	58.5	60.0	63.8	67.6	69.2	2.33
Eastern	59.8	61.2	64.8	68.3	69.7	2.15
Midwest	58.8	60.5	64.4	68.4	70.0	2.36
Canadians						
18–19 years old	57.4	59.0	62.7	66.3	67.8	----
20–24 years old	57.2	58.8	62.9	66.8	68.4	----
25–29 years old	56.9	58.5	62.6	66.5	68.2	----
30–34 years old	57.0	58.7	62.6	66.5	68.2	----
35–44 years old	56.6	58.2	62.4	66.4	68.0	----
45–54 years old	56.6	58.1	61.8	65.4	67.0	----
55–64 years old	55.9	57.4	61.0	64.9	66.4	----
More than 64 years old	54.8	56.4	60.6	64.7	66.4	----
General	57.4	59.1	63.2	67.2	68.8	2.48
Civilian women	57.1	59.0	62.9	67.1	68.8	----

* Includes shoes. Subtract 2 inch for nude height.

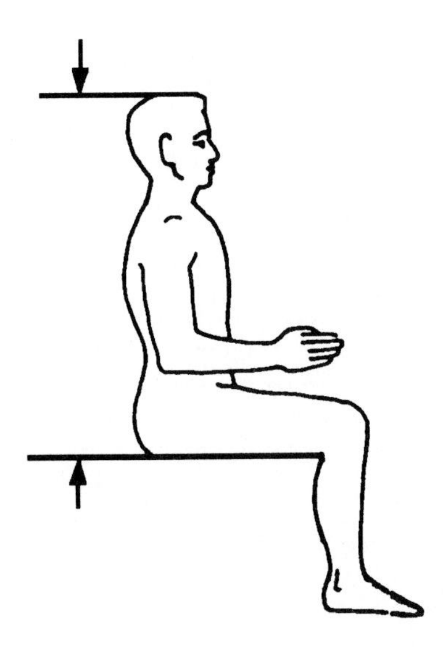

Figure ANTH-C2: Seated Height

Table ANTH-C8: Nude Sitting Height of US Female Military and Civilians[1]

Population	Percentiles (in.)					
	1st	5th	50th	95th	99th	S.D.
Air Force personnel						
Pilots	31.8	32.4	34.1	35.8	36.3	----
Flight nurses	31.1	31.9	33.7	35.7	36.6	----
Working women	30.9	31.7	33.7	35.7	36.5	1.15
College students:						
Eastern	31.6	32.4	34.2	36.0	36.7	1.10
Southern	31.0	31.7	33.6	35.4	36.2	1.06
Civilian women	29.5	30.9	33.4	35.7	36.6	----

Table ANTH-C9: Weight of Nude Male Military Personnel[1]

Population	Percentiles (lb)					
	1st	5th	50th	95th	99th	S.D.
Air Force personnel	123	133	162	201	216	20.9
Pilots, multi-engine	123	----	166	----	217	20.5
Pilots, fighter	123	----	159	----	225	20.7
Cadets	123	----	159	----	199	17.4
Bombardiers	126	----	169	----	211	20.6
Navigators	125	----	165	----	214	20.6
Observers	113	----	166	----	217	22.4
Flight engineers	124	----	166	----	222	23.3
Gunners	121	----	158	----	214	21.3
Radio operators	115	----	157	----	199	19.0
Basic trainees	109	118	145	186	208	21.0
Army personnel						
Inductees less than 20 years old	----	(111)	159	(206)	----	29.4
Inductees more than 20 years old	----	(122)	162	(202)	----	23.9
Separatees, white	114	124	153	192	215	20.6
Separatees, Negro	----	(120)	152	(183)	----	19.2
Marine Corps personnel	130	139	170	212	228	22.4
Recruits	----	(112)	143	(174)	----	18.7
Navy personnel	----	(126)	162	(197)	----	21.5
Recruits, 18 yr old	----	(110)	140	(171)	----	18.5
Recruits, 17–25 yr old	----	(119)	152	(185)	----	20.6
Enlisted men, general	----	132	160	197	----	19.9
Pilots, aircraft	129	140	171	203	221	19.1
Cadets, aviation	----	135	166	196	----	----
Army aviators	124	136	167	200	213	----

* Percentiles in parentheses were computed from the 50th percentile using the S.D.. Because of the skewed distribution of weight, these values might differ somewhat from the true values and should be used with caution.

Table ANTH-C10: Weight of Nude US Female Military Personnel[1]

Population	Percentiles* (lb)					
	1st	5th	50th	95th	99th	S.D.
Air Force personnel:						
WAF basic trainees	95	102	122	148	162	14.5
Pilots	102	106	129	155	169	----
Flight nurses	104	107	122	135	143	----
Army personnel	97	105	129	170	192	20.0
WAC enlisted women	----	(97)	130	(163)	----	20.6
WAC officers	----	(105)	132	(158)	----	16.1
Nurses	----	(95)	129	(162)	----	20.2

* Percentiles in parentheses were computed from the 50th percentile using the S.D.. Because of the skewed distribution of weight, these values might differ somewhat and should be used with caution.

Table ANTH-C11: Weight of Nude Female Civilians (US, Canada)[1]

Population	Percentiles* (lb)					
	1st	5th	50th	95th	99th	S.D.
Railroad travelers	----	104	133	179	----	----
Working women	----	(110)	136	(163)	----	16.0
City women:						
Eastern, white	95	108	140	200	229	27.2
Eastern, Negro	85	104	143	193	210	34.5
College students	----	(94)	121	(149)	----	17.1
Eastern	----	(101)	125	(149)	----	15.2
Midwest	----	(99)	126	(154)	----	16.9
Canadians						
18–19 years old	----	----	120	----	----	----
20–24 years old	----	----	122	----	----	----
25–29 years old	----	----	123	----	----	----
30–34 years old	----	----	126	----	----	----
35–44 years old	----	----	132	----	----	----
45–54 years old	----	----	142	----	----	----
55–64 years old	----	----	145	----	----	----
More than 64 years old	----	----	136	----	----	----
General	91	100	129	184	213	26.0
Civilian women	93	104	137	199	236	----

* Percentiles in parentheses were computed from the 50th percentile using the S.D.. Because of the skewed distribution of weight, these values might differ somewhat and should be used with caution.

Table ANTH-C12: Weight of Nude Male Civilians (US, Canada)[1]

Population	Percentiles* (lb)					
	1st	5th	50th	95th	99th	S.D.
Railroad travelers	----	132	167	218	----	----
Truck and bus drivers	----	129	164	213	247	----
Airline pilots	----	(134)	168	(201)	----	20.3
Industrial workers	----	(130)	170	(210)	----	24.5
College students	----	(112)	142	(172)	----	18.1
Eastern, 18 years old	----	(122)	150	(178)	----	17.2
Eastern, 19 years old	----	(132)	159	(187)	----	16.2
Midwest, 18 years old	----	(115)	148	(180)	----	19.7
Midwest, 18–22 years old	----	(118)	156	(195)	----	23.5
Draft registrants:						
18–19 years old	----	(106)	141	(176)	----	21.1
20–24 years old	----	(109)	146	(183)	----	22.4
25–29 years old	----	(110)	151	(192)	----	24.8
30–34 years old	----	(110)	153	(195)	----	25.8
35–37 years old	----	(111)	154	(197)	----	26.1
Civilian men	112	126	166	217	241	----
Canadians						
18–19 years old	----	----	140	----	----	----
20–24 years old	----	----	151	----	----	----
25–29 years old	----	----	157	----	----	----
30–34 years old	----	----	168	----	----	----
35–44 years old	----	----	165	----	----	----
45–54 years old	----	----	161	----	----	----
55–64 years old	----	----	159	----	----	----
More than 64 years old	----	----	156	----	----	----

* Percentiles in parentheses were computed from the 50th percentile using the S.D.. Because of the skewed distribution of weight, these values might differ somewhat and should be used with caution.

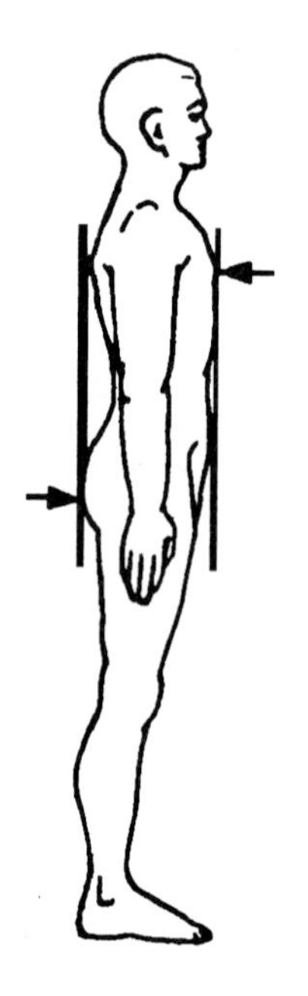

Figure ANTH-C3
Maximum Body Depth

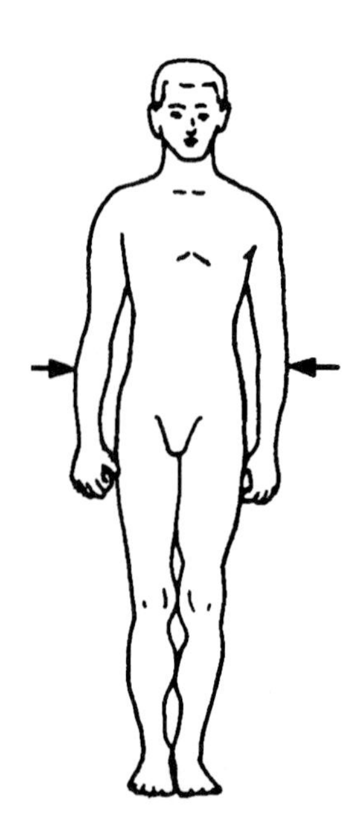

Figure ANTH-C4
Maximum Body Breadth

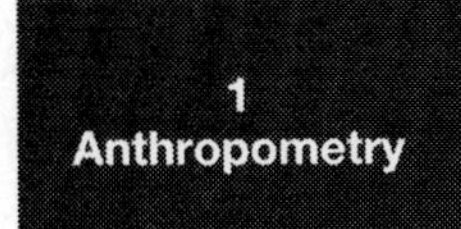

Table ANTH-C13: Maximum Body Depth and Breadth of US Air Force Personnel and College Students[1]

Dimension	Percentiles (in.) 5th	50th	95th	S.D.
Body depth	10.1	11.5	13.0	0.88
Body breadth	18.8	20.9	22.8	1.19

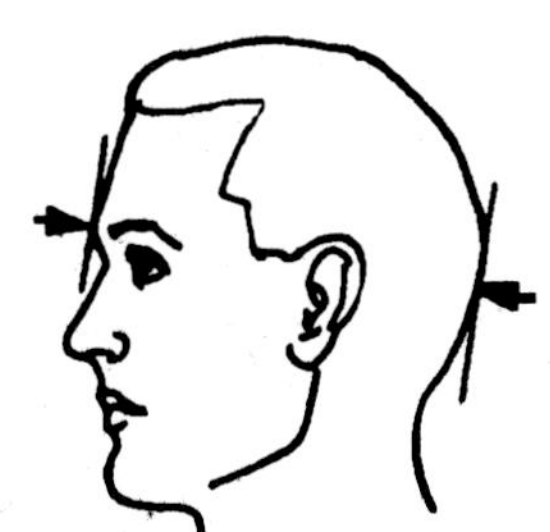

Figure ANTH-C5: Head Length

Table ANTH-C14: Head Length of US Military and Civilian Males[1]

Population	Percentiles (in.) 1st	5th	50th	95th	99th	S.D.
Air Force personnel	7.2	7.3	7.7	8.2	8.3	0.25
Cadets	7.2	7.4	7.8	8.2	8.4	.26
Basic trainees	7.0	7.2	7.6	8.1	8.3	.28
Army separatees	7.0	7.2	7.7	8.1	8.3	.28
Army aviators	7.2	7.3	7.8	8.2	8.5	.27
College students:						
Eastern	7.1	7.3	7.7	8.2	8.4	.27
Midwest	7.1	7.3	7.7	8.2	8.4	.27
Naval aviators	7.2	7.4	7.8	8.2	8.4	.26

Table ANTH-C15: Head Length of US Military and Civilian Females[1]

Population	Percentiles (in.)					S.D.
	1st	5th	50th	95th	99th	
WAF basic trainees	6.1	6.4	6.9	7.3	7.5	0.30
WAC personnel and Army nurses	6.7	6.8	7.2	7.7	7.8	.26
Working women	6.8	7.0	7.4	7.7	7.9	.23
College students:						
Eastern	6.8	7.0	7.4	7.7	7.9	.24
Southern	6.8	7.0	7.4	7.8	7.9	.23

Table ANTH-C16: Head Length of International Military Males[1]

Population	Personnel	N	Percentiles (in.)					S.D.
			1	5	50	95	99	
Canada	Pilots	314	6.9	7.05	7.6	8.15	8.43	0.31
Greece	Mixed military	1084	6.85	7.0	7.45	7.9	8.04	.27
Italy	Mixed military	1358	7.0	7.16	7.6	8.04	8.2	.26
Japan	JASDF pilots	236	6.58	6.78	7.37	7.84	7.95	.31
South Korea	ROKAF pilots	264	6.65	6.85	7.25	7.88	7.8	.26
Thailand	Mixed military	2950	6.42	6.63	7.0	7.45	7.65	.24
Turkey	Mixed military	915	6.7	6.9	7.33	7.75	7.8	.26
Vietnam	Mixed military	2129	6.5	6.7	7.23	7.56	7.75	.28

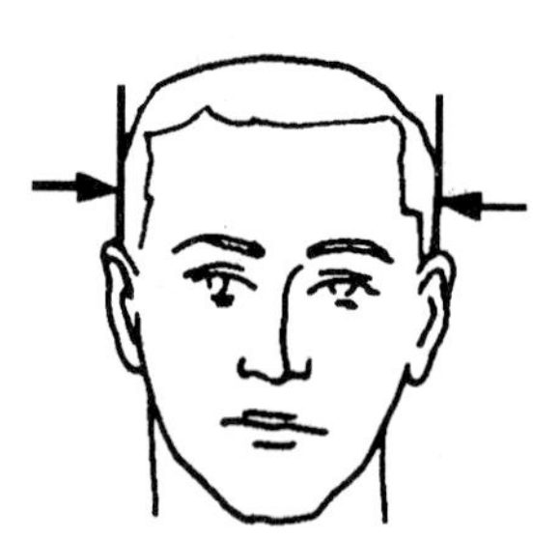

Figure ANTH-C6: Head Breadth

Table ANTH-C17: Head Breadth of US Military and Civilian Males[1]

Population	Percentiles (in.) 1st	5th	50th	95th	99th	S.D.
Air Force personnel	5.6	5.7	6.1	6.4	6.6	0.20
Cadets	5.6	5.7	6.1	6.4	6.6	.21
Basic trainees	5.4	5.6	5.9	6.3	6.5	.23
Army separatees	5.4	5.6	6.0	6.4	6.6	.23
Army aviators	5.6	5.7	6.1	6.5	6.8	.21
College students:						
Eastern	5.5	5.7	6.0	6.4	6.5	.22
Midwest	5.6	5.8	6.1	6.5	6.6	.20
Naval aviators	5.6	5.8	6.1	6.5	6.6	.21

Table ANTH-C18: Head Breadth of US Military and Civilian Females[1]

Population	Percentiles (in.) 1st	5th	50th	95th	99th	S.D.
WAF basic trainees	5.3	5.4	5.7	6.1	6.2	0.20
WAC personnel and Army nurses	5.2	5.4	5.7	6.1	6.2	.22
Working women	5.5	5.6	5.9	6.1	6.3	.17
College students:						
Eastern	5.4	5.5	5.8	6.2	6.3	.20
Southern	5.4	5.5	5.8	6.1	6.2	.18

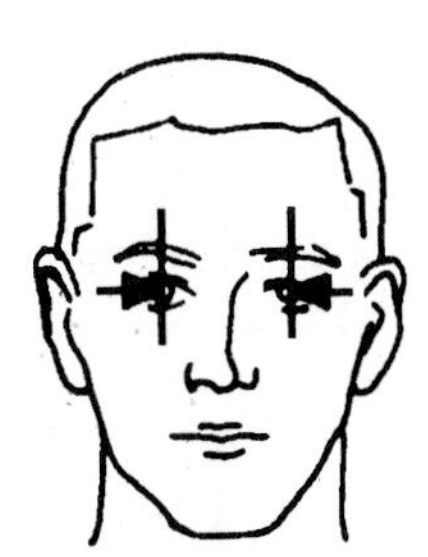

Figure ANTH-C7: Interpupilary Distance

Table ANTH-C19: Intrerpupillary Distance and Eye Height of Male USAF Personnel[1]

Dimension	Percentiles (in.) 1st	5th	50th	95th	99th	S.D.
Interpupillary distance	2.19	2.27	2.49	2.74	2.84	0.14
Eye height	59.2	60.8	64.7	68.6	70.3	2.38

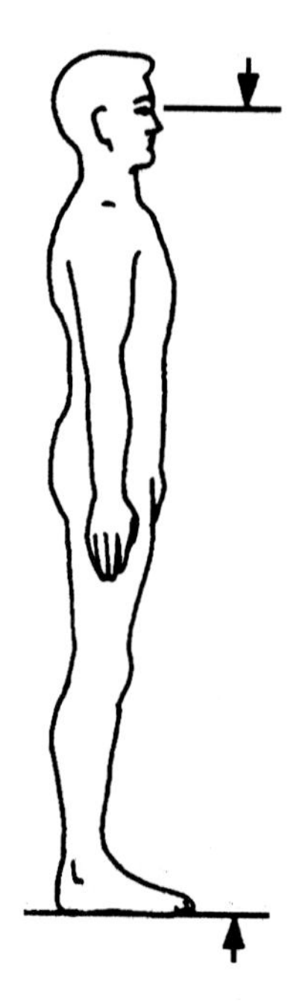

Figure ANTH-C8: Eye Height

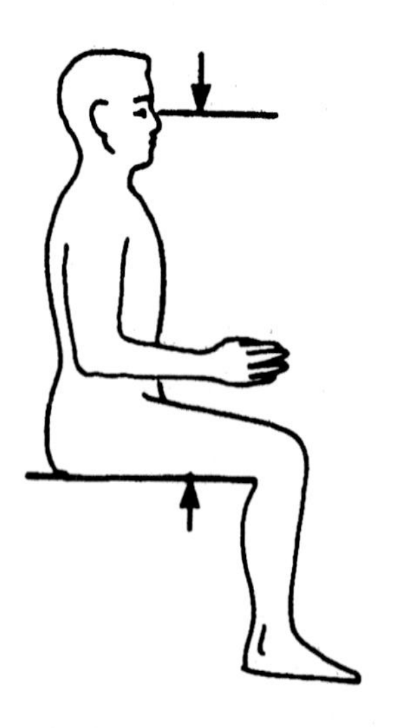

Figure ANTH-C9: Seated Eye Height

Table ANTH-C20: Seated Eye Height of US Military Males and Females [1]

Population	Percentiles (in.)					
	1st	5th	50th	95th	99th	S.D.
Male flight personnel	28.5	29.4	31.5	33.5	34.4	1.27
Female pilots	27.9	28.5	30.0	31.6	32.4	----
Female flight nurses	26.3	27.3	29.3	31.1	32.2	----
Army aviators	28.1	28.8	30.9	33.1	34.5	1.28
Naval aviators	28.8	29.7	31.5	33.6	34.5	1.18

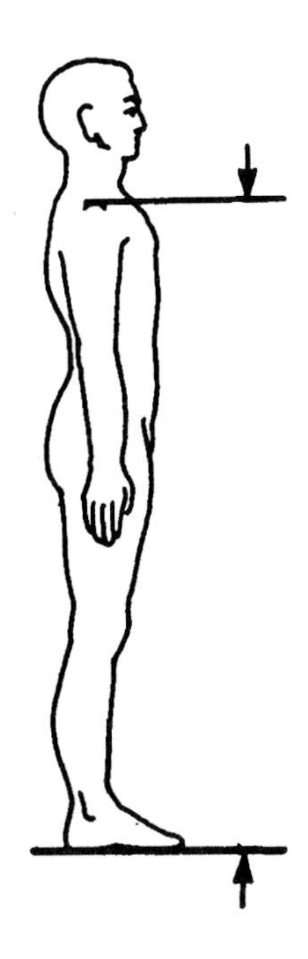

Figure ANTH-C10: Shoulder Height

Table ANTH-C21: Shoulder Height of USAF Males and Females [1]

Population	Percentiles (in.)					
	1st	5th	50th	95th	99th	S.D.
Male flight personnel	51.2	52.8	56.6	60.2	61.9	2.28
Male basic trainees	50.3	52.0	55.9	59.9	61.8	2.41
Female basic trainees	46.9	48.2	51.9	55.4	57.3	2.18

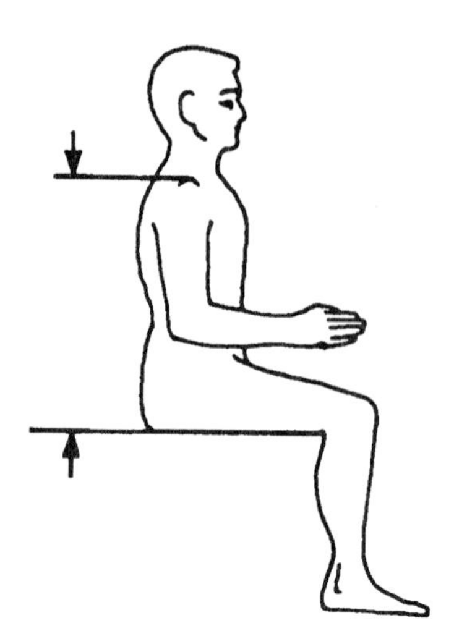

Figure ANTH-C11: Seated Shoulder Height

Table ANTH-C22: Seated Shoulder Height of US Military Males and Females [1]

Population	Percentiles (in.)					S.D.
	1st	5th	50th	95th	99th	
Male flight personnel	20.6	21.3	23.3	25.1	25.8	1.14
Female pilots	21.8	22.4	23.8	25.2	25.9	----
Female flight nurses	20.4	21.1	23.1	24.8	25.9	----
Naval aviators	21.5	22.0	23.8	25.5	26.4	1.06

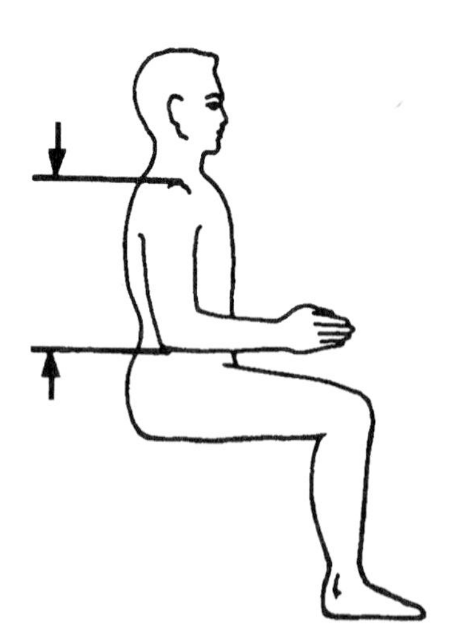

Figure ANTH-C12: Shoulder-Elbow Length

Table ANTH-C23: Shoulder-Elbow Length of US Military and Civilian Males [1]

Population	Percentiles (in.)					
	1st	5th	50th	95th	99th	S.D.
Air Force personnel	12.8	13.2	14.3	15.4	15.9	0.69
Cadets	13.2	13.6	14.7	15.8	16.3	----
Gunners	12.9	13.3	14.5	15.6	16.1	----
Army personnel:						
Army aviators	13.4	13.9	15.0	16.1	16.5	.70
Separatees, white	12.3	12.9	14.3	15.6	16.3	.81
Separatees, Negro	12.4	13.0	14.3	15.6	16.1	.80
Truck and bus drivers	13.3	13.8	14.8	15.9	16.3	.81
College students	12.8	13.3	14.5	15.7	16.1	.66
Naval aviators	13.0	13.4	14.5	15.6	16.1	.67

Table ANTH-C24: Shoulder-Elbow Length of US Military Females [1]

Population	Percentiles (in.)					
	1st	5th	50th	95th	99th	S.D.
Air Force personnel:						
Pilots	12.3	12.7	13.7	14.7	15.2	----
Flight nurses	12.3	12.7	13.6	14.8	15.3	----
Army personnel	11.3	11.9	13.1	14.3	14.9	0.74

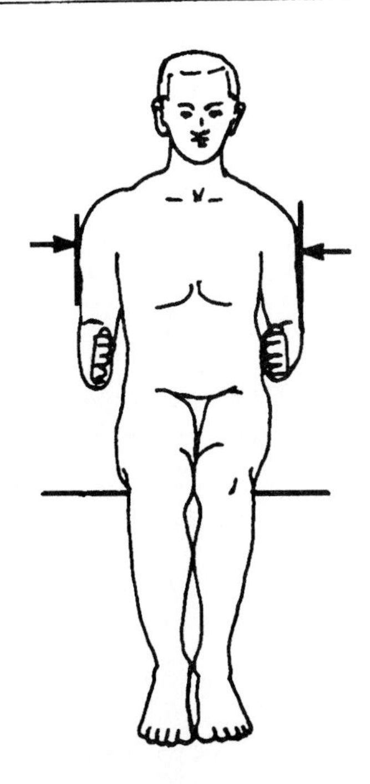

Figure ANTH-C13: Shoulder Breadth

Table ANTH-C25: Shoulder Breadth of US Military and Civilian Males [1]

Population	Percentiles (in.)					S.D.
	1st	5th	50th	95th	99th	
Air Force personnel	15.9	16.5	17.9	19.4	20.1	0.91
Cadets	16.1	16.7	18.0	19.3	19.9	----
Gunners	16.0	16.5	17.7	19.0	19.5	----
Army personnel:						
Aviators	16.4	16.8	18.2	20.0	20.5	.88
Separatees, white	15.8	16.4	17.9	19.6	20.6	.99
Separatees, Negro	15.8	16.4	17.9	19.4	20.0	.89
Navy personnel	15.1	15.8	17.6	19.4	20.2	1.09
Railroad travelers	*15.7	*16.4	*17.6	*19.2	*19.8	----
Truck and bus drivers	16.2	16.9	18.3	19.9	20.5	----
College students	15.1	15.7	17.2	18.7	19.3	.86
Naval aviators	16.6	17.3	18.8	20.3	20.9	.91

*Including light clothing (subtract 0.3 in. for nude)

Table ANTH-C26: Shoulder Breadth of US Military and Civilian Females [1]

Population	Percentiles (in.)				
	1st	5th	50th	95th	99th
Air Force personnel:					
Pilots	14.3	14.9	16.1	17.6	18.0
Flight nurses	14.1	14.5	15.7	16.8	17.2
Railroad travelers	*13.7	*14.4	*15.7	*17.6	*18.2

*Including light clothing (subtract 0.3 in. for nude)

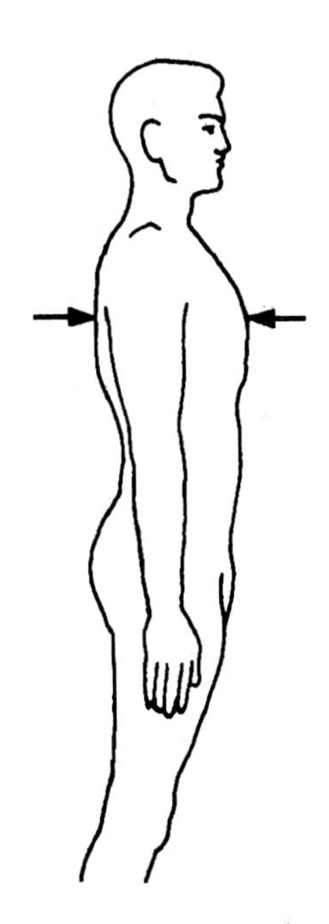

Figure ANTH-C14: Chest Depth

Table ANTH-C27: Chest Depth of US Military and Civilian Males [1]

Population	Percentiles (in.)					
	1st	5th	50th	95th	99th	S.D.
Air Force personnel	7.6	8.0	9.0	10.4	11.1	0.75
Cadets	6.8	7.2	8.2	9.3	9.7	----
Gunners	6.7	7.1	8.2	9.2	9.6	----
Army separatees	6.7	7.2	8.3	9.6	10.5	.75
Aviators	7.4	7.9	8.9	10.4	11.0	.79
Truck and bus drivers	7.1	7.6	8.9	10.5	11.1	----
College students:						
Eastern	6.5	6.9	7.9	8.9	9.3	.55
Midwest	6.4	6.9	8.0	9.2	9.7	.71
Naval aviators	7.8	8.3	9.4	10.6	11.1	.71

Table ANTH-C28: Chest Depth of US Female College Students [1]

Population	Percentiles (in.)					
	1st	5th	50th	95th	99th	S.D.
Eastern	5.8	6.3	7.4	8.6	9.0	0.68
Midwest	6.0	6.4	7.3	8.2	8.6	.56

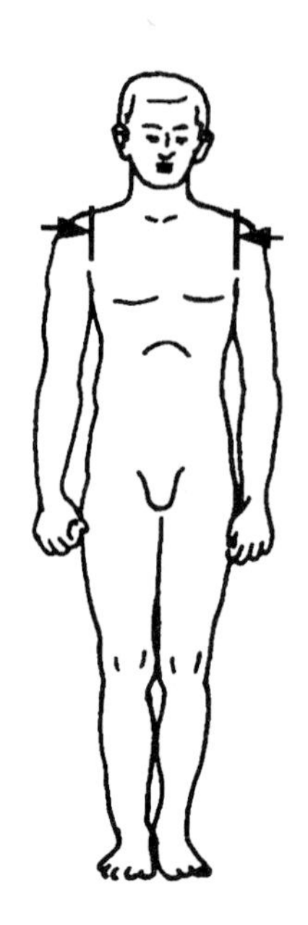

Figure ANTH-C15: Chest Breadth

Table ANTH-C29: Chest Breadth of US Military and Civilian Males [1]

Population	Percentiles (in.)					S.D.
	1st	5th	50th	95th	99th	
Air Force personnel	10.4	10.8	12.0	13.4	14.1	0.80
Cadets	9.8	10.3	11.3	12.4	12.8	----
Gunners	9.7	10.1	11.1	12.1	12.5	----
Basic trainees	9.7	10.2	11.4	13.0	14.3	.91
Army separatees	9.3	10.0	11.1	12.4	13.2	.77
Truck and bus drivers	9.6	10.2	11.8	13.5	13.9	----
College students:						
Eastern	9.9	10.4	11.5	12.7	13.1	.67
Midwest	9.3	9.9	11.1	12.4	12.9	.79

Table ANTH-C30: Chest Breadth of US Female College Students [1]

Population	Percentiles (in.)					S.D.
	1st	5th	50th	95th	99th	
WAF basic trainees	8.9	9.1	9.9	10.9	11.3	0.55
College students:						
Eastern	8.3	8.7	9.7	10.7	11.1	.59
Midwest	8.6	9.0	10.1	11.1	11.5	.64

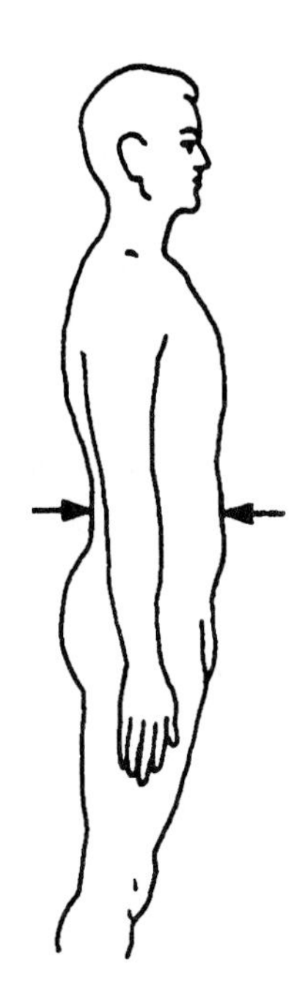

Figure ANTH-C16: Waist Depth

Table ANTH-C31: Waist Depth of US Military and Civilian Males [1]

Population	Percentiles (in.)					S.D.
	1st	5th	50th	95th	99th	
Air Force personnel	6.3	6.7	7.9	9.5	10.3	0.88
Cadets	6.7	7.2	8.2	9.3	9.8	----
Gunners	6.7	7.2	8.2	9.3	9.8	----
Army separatees	7.5	7.9	9.0	10.5	11.5	.81
Truck and bus drivers	7.3	7.9	9.5	12.1	13.1	----
Naval aviators	6.9	7.3	8.5	9.8	10.5	.76

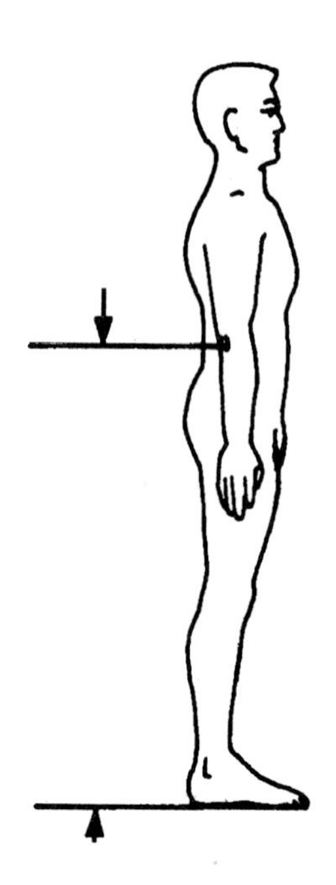

Figure ANTH-C17: Standing Elbow Height

Table ANTH-C32: Standing Elbow Height of US Military Males [1]

Posture	Percentiles (in.)					S.D.
	1st	5th	50th	95th	99th	
Standing, USAF	39.5	40.6	43.5	46.4	47.7	1.77
Sitting, USAF	6.6	7.4	9.1	10.8	11.5	1.04
Naval aviators	7.0	7.6	9.3	10.9	11.7	.99

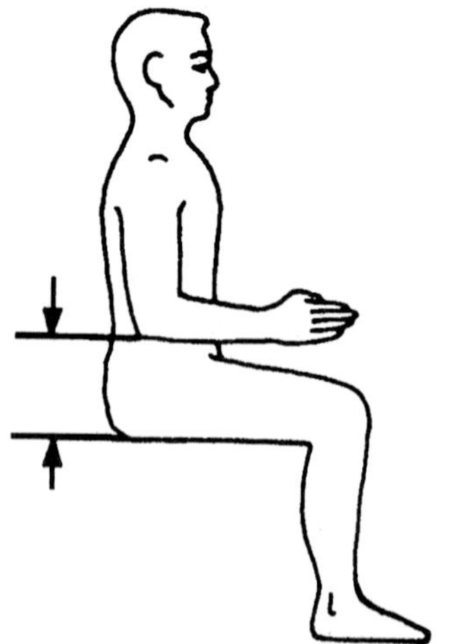

Figure ANTH-C18: Seated Elbow Height

Table ANTH-C33: Seated Elbow Height of US Civilians [1]

Sex	Percentiles (in.)				
	1st	5th	50th	95th	99th
Male	6.3	7.4	9.5	11.6	12.5
Female	6.1	7.1	9.2	11.0	11.9

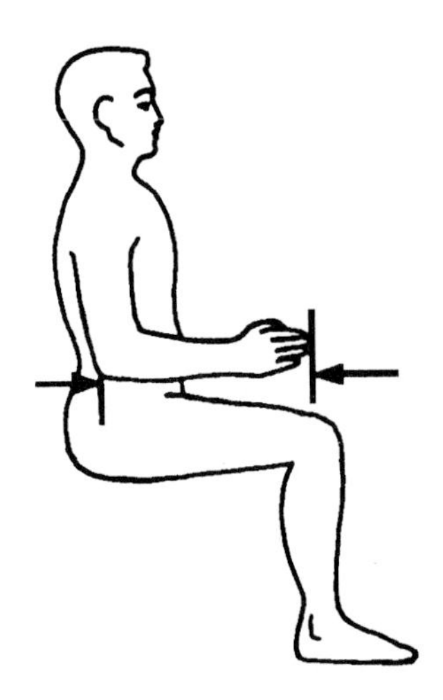

Figure ANTH-C19: Forearm-Hand Length

Table ANTH-C34: Forearm-Hand Length of US Military and Civilian Males[1]

Population	Percentiles (in.)					S.D.
	1st	5th	50th	95th	99th	
Air Force personnel	17.0	17.6	18.9	20.2	20.7	0.81
Army personnel:						
Aviators	16.1	17.6	19.1	20.4	21.5	.86
Separatees, white	16.6	17.3	18.7	20.1	20.8	.88
Separatees, Negro	17.3	18.0	19.6	21.4	22.1	.94
Truck and bus drivers	16.7	17.3	18.8	20.2	20.8	----
College students	17.0	17.6	18.9	20.2	20.7	.75
Naval aviators	17.5	17.9	19.1	20.4	20.8	.75

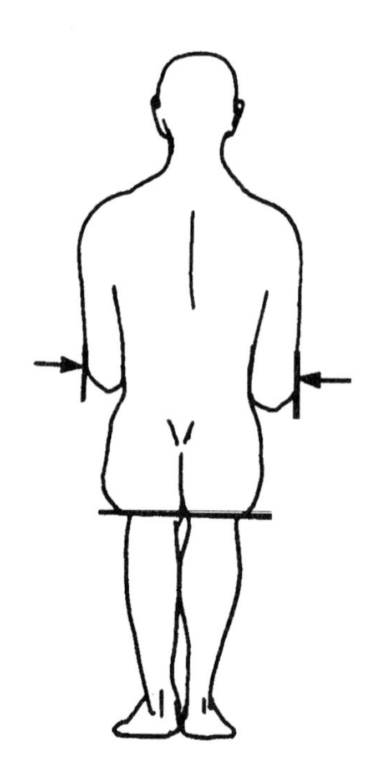

Figure ANTH-C20: Elbow-to-Elbow Breadth

Table ANTH-C35: Elbow-to-Elbow Breadth of US Military and Civilian Males [1]

Population	Percentiles (in.)					
	1st	5th	50th	95th	99th	S.D.
Air Force personnel	14.5	15.2	17.2	19.8	20.9	1.42
Cadets	14.4	15.1	16.7	18.4	19.1	----
Gunner	13.9	14.6	16.4	18.2	18.9	----
Army personnel:						
Separatees, white	14.4	15.3	17.4	20.3	21.8	1.54
Separatees, Negro	14.4	15.1	16.9	19.3	20.4	1.28
Truck and bus drivers	13.8	14.9	17.5	20.7	22.2	----

Table ANTH-C36: Elbow-to-Elbow Breadth of USAF Females[1]

Population	Percentiles (in.)				
	1st	5th	50th	95th	99th
Pilots	12.8	13.3	15.1	17.1	18.5
Flight nurses	13.0	13.5	14.9	16.7	17.3

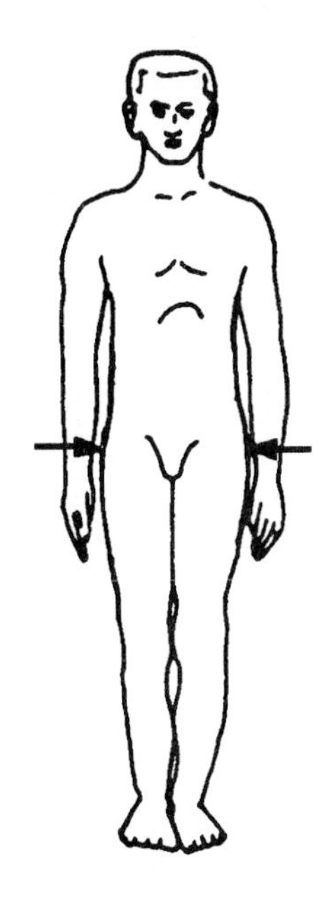

Figure ANTH-C21: Standing Hip Breadth

Table ANTH-C37: Hip Breadth of US Male and Female US Air Force and Navy Personnel, and College Students [1]

Population	Percentiles (in.)					S.D.
	1st	5th	50th	95th	99th	
Male USAF flight personnel	11.3	12.1	13.2	14.4	15.2	0.73
Male basic trainees	11.5	12.1	13.3	15.0	15.0	.94
Naval aviators	12.1	12.6	13.8	14.9	15.4	.70
Male college students	11.4	11.8	13.0	14.2	14.7	.67
Female basic trainees	12.2	12.5	13.5	15.4	16.9	.95

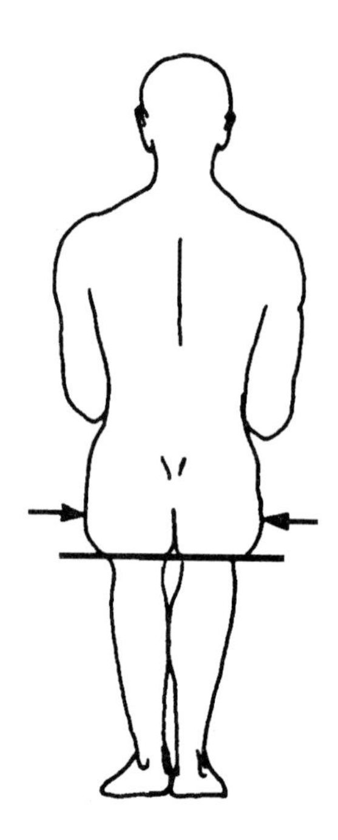

Figure ANTH-C22: Seated Hip Breadth

Table ANTH-C38: Seated Hip Breadth of US Military and Civilian Males [1]

Population	Percentiles (in.)					S.D.
	1st	5th	50th	95th	99th	
Air Force personnel	12.2	12.7	13.9	15.4	16.2	0.87
Cadets	12.6	13.1	14.2	15.5	15.9	----
Gunners	12.1	12.7	13.8	15.1	15.5	----
Army personnel:						
Separatees, white	12.2	12.7	13.9	15.5	16.7	.90
Separatees, Negro	11.6	12.1	13.4	15.0	15.8	.84
Aviators	12.4	12.8	14.2	15.7	16.3	.87
Navy personnel:						
Enlisted men	12.4	13.0	14.8	16.4	17.2	1.05
Cadets, aviation	13.4	14.0	15.4	16.8	17.3	----
Aviators	12.7	13.1	14.5	15.9	16.6	.85
Railroad travelers	*12.9	*13.7	*15.3	*17.4	*18.1	----
Truck and bus drivers	12.4	13.2	14.5	16.3	16.8	----
Civilian men	11.5	12.2	14.0	15.9	17.0	----

*Including Light Clothing (subtract 0.5 in. for nude)

Table ANTH-C39: Seated Hip Breadth of US Military and Civilian Females [1]

Population	Percentiles (in.)				
	1st	5th	50th	95th	99th
Air Force personnel:					
Pilots	13.0	13.5	15.0	16.9	18.1
Flight nurses	13.1	13.5	15.1	16.6	17.1
Railroad travelers	*12.2	*13.1	*14.6	*17.2	*17.8
Civilian women	11.7	12.3	14.3	17.1	18.8

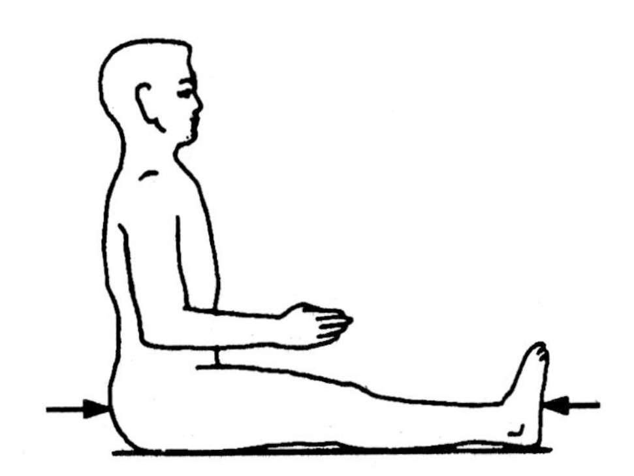

Figure ANTH-C23: Buttock-Leg Length

Table ANTH-C40: Buttock-Leg Length of US Military Males [1]

Population	Percentiles (in.)					S.D.
	1st	5th	50th	95th	99th	
Air Force personnel	38.2	39.4	42.7	46.1	47.7	2.04
Navy personnel:						
General	36.5	38.0	41.5	44.9	46.4	2.07
Pilots, aircraft	36.8	38.3	42.3	46.3	48.8	----

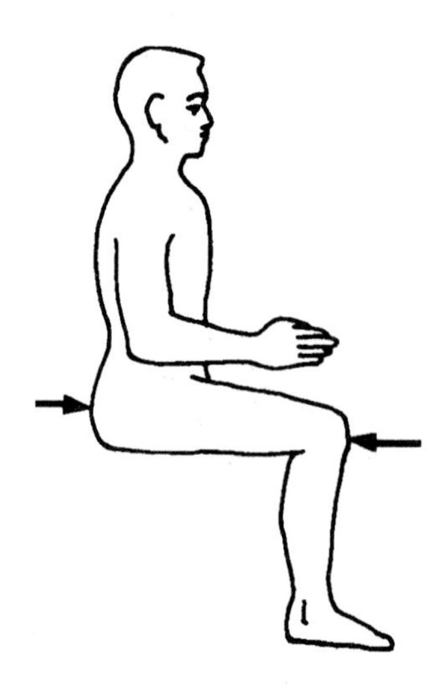

Figure ANTH-C24: Buttock-Knee Length

Table ANTH-C41: Buttock-Knee Length of US Military and Civilian Males [1]

Population	Percentiles (in.)					S.D.
	1st	5th	50th	95th	99th	
Air Force personnel	21.2	21.9	23.6	25.4	26.2	1.06
Cadets	21.2	22.0	23.6	25.6	26.2	----
Gunners	20.5	21.1	23.1	24.7	25.6	----
Army personnel:						
Aviators	21.4	22.1	23.8	25.8	26.7	1.08
Separatees, white	20.7	21.5	23.4	25.2	26.0	1.12
Separatees, Negro	21.1	21.9	23.8	25.8	26.6	1.17
Navy personnel	20.6	21.4	23.4	25.0	25.8	1.18
Enlisted men	21.7	22.5	24.5	26.5	27.3	1.23
Cadets, aviation	21.8	22.6	24.3	26.2	26.9	----
Truck and bus drivers	21.3	22.1	23.8	25.8	26.5	----
Naval aviators	21.8	22.5	24.1	25.8	26.5	1.00
Civilian men	20.3	21.3	23.3	25.2	26.3	----

Table ANTH-C42: Buttock-Knee Length of USAF Females [1]

Population	Percentiles (in.)				
	1st	5th	50th	95th	99th
Pilots	20.4	21.1	22.6	24.2	25.0
Flight nurses	20.2	20.9	22.4	24.0	24.8

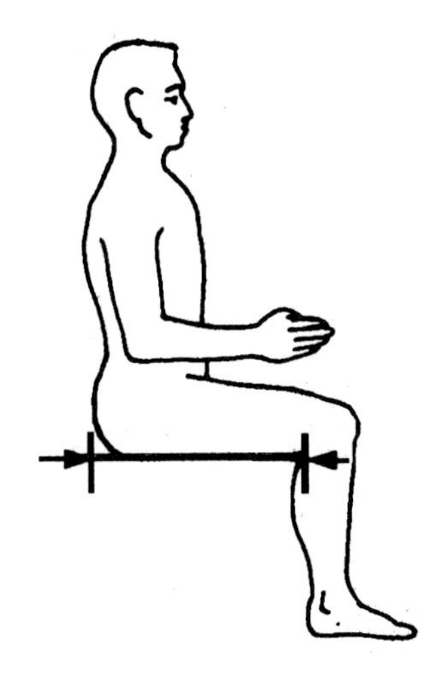

Figure ANTH-C25: Buttock-Popliteal Length

Table ANTH-C43: Buttock-Popliteal Length of US Civilians [1]

Sex	Percentiles (in.) 1st	5th	50th	95th	99th
Male	16.6	17.4	18.9	20.8	21.5
Female	16.0	16.8	18.2	20.0	20.6
Male	16.5	17.3	19.5	21.6	22.7
Female	16.1	17.0	18.9	21.0	22.0

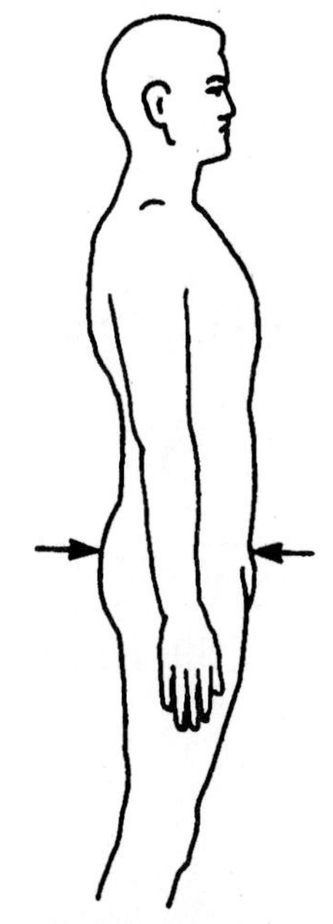

Figure ANTH-C26: Buttock Depth

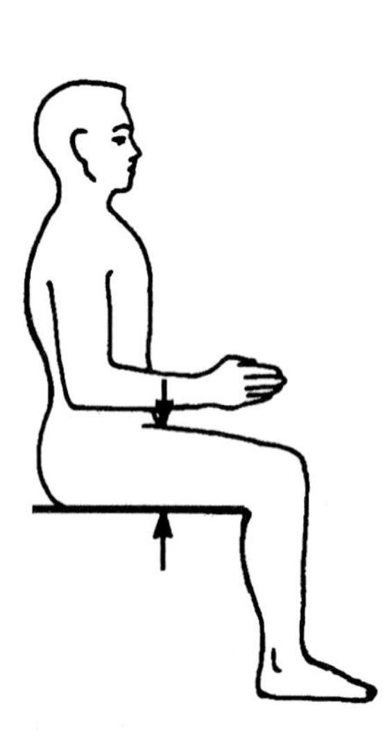

Figure ANTH-C27:
Thigh Clearance Height, Sitting

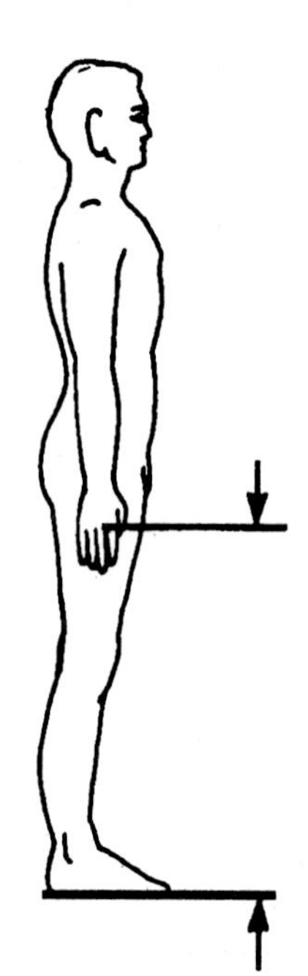

Figure ANTH-C28:
Knuckle Height

Table ANTH-C44: Buttock-Depth, Thigh Clearance Height, Sitting, and Knuckle Height of USAF Males [1]

Dimension	Percentiles (in.)					
	1st	5th	50th	95th	99th	S.D.
Buttock depth	7.2	7.6	8.8	10.2	10.9	0.82
Thigh clearance height, sitting	4.5	4.8	5.6	6.5	6.8	.52
Knuckle height	26.7	27.7	30.0	32.4	33.5	1.45

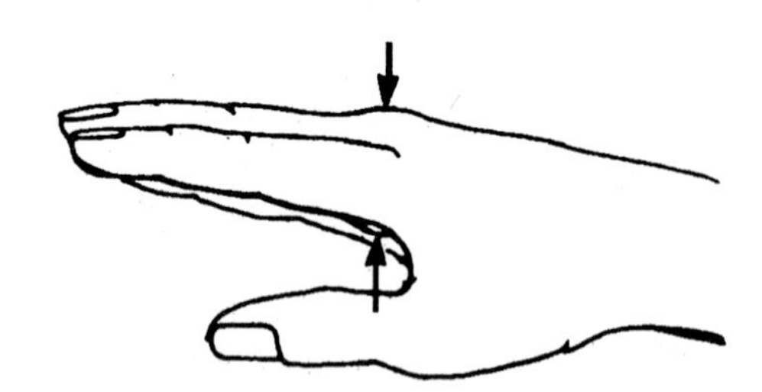

Figure ANTH-C29: Hand Thickness

Table ANTH-C45: Hand Thickness of USAF and Navy Males and Females[1]

Population	Percentiles (in.) 1st	5th	50th	95th	99th	S.D.
Male flight personnel	1.0	1.1	1.2	1.3	1.4	0.07
Male basic trainees	1.0	1.1	1.2	1.4	1.4	.09
Female basic trainees	0.8	0.8	1.0	1.1	1.2	.09
Naval aviators	1.0	1.1	1.2	1.4	1.5	.08

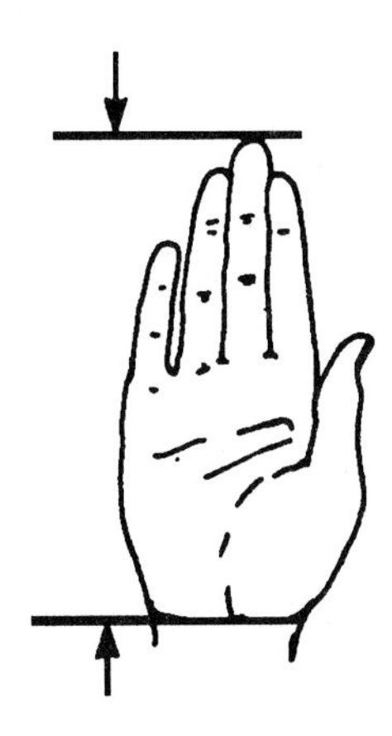

Figure ANTH-C30: Hand Length

Table ANTH-C46: Hand Length of US Military and Civilian Males [1]

Population	Percentiles (in.) 1st	5th	50th	95th	99th	S.D.
Air Force personnel	6.7	6.9	7.5	8.0	8.3	0.34
Cadets	6.8	7.1	7.6	8.2	8.4	----
Gunners	6.6	6.9	7.5	8.1	8.4	----
Basic trainees	6.7	6.9	7.5	8.2	8.5	.38
Army personnel:						
Aviators	6.7	6.9	7.5	8.1	8.3	.34
Separatees, white	6.7	7.0	7.6	8.2	8.5	.36
Separatees, Negro	7.0	7.3	8.0	8.7	9.0	.42
Truck and bus drivers	6.9	7.1	7.6	8.1	8.3	----
Naval aviators	6.8	7.0	7.5	8.1	8.3	.34

Table ANTH-C47: Hand Length of US Military and Civilian Females [1]

Population	Percentiles (in.) 1st	5th	50th	95th	99th	S.D.
Air Force personnel:						
Pilots	6.2	6.4	6.9	7.5	7.7	----
Flight nurses	6.3	6.5	6.9	7.4	7.6	----
Basic trainees	6.0	6.2	6.8	7.3	7.6	0.34
Army personnel	6.1	6.4	6.9	7.4	7.7	.33
College students	6.0	6.2	6.7	7.2	7.4	.31

Table ANTH-C47: Hand Length of US Military and Civilian Females [1]

Population	Percentiles (in.)					S.D.
	1st	5th	50th	95th	99th	
Air Force personnel:						
Pilots	6.2	6.4	6.9	7.5	7.7	----
Flight nurses	6.3	6.5	6.9	7.4	7.6	----
Basic trainees	6.0	6.2	6.8	7.3	7.6	0.34
Army personnel	6.1	6.4	6.9	7.4	7.7	.33
College students	6.0	6.2	6.7	7.2	7.4	.31

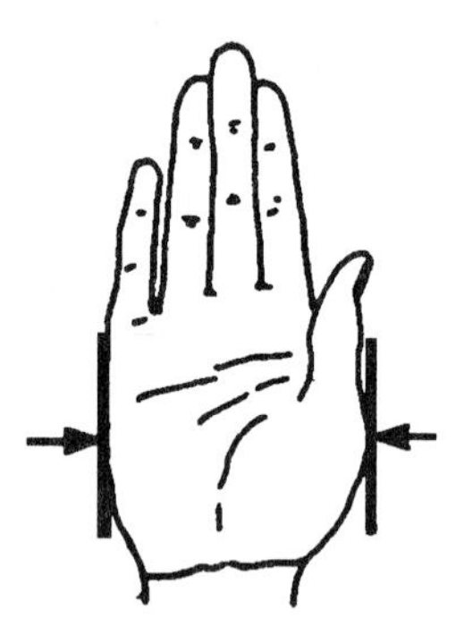

Figure ANTH-C31: Hand Breadth at Thumb

Table ANTH-C48: Hand Breadth at Thumb of US Air Force and Navy Males and Females [1]

Population	Percentiles (in.)					S.D.
	1st	5th	50th	95th	99th	
Male flight personnel	3.6	3.7	4.1	4.4	4.6	0.21
Male basic trainees	3.5	3.7	4.1	4.5	4.7	.25
Female basic trainees	3.1	3.2	3.6	4.0	4.1	.23
Naval aviators	3.7	3.9	4.2	4.5	4.6	.19

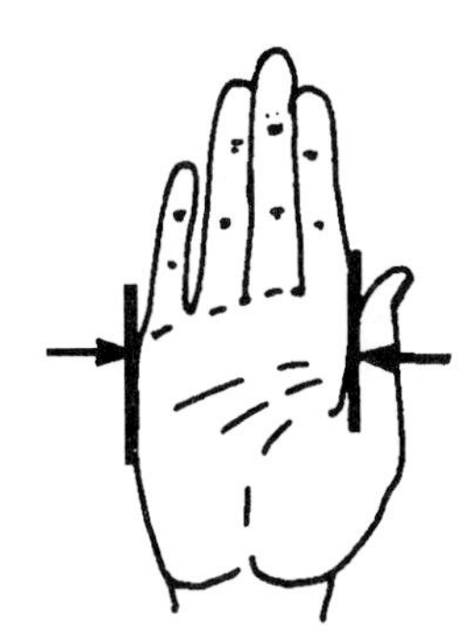

Figure ANTH-C32: Hand Breadth at Metacarpal

Table ANTH-C49: Hand Breadth at Metacarpal of US Military and Civilian Males [1]

Population	Percentiles (in.)					S.D.
	1st	5th	50th	95th	99th	
Air Force personnel	3.1	3.2	3.5	3.7	3.9	0.16
Cadets	3.0	3.1	3.4	3.7	3.8	----
Gunners	3.0	3.1	3.4	3.6	3.7	----
Basic trainees	3.0	3.2	3.5	3.7	3.9	.18
Army personnel:						
Separatees, white	2.9	3.1	3.4	3.8	3.9	.19
Separatees, Negro	3.0	3.2	3.5	3.8	4.0	.20
Truck and bus drivers	3.1	3.2	3.5	3.8	4.0	----
Army aviators	3.1	3.2	3.5	3.8	3.9	.16
Naval aviators	3.2	3.3	3.5	3.8	3.9	.17

Table ANTH-C50: Hand Breadth at Metacarpal of US Military Females [1]

Population	Percentiles (in.)					S.D.
	1st	5th	50th	95th	99th	
Air Force personnel:						
Pilots	2.8	2.8	3.0	3.3	3.4	----
Flight nurses	2.7	2.8	3.0	3.2	3.3	----
Basic trainees	2.6	2.7	3.0	3.4	3.6	0.19
Army personnel	2.6	2.7	3.0	3.4	3.6	.20

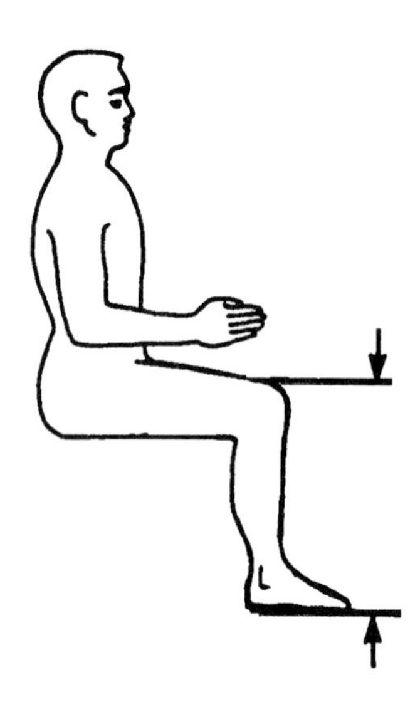

Figure ANTH-C33: Seated Knee Height

Table ANTH-C51: Seated Knee Height of US Military and Civilian Males [1]

Population	Percentiles (in.)					S.D.
	1st	5th	50th	95th	99th	
Air Force personnel	19.5	20.1	21.7	23.3	24.0	0.99
Cadets	19.7	20.4	22.0	23.6	24.3	----
Gunners	19.2	19.8	21.5	23.0	2.37	----
Army personnel:						
Separatees, white	19.0	19.8	21.6	23.5	24.3	1.09
Separatees, Negro	19.6	20.3	22.2	24.0	24.7	1.14
Naval aviators	19.7	20.3	21.8	23.5	24.2	.98
Truck and bus drivers	19.3	20.1	21.7	23.5	24.2	----
Civilian men	18.3	19.3	21.4	23.4	24.1	----

Table ANTH-C52: Seated Knee Height of US Military and Civilian Females [1]

Population	Percentiles (in.)					S.D.
	1st	5th	50th	95th	99th	
Air Force personnel						
Pilots	18.3	18.7	20.1	21.5	22.2	----
Flight nurses	17.7	18.1	19.5	20.8	21.5	----
Army personnel	16.6	17.2	18.8	20.3	21.1	0.95
Civilian women	17.1	17.9	19.6	21.5	22.4	----

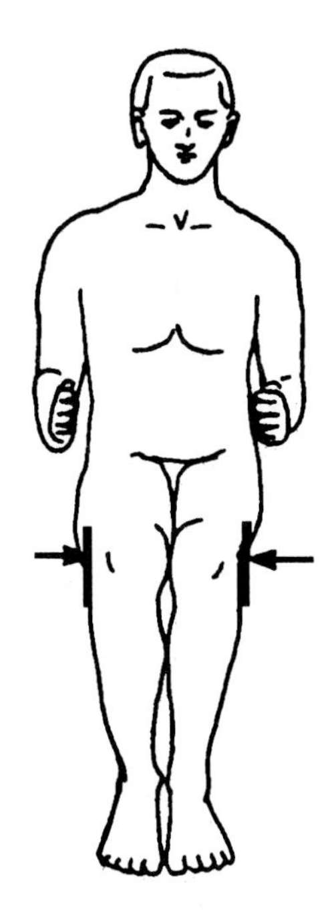

Figure ANTH-C34: Knee-to-Knee Breadth

Table ANTH-C53: Knee-to-Knee Breadth of US Military and Civilian Males [1]

Population	Percentiles (in.)					S.D.
	1st	5th	50th	95th	99th	
Air Force personnel	7.0	7.2	7.9	8.8	9.4	0.52
Cadets	6.8	7.1	7.7	8.4	8.7	----
Gunners	6.7	6.9	7.6	8.2	8.5	----
Truck and bus drivers	6.8	7.3	8.1	9.2	9.5	----

Table ANTH-C54: Knee-to-Knee Breadth of USAF Females [1]

Population	Percentiles (in.)				
	1st	5th	50th	95th	99th
Pilots	6.5	6.7	7.6	8.6	9.6
Flight nurses	6.6	6.8	7.5	8.4	9.6

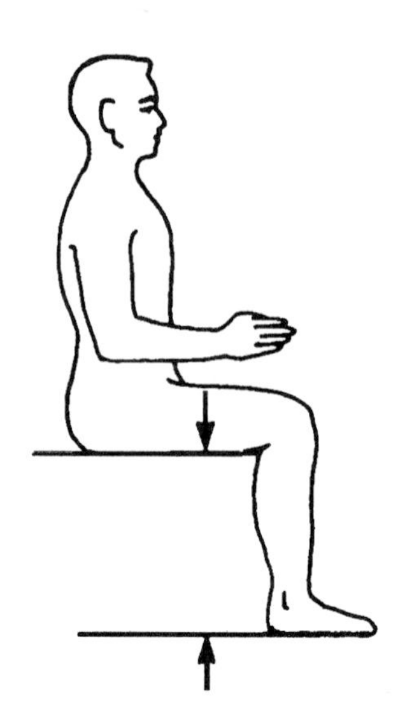

Figure ANTH-C35: Seated Popliteal Height

Table ANTH-C55: Seated Popliteal Height of US Military and Civilian Males and Females[1]

Population	Percentiles (in.)					S.D.
	1st	5th	50th	95th	99th	
Male Air Force personnel	15.3	15.7	17.0	18.2	18.8	0.77
Male railroad travelers	*16.9	*17.6	*19.0	*20.6	*21.1	----
Female railroad travelers	*16.2	*16.7	*18.1	*19.5	*20.1	----
Naval aviators	15.4	15.9	17.3	18.8	19.3	.86

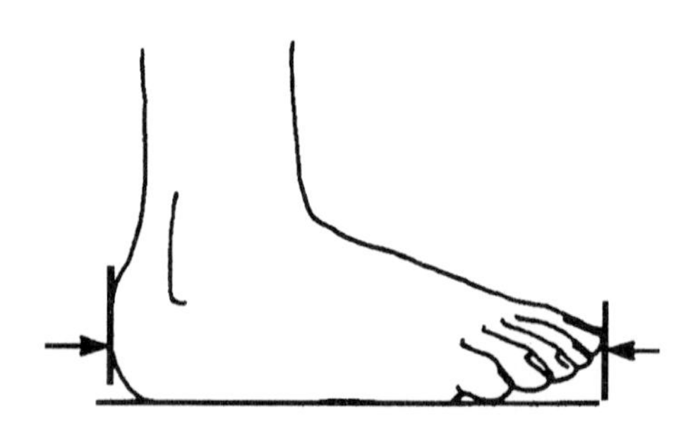

Figure ANTH-C36: Foot Length

Table ANTH-C56: Foot Length of US Military and Civilian Males [1]

Population	Percentiles (in.)					S.D.
	1st	5th	50th	95th	99th	
Air Force personnel	9.5	9.8	10.5	11.3	11.6	0.45
Cadets	9.5	9.8	10.5	11.3	11.6	----
Gunners	9.3	9.6	10.4	11.1	11.4	----
Basic trainees	9.2	9.5	10.3	11.2	11.5	.50
Army personnel	9.3	9.6	10.4	11.1	11.5	.47
Separatees, white	9.3	9.7	10.4	11.2	11.5	.48
Separatees, Negro	9.6	9.9	10.8	11.6	12.0	.50
Aviators	9.5	9.9	10.6	11.5	11.9	.49
Truck and bus drivers	9.2	9.6	10.4	11.3	11.6	----
College students	9.2	9.4	10.3	11.1	11.4	.48
Naval aviators	9.5	9.7	10.5	11.3	11.6	.47

Table ANTH-C57: Foot Length of US Military and Civilian Females [1]

Population	Percentiles (in.)					S.D.
	1st	5th	50th	95th	99th	
Air Force personnel:						
Pilots	8.6	8.9	9.6	10.2	10.5	----
Flight nurses	8.7	8.9	9.6	10.3	10.5	----
Basic trainees	8.4	8.7	9.4	10.2	10.5	0.46
Army personnel	8.4	8.7	9.4	10.2	10.5	.44
College students	8.3	8.7	9.5	10.3	10.7	.45

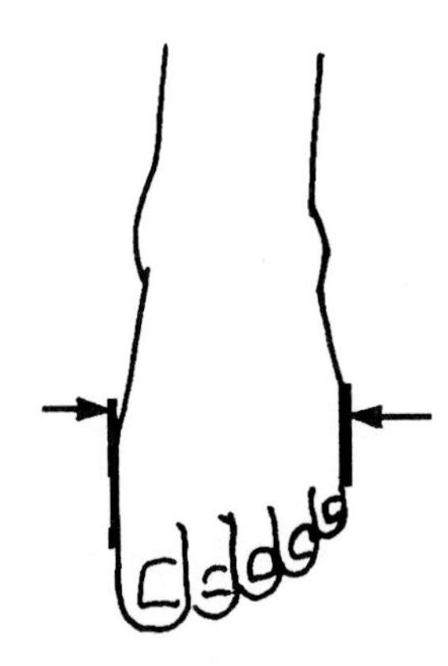

Figure ANTH-C36: Foot Breadth

Table ANTH-C58: Foot Breadth of US Military and Civilian Males [1]

Population	Percentiles (in.)					S.D.
	1st	5th	50th	95th	99th	
Air Force personnel	3.4	3.5	3.8	4.1	4.4	0.19
Cadets	3.5	3.6	3.9	4.2	4.3	----
Gunners	3.3	3.5	3.8	4.2	4.3	----
Basic trainees	3.5	3.6	4.0	4.4	4.7	.25
Army personnel	3.4	3.5	3.9	4.2	4.3	.20
Separatees, white	3.3	3.5	3.9	4.3	4.4	.25
Separatees, Negro	3.4	3.6	4.0	4.4	4.6	.25
Aviators	3.5	3.6	4.0	4.4	4.5	.21
Truck and bus drivers	3.6	3.7	4.0	4.3	4.4	----
Naval aviators	3.5	3.6	4.0	4.6	4.9	.30

Table ANTH-C59: Foot Beadth of US Military Females [1]

Population	Percentiles (in.)					S.D.
	1st	5th	50th	95th	99th	
Air Force personnel:						
Pilots	3.2	3.3	3.6	3.9	4.1	----
Flight nurses	3.2	3.3	3.6	3.9	4.1	----
Basic trainees	3.1	3.2	3.6	3.9	4.0	0.20
Army personnel	3.1	3.2	3.6	4.0	4.1	.22

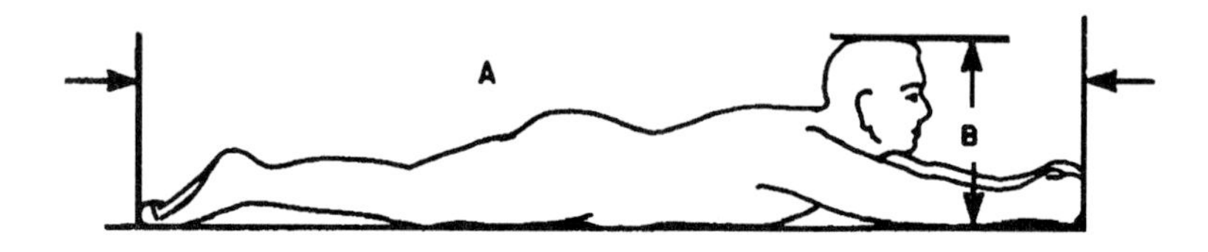

Figure ANTH-C36: Prone Length (A) and Prone Height (B)

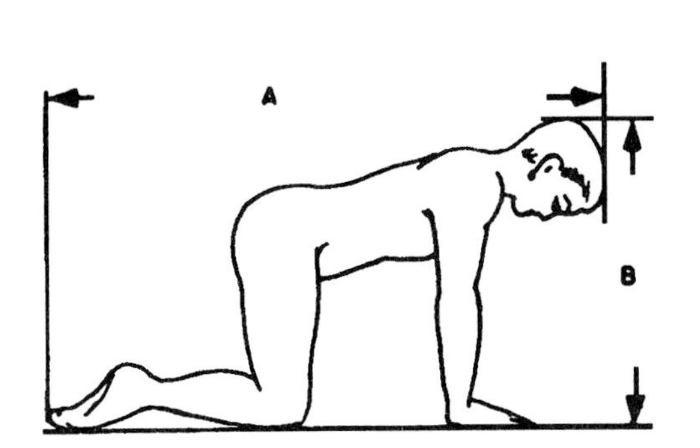

Figure ANTH-C37:
Crawling Length (A) and
Crawling Height (B)

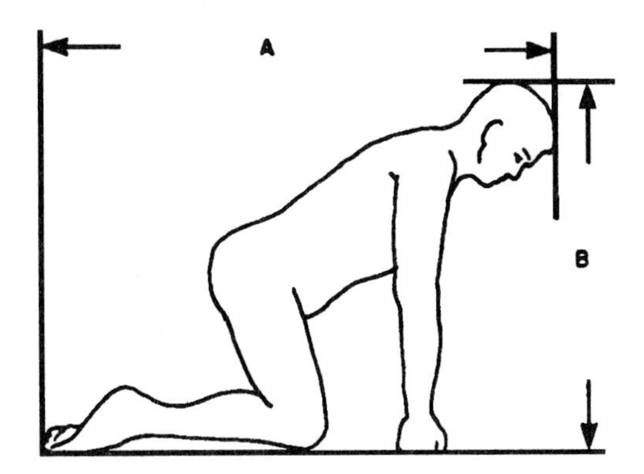

Figure ANTH-C38:
Kneeling Length (A) and Kneeling
Height, Crouched (B)

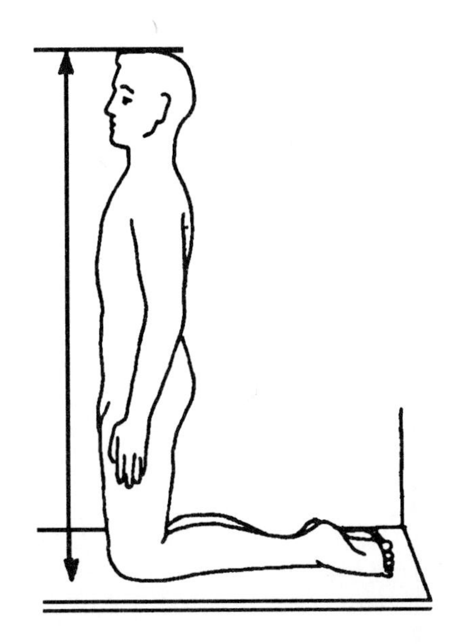

Figure ANTH-C39: Kneeling Height, Upright

Table ANTH-C60: Working Position Dimensions of USAF Males [1]

Dimensions	Percentiles (in.)			S.D.
	5th	50th	95th	
Kneeling:				
Height, crouched	29.7	32.0	34.5	1.57
Length	37.6	43.0	48.1	3.26
Crawling:				
Height	26.2	28.4	30.5	1.30
Length	49.3	53.2	58.2	2.61
Prone:				
Height	12.3	14.5	16.4	1.28
Length	84.7	90.1	95.8	3.41

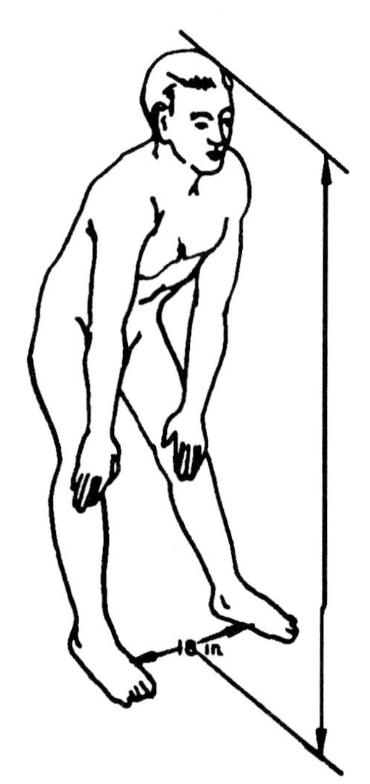

Figure ANTH-C40: Bent Torso Height

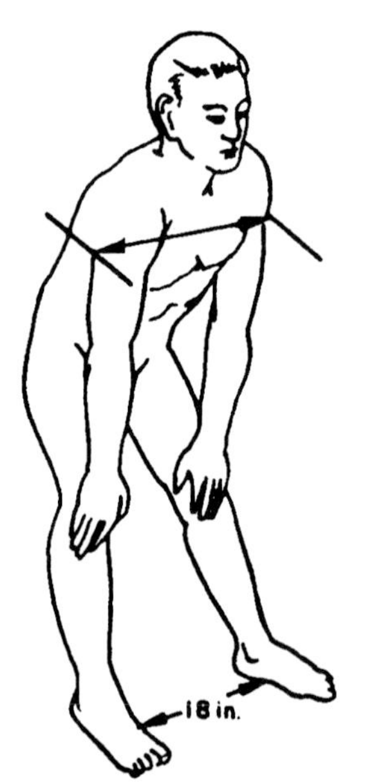

Figure ANTH-C41: Bent Torso Breadth

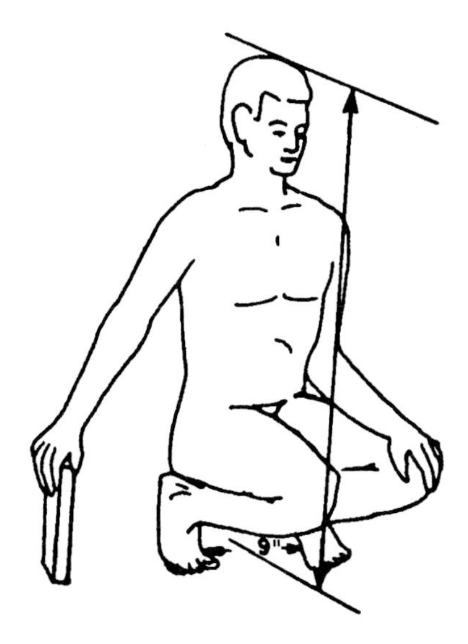

Figure ANTH-C42:
Squatting Height

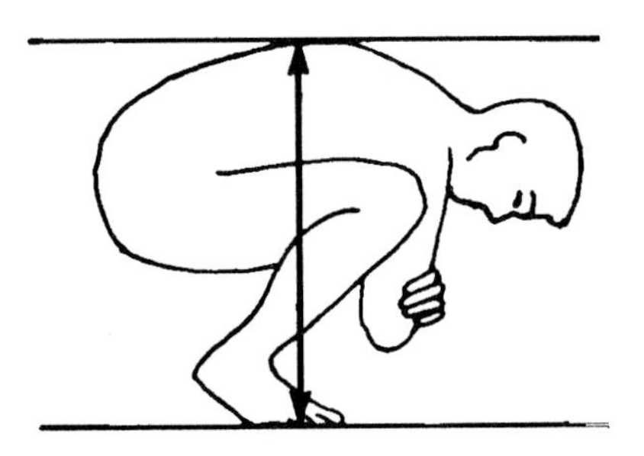

Figure ANTH-C43:
Minimum Squatting Height

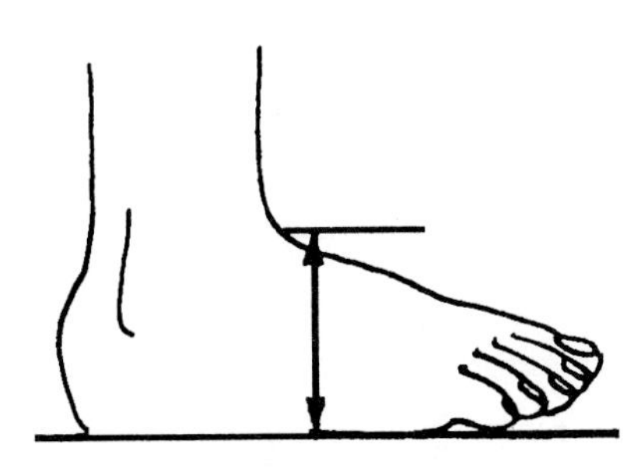

Figure ANTH-C44:
Functional Foot Height

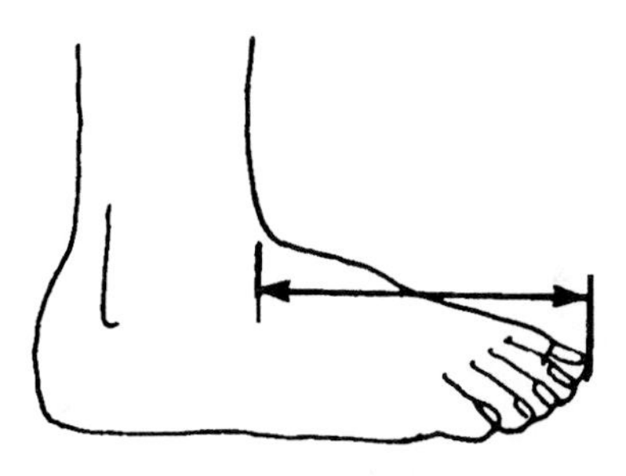

Figure ANTH-C45:
Functional Foot Length

Table ANTH-C61: Working Position Dimensions [1]

Dimension	Percentiles (in.)			S.D.
	5th	50th	95th	
Bent torso height	46.3	52.0	55.9	2.76
Bent torso breadth	16.3	17.5	19.1	.88
Kneeling height, upright	48.2	51.0	54.4	1.77
Squatting height	40.8	43.6	47.0	1.44
Functional foot length	5.48	6.11	6.72	.37
Functional foot height	2.72	2.97	3.33	.21
Vertical reach height, sitting	51.6	55.0	59.0	2.24
Minimum squatting height	21.5	24.8	28.0	2.04

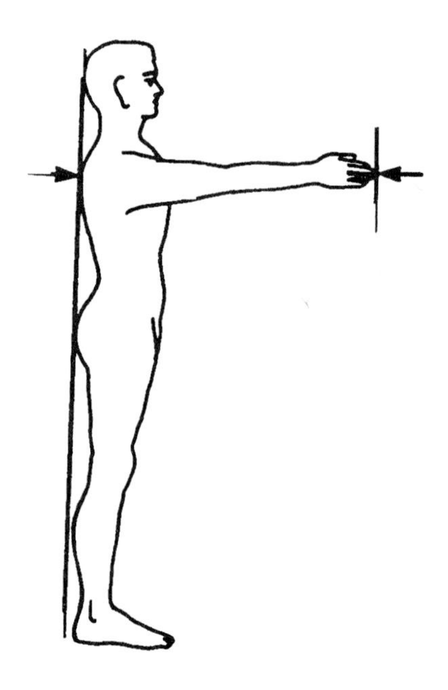

Figure ANTH-C46: Arm Reach

Table ANTH-C62: Arm Reach of US Military and Civilian Males [1]

Population	Percentiles (in.) 1st	5th	50th	95th	99th	S.D.
Air Force personnel	30.9	31.9	34.6	37.3	38.6	1.70
Cadets	31.6	32.7	35.2	37.8	38.8	----
Gunners	30.9	31.9	34.8	37.4	38.6	----
Navy personnel	30.0	31.1	33.7	36.3	37.4	1.57
Enlisted men	31.6	32.7	35.7	38.2	39.5	1.70
Cadets, aviation	31.7	32.8	35.4	38.1	39.2	----
Truck and bus drivers	31.9	32.9	35.7	38.4	39.5	----
Army aviators	32.3	33.5	36.0	38.5	39.6	1.47

Table ANTH-C63: Arm Reach of USAF Females [1]

Population	Percentiles (in.) 1st	5th	50th	95th	99th
Pilots	29.2	29.7	31.8	34.1	34.9
Flight nurses	27.9	28.7	31.0	33.5	34.4

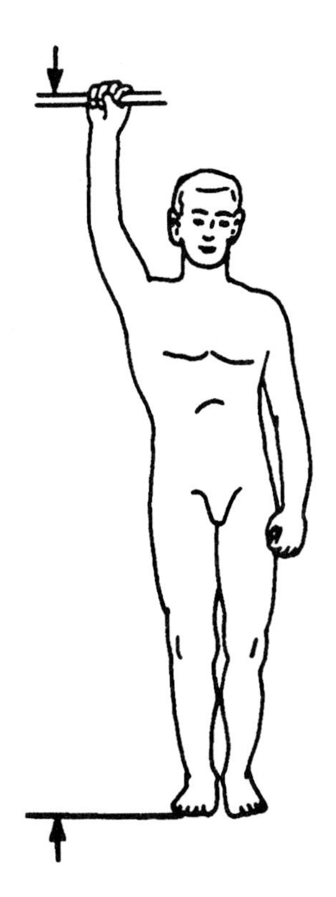

Figure ANTH-C47: Overhead Reach

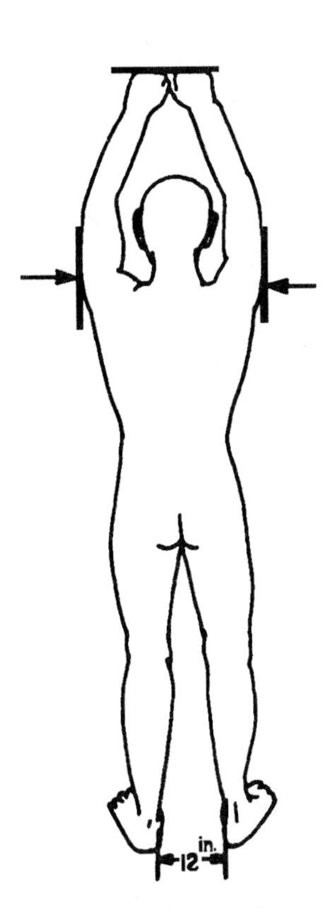

Figure ANTH-C48: Overhead Reach Breadth

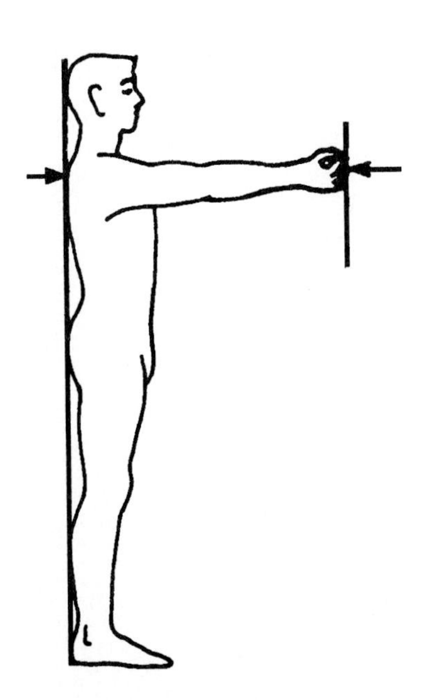

Figure ANTH-C49:Thumb-tip Reach

Figure ANTH-C50: Seated Vertical Reach Height

Table ANTH-C64: Various Arm Reaches, and Reach Breadth of US Military Males [1]

Dimension	Percentiles (in.)					
	1st	5th	50th	95th	99th	S.D.
Thumb-tip reach:*						
USAF	28.8	29.7	32.3	35.0	36.4	1.60
Naval aviators	28.6	29.3	31.4	34.0	35.1	1.42
Overhead reach (1-arm)	----	76.8	82.5	88.5	----	3.35
Overhead reach breadth (2-arm)	----	13.6	14.9	15.9	----	.68

*Formerly called "functional reach"

Note[1]: The tables on the pages that follow make reference to SRV and SRP. SRP is the Seated Reference Point and refers to the commonly used starting point for determining the reach dimensions of seated operators.

The tables that follow do not represent measured reach dimensions, but rather measurements that have been plotted and converted into horizontal contours that center about an imaginary vertical line (SRV) erected from the midpoint of the SRP. The data which follows was computed based upon USAF and cannot be generalized to all populations, thus care must be used when refering to this data. In addition, the data is specific to the angles listed. Thus the values will change if the angles change or if heavy clothing is worn.

Table ANTH-C65: Shirt-Sleeved Grasping Reach, Horizontal Boundaries, SRP Level[1]

Angle (deg)	N	Min	Percentiles (in.) 5th	50th	95th
L165					
L150					
L135					
L120					
L105					
L 90					
L 75					
L 60					
L 45					
L 30					
L 15					
0					
R 15					
R 30	19		17.50	20.75	25.00
R 45	20	16.25	19.50	21.75	26.00
R 60	20	17.50	20.50	22.25	26.25
R 75	20	17.25	20.00	22.25	26.00
R 90	20	17.00	19.50	22.25	25.50
R105	20	16.25	18.75	22.00	25.25
R120	20	15.00	18.25	20.75	24.50
R135	20	13.00	16.50	19.00	23.50
R150	19		14.00	16.50	20.25
R165	13			13.00	17.00
180					

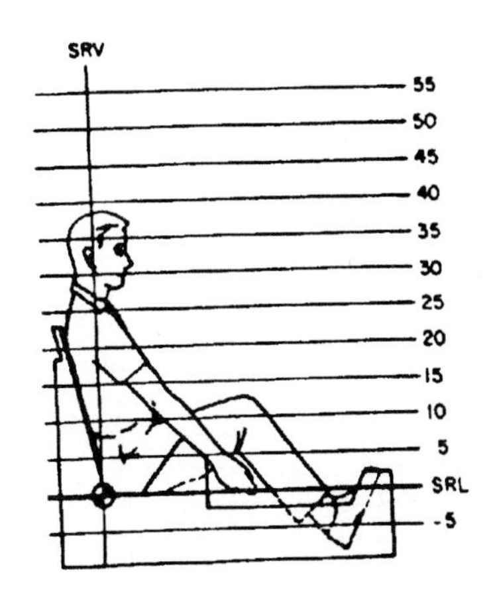

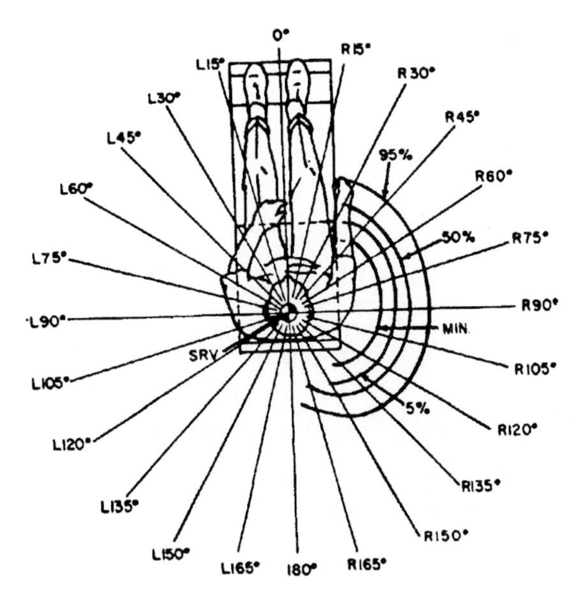

Angular Reach from SRV at the SRP Level

Minimum	R37°* to R140°
5th %ile	R30°* to R151°
50th %ile	R26°* to R166°
95th %ile	R21°* to R175°

* Right side of Leg Support

Horizontal Boundaries, SRP Level

Table ANTH-C66: Shirt-Sleeved Grasping Reach, Horizontal Boundaries, 10-IN Level[1]

Angle (deg)	N	Min	Percentiles (in.) 5th	50th	95th
L165					
L150					
L135					
L120					
L105					
L 90	4				13.50
L 75	4				17.25
L 60	14			16.50	21.00
L 45	15			19.50	23.25
L 30	15			21.00	24.75
L 15	10			22.00	26.25
0					
R 15	20				
R 30	20	26.25	27.00	29.25	33.00
R 45	20	27.25	28.25	30.50	33.75
R 60	20	28.00	29.00	30.75	33.50
R 75	20	28.25	29.25	30.75	33.50
R 90	20	28.25	29.25	31.00	33.50
R105	20	27.75	28.75	30.50	32.75
R120	20	26.75	27.75	29.75	31.50
R135	19		26.25	28.25	30.75
R150	14			25.25	28.75
R165	1				
180					

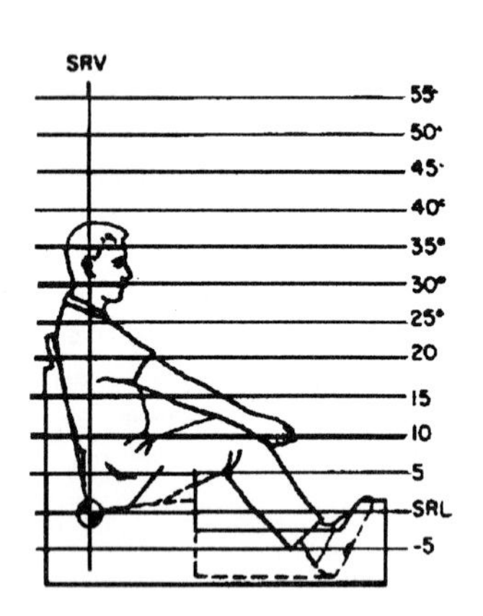

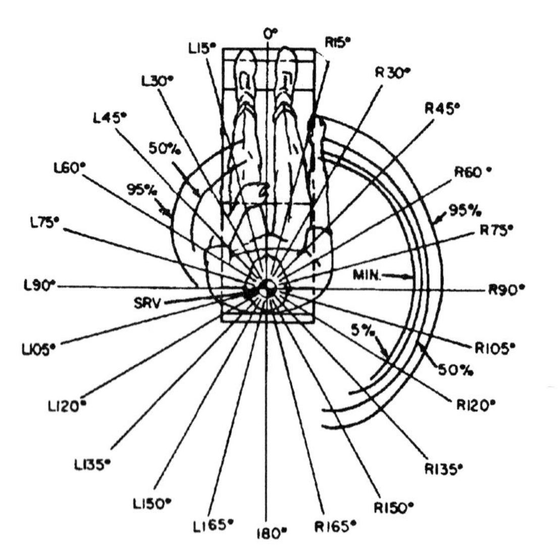

Angular Reach from SRV at the 10-inch Level

Minimum	----	R21°** to R130°
5th %ile	----	R20°** to R141°
50th %ile	L65° to L 7°* and	R18°** to R155°
95th %ile	L90° to L11°* and	R16°** to R157°

* Left side of knees
** Right side of knees

Shirt-Sleeved Grasping Reach: Horizontal Boundaries, 10-IN Level

Table ANTH-C67: Shirt-Sleeved Grasping Reach, Horizontal Boundaries, 20-IN Level[1]

Angle (deg)	N	Min	Percentiles (in.) 5th	50th	95th
L165					
L150					
L135					
L120					
L105					
L 90	11			14.00	18.75
L 75	16			18.00	21.50
L 60	20	17.00	17.50	20.50	24.50
L 45	20	18.25	19.50	22.75	26.75
L 30	20	20.25	21.50	24.75	28.25
L 15	20	22.50	23.50	26.75	29.75
0	20	25.00	25.50	28.75	31.75
R 15	20	27.25	28.00	30.50	34.00
R 30	20	29.00	30.00	32.00	35.75
R 45	20	30.50	31.00	33.50	36.25
R 60	20	31.50	32.00	33.75	36.25
R 75	20	31.50	32.25	34.00	36.50
R 90	20	31.75	32.25	34.00	36.00
R105	20	31.50	31.75	33.50	35.75
R120	19		30.50	33.00	35.50
R135	9				34.50
R150					
R165					
180					

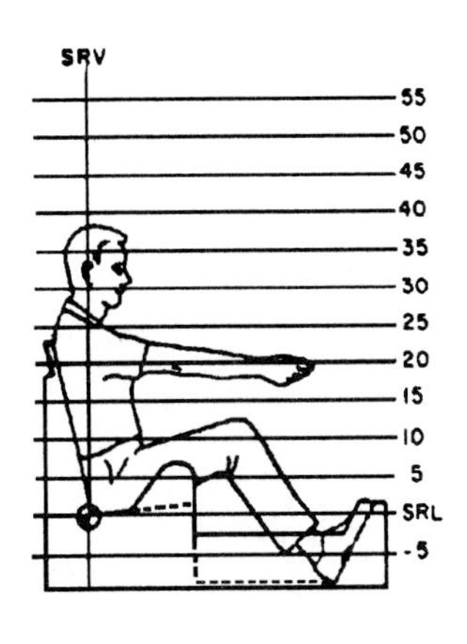

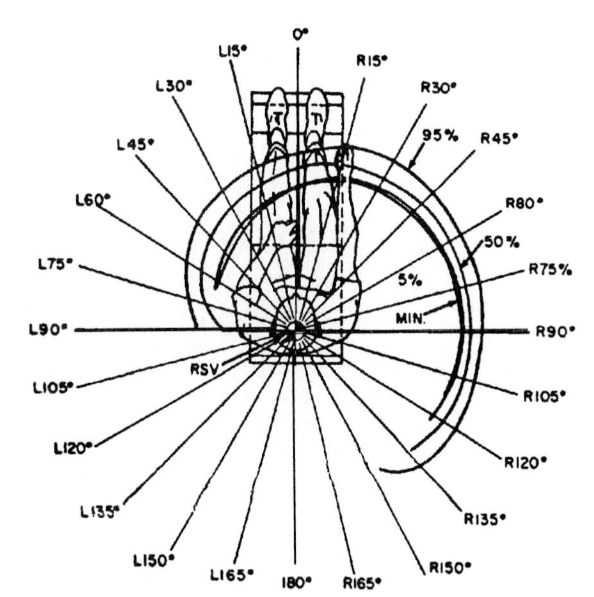

Angular Reach from SRV at the 20-inch Level

Minimum	L65° to R110°
5th %ile	L66° to R122°
50th %ile	L90° to R134°
95th %ile	L90° to R146°

Shirt-Sleeved Grasping Reach: Horizontal Boundaries, 20-IN Level

Table ANTH-C68: Shirt-Sleeved Grasping Reach, Horizontal Boundaries, 30-IN Level[1]

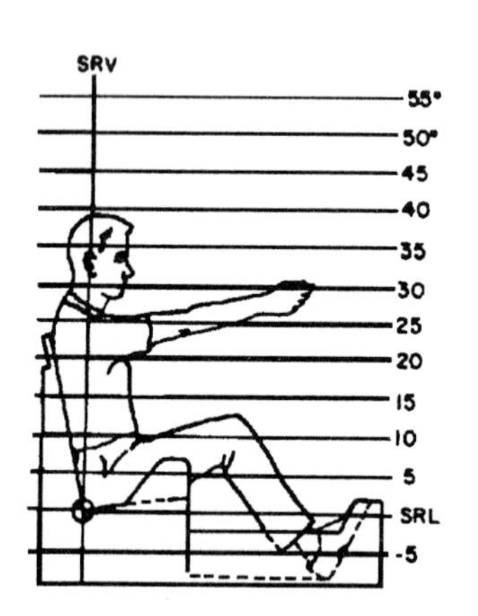

Angle (deg)	N	Min	Percentiles (in.) 5th	50th	95th
L165	4				18.75
L150	4				19.25
L135	6				20.00
L120	7				18.75
L105	9				19.00
L 90	16			16.75	20.75
L 75	18			18.75	22.50
L 60	20	17.00	17.25	20.75	24.50
L 45	20	18.25	19.00	22.50	26.50
L 30	20	19.75	21.50	24.50	28.25
L 15	20	22.00	23.75	26.75	29.50
0	20	23.75	25.50	28.50	31.00
R 15	20	26.00	27.25	29.75	33.00
R 30	20	27.75	29.00	31.50	34.25
R 45	20	28.75	30.25	32.25	34.75
R 60	20	30.00	31.00	32.75	35.75
R 75	20	30.75	31.25	33.00	35.50
R 90	20	31.00	31.25	33.25	35.75
R105	20	30.75	31.00	33.00	35.25
R120	19		30.25	32.50	34.75
R135	9				34.50
R150	1				
R165	2				19.50
180	2				20.25

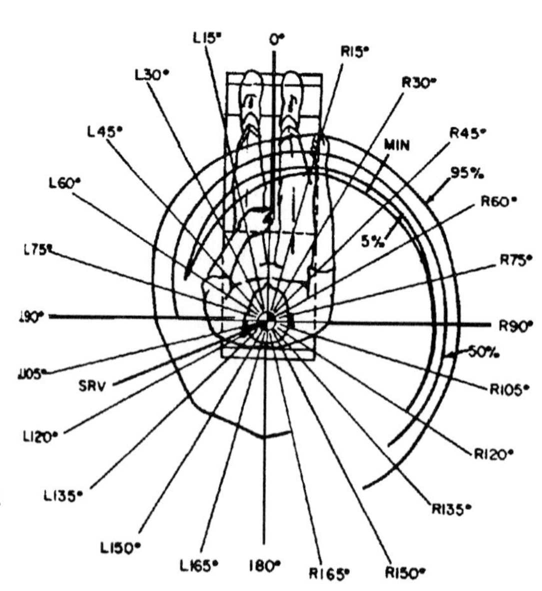

Angular Reach from SRV at the 30-inch Level

Minimum	L 67° to R111°
5th %ile	L 67° to R122°
50th %ile	L 90° to R134°
95th %ile	R165° to R149°

Shirt-Sleeved Grasping Reach: Horizontal Boundaries, 30-IN Level

Table ANTH-C69: Shirt-Sleeved Grasping Reach, Horizontal Boundaries,

40-IN Level[1]

Angle (deg)	N	Min	Percentiles (in.) 5th	50th	95th
L165	14			15.50	21.50
L150	13			14.75	20.00
L135	16			14.00	19.25
L120	19		11.25	13.25	18.50
L105	19		11.75	13.25	18.25
L 90	20	12.00	12.25	12.75	18.25
L 75	20	12.25	12.50	15.00	18.75
L 60	20	12.50	13.25	16.25	20.00
L 45	20	13.00	14.00	17.75	21.50
L 30	20	13.75	15.50	19.50	23.50
L 15	20	15.25	17.00	21.25	24.50
0	20	17.00	19.00	23.00	25.75
R 15	20	18.75	21.00	24.50	28.50
R 30	20	21.00	22.75	22.75	30.50
R 45	20	23.25	24.75	27.75	31.50
R 60	20	24.25	25.50	28.00	31.25
R 75	20	25.00	26.00	28.00	31.50
R 90	20	25.00	26.25	28.25	31.50
R105	20	25.75	26.75	28.50	31.75
R120	19		26.25	28.75	31.50
R135	16			27.00	31.00
R150	8				29.25
R165	10			16.75	23.75
180	10			17.75	23.50

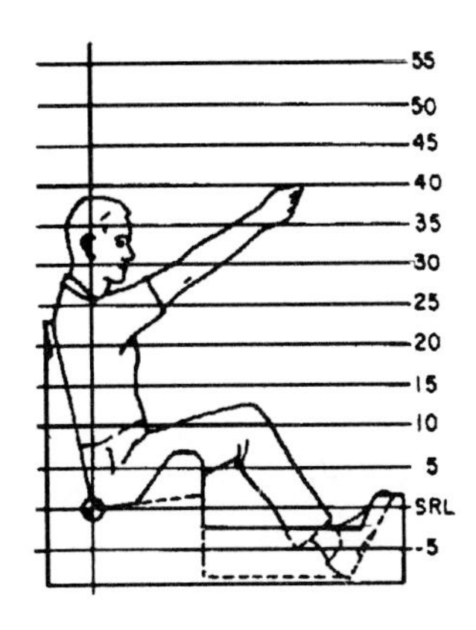

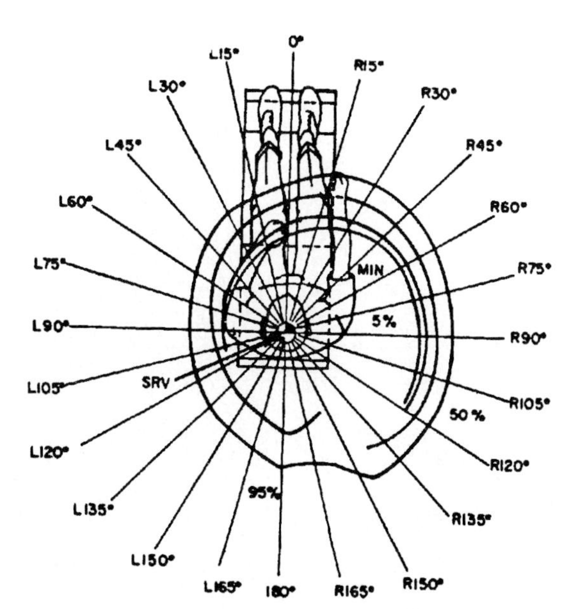

Angular Reach from SRV at the 40-inch Level

Minimum	L 90° to R119°
5th %ile	L120° to R120°
50th %ile	R156° to R143°
95th %ile	360°

Shirt-Sleeved Grasping Reach: Horizontal Boundaries, 40-IN Level

Table ANTH-C70: Shirt-Sleeved Grasping Reach, Horizontal Boundaries,

50-IN Level[1]

Angle (deg)	N	Min	Percentiles (in.) 5th	50th	95th
L165	17			7.50	13.75
L150	17			6.00	13.00
L135	17			5.00	12.00
L120	17			4.50	10.75
L105	17			4.25	9.75
L 90	17			4.25	9.50
L 75	17			4.25	9.75
L 60	17			4.74	10.25
L 45	17			5.25	11.50
L 30	17			6.50	13.25
L 15	17			7.75	15.00
0	17			9.50	17.25
R 15	17			11.75	18.75
R 30	17			14.00	20.00
R 45	17			15.75	21.25
R 60	17			16.75	21.75
R 75	17			16.75	21.75
R 90	17			17.25	22.25
R105	17			17.50	22.25
R120	17			17.50	22.00
R135	17			16.50	20.75
R150	17			14.25	19.00
R165	17			11.50	17.25
180	17			8.75	15.50

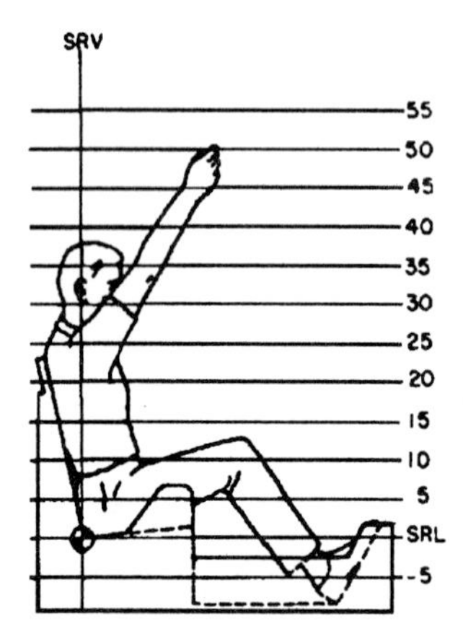

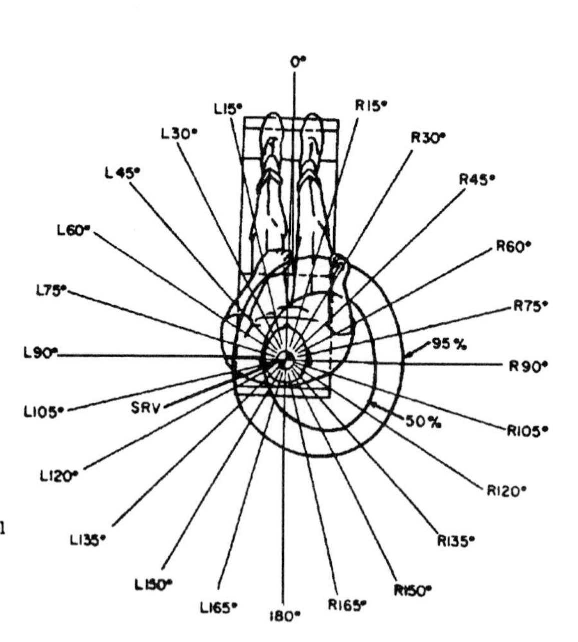

Angular Reach from SRV at the 50-inch Level

Minimum	------------
5th %ile	------------
50th %ile	360°
95th %ile	360°

Shirt-Sleeved Grasping Reach: Horizontal Boundaries, 50-IN Level

Table ANTH-C71: Standing Forward Reach, Both Arms[1]

	Percentiles (in.)				
	5th	25th	50th	75th	95th
A. Depth of reach Range: 17.50 to 25.25 SD: 1.50	19.25	21.00	22.25	22.75	24.50
B. Breadth of aperture Range: 15.00 to 20.25 Mean: 17.69 SD: 1.19	15.50	17.00	17.75	18.50	19.50
C. Floor to top of aperture Range: 58.75 to 70.50 SD: 2.34	61.00	63.50	65.25	66.50	69.00
D. Floor to bottom of aperture Range: 51.25 to 61.75 Mean: 56.09 SD: 2.05	52.25	54.75	56.00	57.25	59.00
E. Vertical dimension of aperture	(1)	(1)	(1)	(1)	(1)

The ranges and all percentiles have been rounded off to the nearest 0.25 in

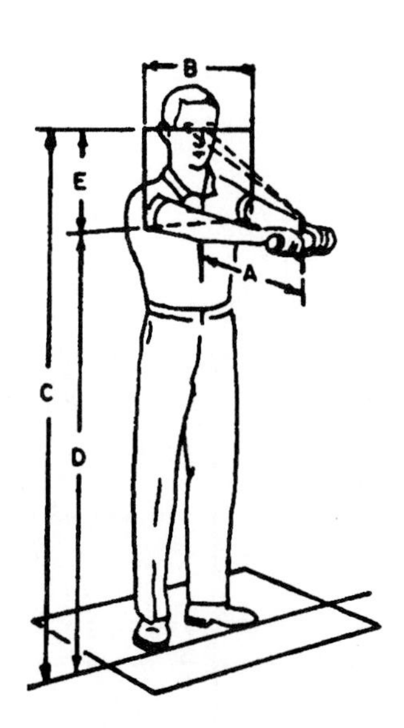

Figure ANTH-C51: Standing Forward Reach, Both Arms

Table ANTH-C72: Standing Forward Reach, Preferred Arm[1]

	Percentiles (in.)[1]				
	5th	25th	50th	75th	95th
A. Depth of reach Range: 19.50 to 27.50 Mean: 23.61 SD: 1.82	20.25	22.25	23.75	25.00	26.75
B. Breadth of aperture 12.00					
C. Floor to top of aperture Range: 58.25 to 70.50 Mean: 64.88 SD: 2.36	61.00	63.25	65.00	66.25	69.00
D. Floor to bottom of aperture Range: 51.25 Mean: 56.09 SD: 2.05	52.25	54.75	56.00	57.25	59.00
E. Vertical dimension of aperture	(1)	(1)	(1)	(1)	(1)

The ranges and all percentiles have been rounded off to the nearest 0.25 in

Note[1]: The measuring technique was as follows: The fixed blade of the beam caliper was placed at the outside of the subject's preferred arm. The subject then closed the corresponding eye and sighted the target grip cylinder with his other eye, while holding his head as straight as possible. The sliding blade of the beam caliper was then moved until it began to cut off the subject's view of the free end of the target grip. The measurement was then taken. A breadth of 12 in will accomodate 95% of the Air Force population.

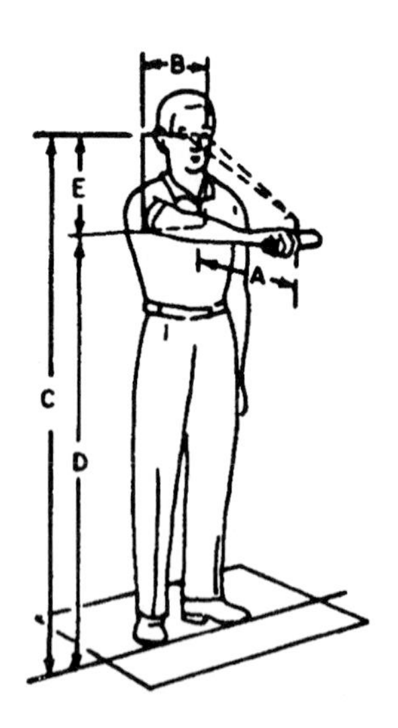

Figure ANTH-C52: Standing Forward Reach, Preferred Arm

Table ANTH-C73: Standing Lateral Reach, Preferred Arm[1]

	Percentiles (in.)				
	5th	25th	50th	75th	95th
A. Depth of reach Range: 21.75 to 28.63 Mean: 24.65 SD: 1.51	22.00	23.50	24.75	25.75	26.75
B. Breadth of aperture 10.00					
C. Floor to top of aperture Range: 58.25 to 70.00 Mean: 64.70 SD: 2.32	60.75	63.25	64.25	66.00	68.75
D. Floor to bottom of aperture Range: 51.25 to 61.75 Mean: 56.09 SD: 2.05	52.25	54.75	56.00	57.25	59.00
E. Vertical dimension of aperture	(1)	(1)	(1)	(1)	(1)

The ranges and all percentiles have been rounded off to the nearest 0.25 in

Note[1]: The measuring technique for aperture breadth for one handed lateral reach is similar to that used to measure aperture breadth for one handed forward reach. A breadth of 10 in will accomodate 95% of the Air Force population.

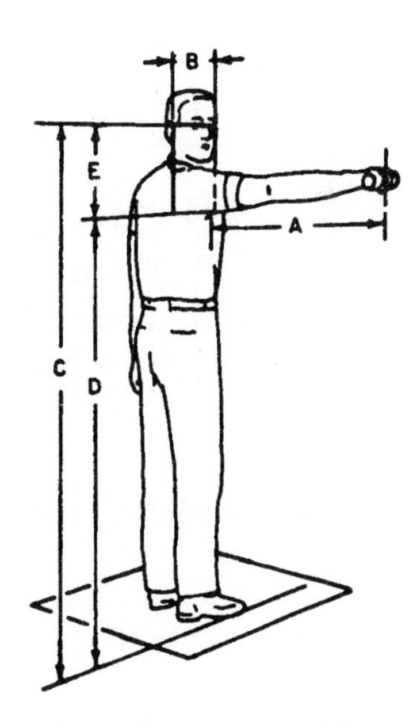

Figure ANTH-C53: Standing Lateral Reach, Preferred Arm

Table ANTH-C74: Seated Forward Reach, Both Arms[1]

	Percentiles (in.)				
	5th	25th	50th	75th	95th
A. Depth of reach Range: 14.00 to 23.50 Mean: 18.26 SD: 2.15	15.00	16.50	17.75	19.50	22.25
B. Breadth of aperture Range: 13.50 to 18.75 Mean: 16.12 SD: 1.25	13.75	15.25	16.00	17.00	18.25
C. Floor to top of aperture † Range: 39.25 to 51.00 Mean: 43.25 SD: 2.05	19.75	41.75	43.00	44.25	46.50
D. Floor to bottom of aperture † Range: 32.50 to 41.75 Mean: 36.59 SD: 1.59	34.25	35.50	36.50	37.50	39.00
E. Vertical dimension of aperture	([1])	([1])	([1])	([1])	([1])

The ranges and all percentiles have been rounded off to the nearest 0.25 in

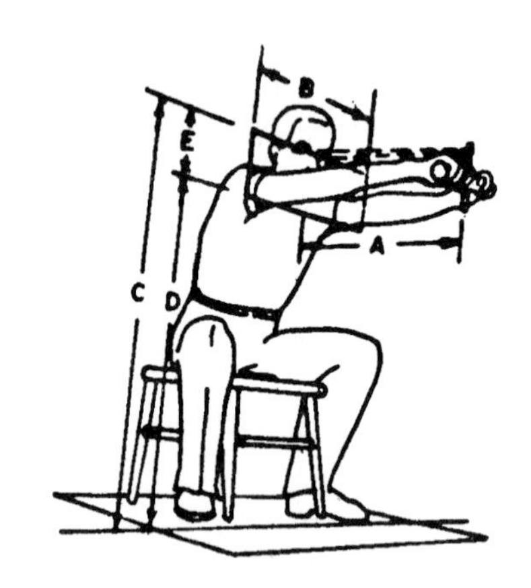

Figure ANTH-C54: Seated Forward Reach, Both Arms

Table ANTH-C75: Cross-Legged Seated, Forward Reach, Both Arms[1]

	Percentiles (in.)				
	5th	25th	50th	75th	95th
A. Depth of reach Range: 13.50 to 22.25 Mean: 17.08 SD: 1.91	13.75	15.75	16.75	18.25	20.00
B. Breadth of aperture Range: 13.50 to 18.50 Mean: 15.89 SD: 1.54	13.75	14.75	16.00	16.75	17.75
C. Floor to top of aperture Range: 22.25 to 30.50 Mean: 25.30 SD: 1.54	22.75	24.25	25.25	26.25	28.00
D. Floor to bottom of aperture Range: 17.00 to 23.25 Mean: 19.23 SD: 1.19	17.00	18.50	19.25	20.00	21.25
E. Vertical dimension of aperture	(1)	(1)	(1)	(1)	(1)

The ranges and all percentiles have been rounded off to the nearest 0.25 in

Figure ANTH-C55: Cross-Legged Seated, Forward Reach, Both Arms

Table ANTH-C76: Recommended Aperture Sizes and Depth of Reach For Shirt-Sleeved Technicians[1]

	Standing positions			Seated positions	
	Forward reach, both arms	Forward reach, preferred arm	Lateral reach, preferred arm	Normal, both arms	Cross-legged, both arms
A. Depth of reach, 5th percentile	19.25	20.25	22.00	15.00	13.75
B. Breadth of aperture, 95th percentile	19.50	12.00	10.00	18.25	17.75
C. Floor to top of aperture, 95th percentile	69.00	69.00	68.75	46.50	28.00
D. Floor to bottom of aperture, 5th percentile	52.25	52.25	52.25	34.25	17.00
E. Vertical dimension of aperture (C minus D; see Fig. 11–66)	16.75	16.75	16.50	12.25	11.00

The ranges and all percentiles have been rounded off to the nearest 0.25 in

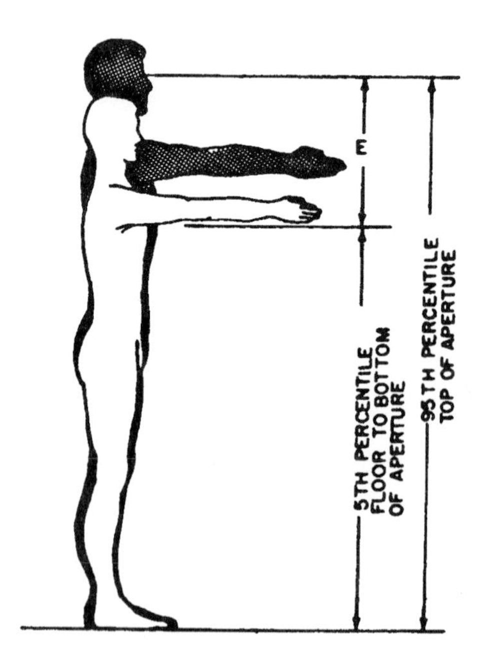

Figure ANTH-C56: Reach Through Verticle Aperture

Table ANTH-C77: Average Increase in Range of Joint Movement of Women Over Men [1]

Movement	Difference (deg)
Wrist flexion and extension	14
Wrist adduction and abduction	11
Elbow flexion and extension	8
Shoulder abduction (rearward)	2
Ankle flexion and extension	4
Knee flexion and extension	0
Hip flexion	3

Table ANTH-C78: Range of Movement at the Neck of Male Civilians [1]

Movement	Range (deg)	
	Avg.	S.D.
Ventral flexion	60	12
Dorsal flexion	61	27
Right or left flexion	41	7
Right or left rotation	79	14

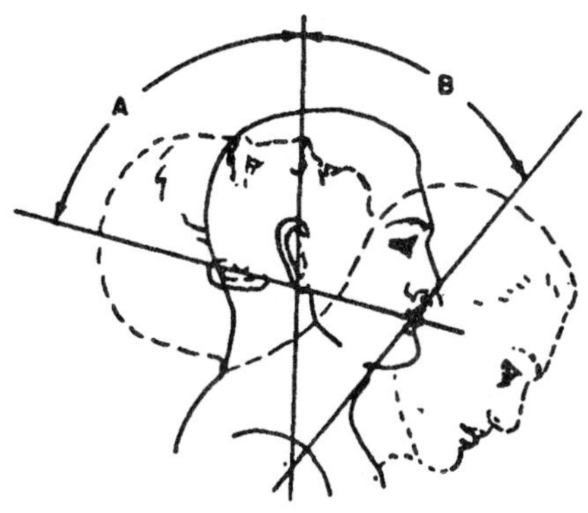

Figure ANTH-C57: Neck Flexion: Dorsal (A) Ventral (B)

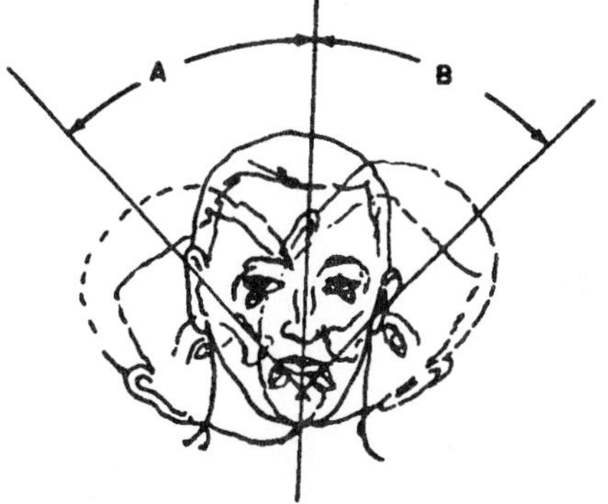

Figure ANTH-C58: Neck Flexion: Right (A) Left (B)

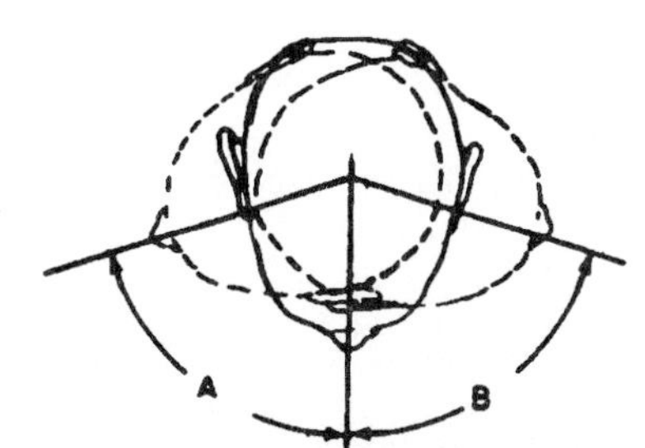

Figure ANTH-C59: Neck Rotation: Right (A) Left (B)

Table ANTH-C79: Range of Movement at the Joints of the Hand and Arm of USAF Males[1]

Movement	Range (deg) Avg.	S.D.
Wrist flexion	90	12
Wrist extension	99	13
Wrist adduction	27	9
Wrist abduction	47	7
Forearm supination	113	22
Forearm pronation	77	24
Elbow flexion	142	10
Shoulder flexion	188	12
Shoulder extension	61	14
Shoulder adduction	48	9
Shoulder abduction	134	17
Shoulder rotation:		
Medial	97	22
Lateral	34	13

Table ANTH-C80: Range of Movement at the Joints of the Foot and Leg of USAF Males[1]

Movement	Range (deg) Avg.	S.D.
Ankle flexion	35	7
Ankle extension	38	12
Ankle adduction	24	9
Ankle abduction	23	7
Knee flexion:		
Standing	113	13
Kneeling	159	9
Prone	125	10
Knee rotation:		
Medial	35	12
Lateral	43	12
Hip flexion	113	13
Hip adduction	31	12
Hip abduction	53	12
Hip rotation (sitting):		
Medial	31	9
Lateral	30	9
Hip rotation (prone):		
Medial	39	10
Lateral	34	10

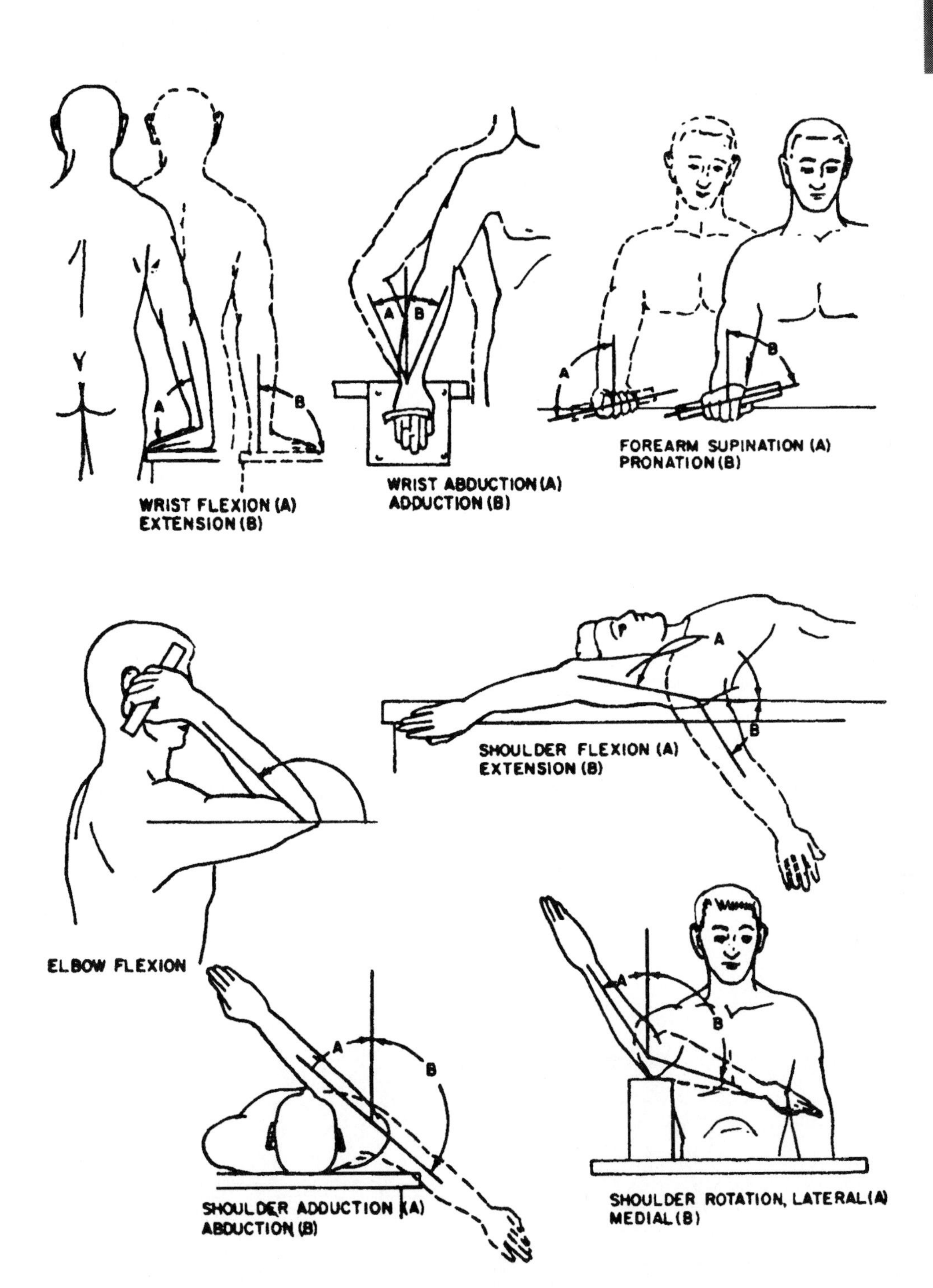

Figure ANTH-C60: Wrist, Shoulder, and Elbow Movements

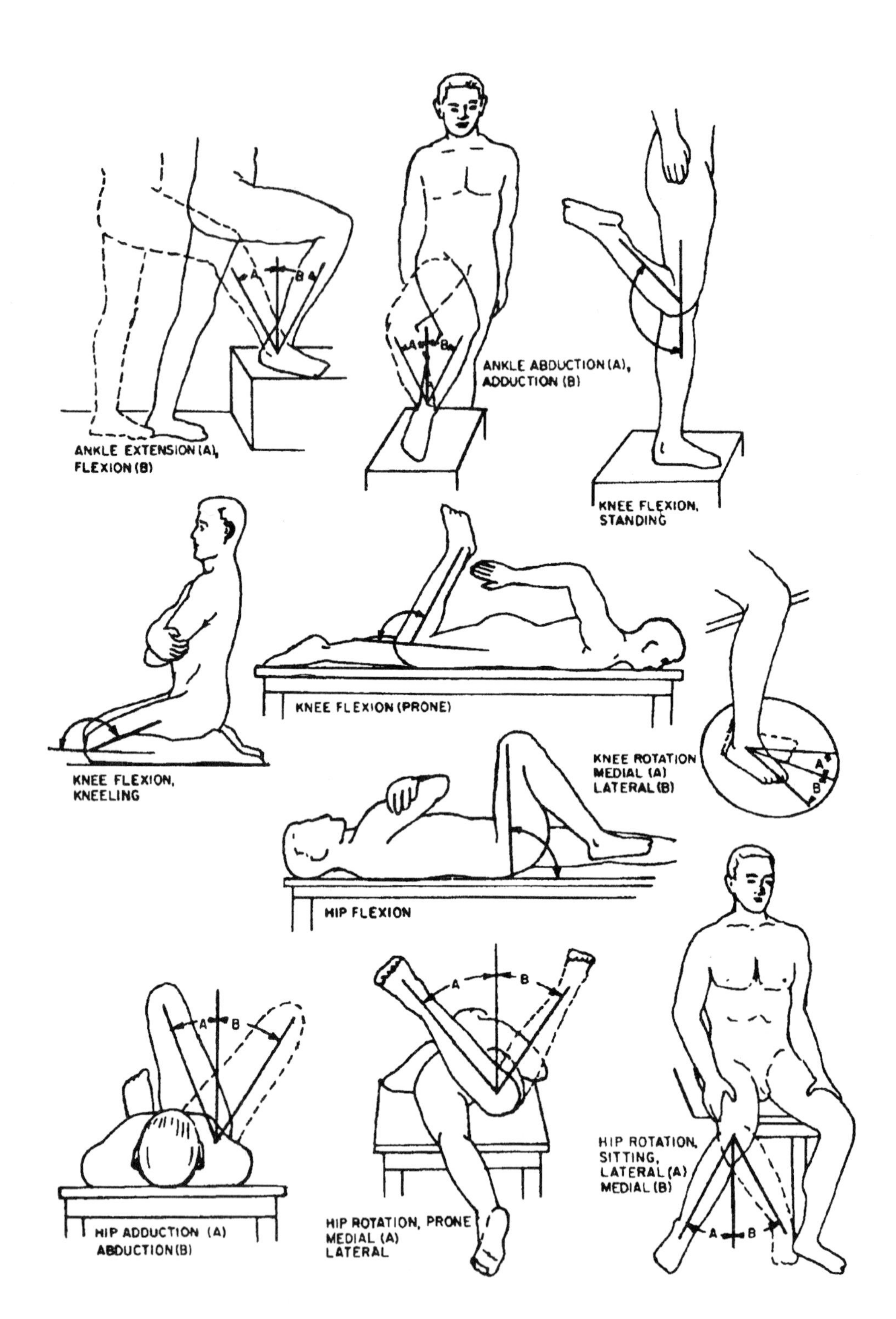

Figure ANTH-C61: Ankle, Knee, and Hip Movements

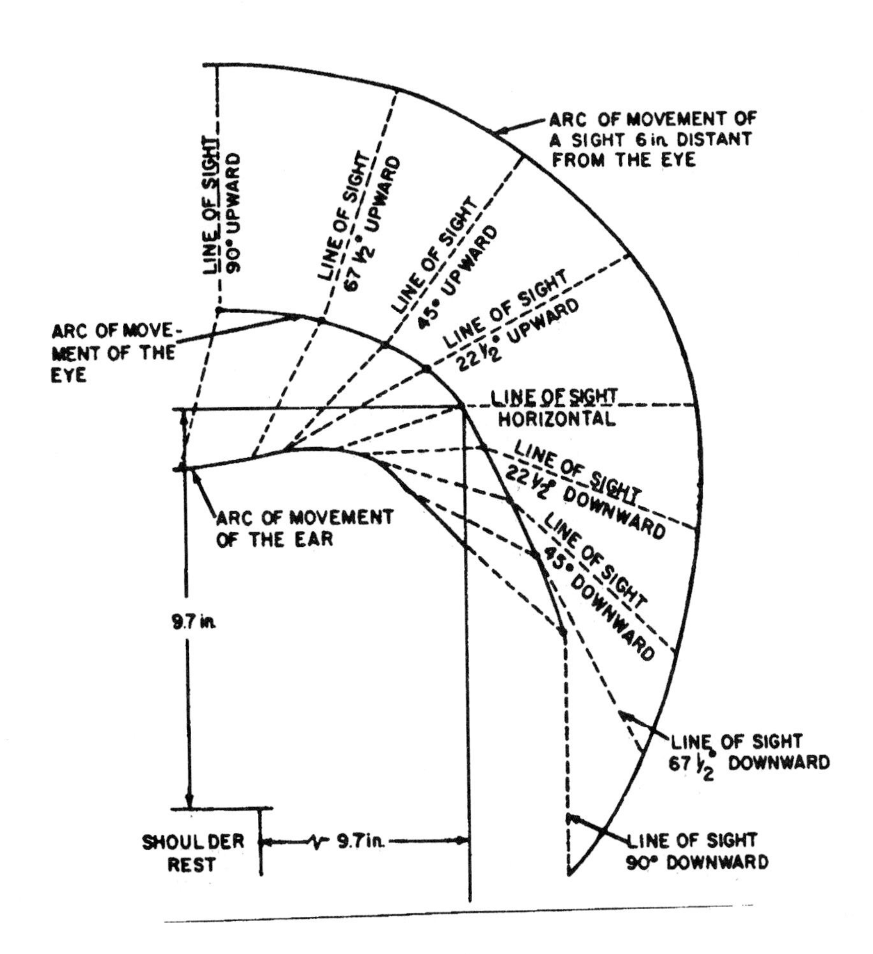

Figure ANTH-C62: Angles of Sight at Various Head Positions

Table ANTH-C81: Maximal, Standing, Right-Handed Static Forces and Torques Exerted on a Verticle Hand-Grip[1]

	Angle (deg)	Location of the handgrip — At percentages of maximal grip distance 50%	75%	100%
		Force in lb[1]		
Push, horizontal	30	16	24	32
	0	30	35	40
	−30	28	30	32
	−60	28	32	36
Pull, horizontal	30	19	22	26
	0	23	26	29
	−30	28	30	31
	−60	23	28	34
To the left, horizontal	30	35	30	24
	0	42	33	24
	−30	42	34	26
	−60	33	30	26
To the right, horizontal	30	24	22	21
	0	30	25	20
	−30	33	27	22
	−60	25	23	21
Up, vertical	30	28	24	19
	0	34	26	18
	−30	50	40	28
	−60	63	51	41
Down, vertical	30	76	58	41
	0	56	40	33
	−30	35	33	30
	−60	39	36	32
		Torque in ft.-lb.[2]		
Clockwise (supination)	30	12.9	12.2	5.8
	0	13.5	11.7	4.3
	−30	11.5	9.8	3.6
	−60	8.5	6.7	4.9
Counterclockwise (pronation)	30	12.4	11.0	8.0
	0	13.2	11.4	8.0
	−30	15.3	13.4	8.7
	−60	16.8	14.8	10.1

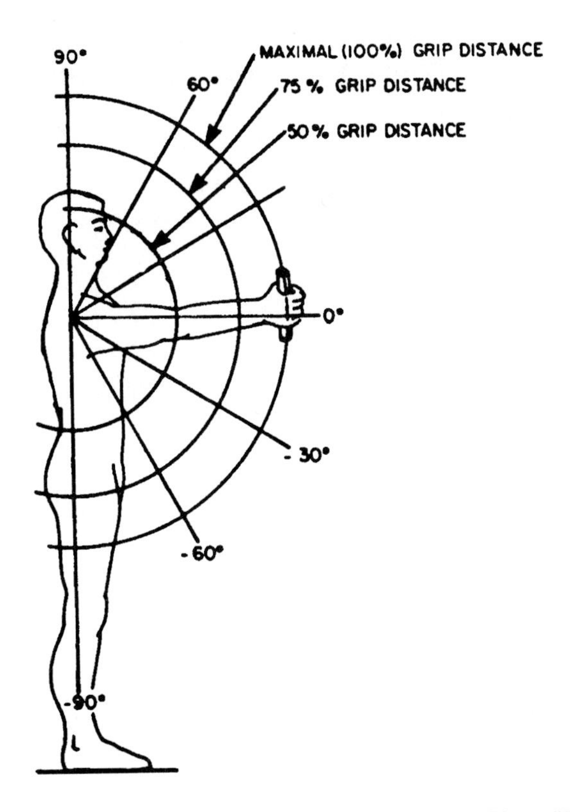

Figure ANTH-C62: Hand Forces in Standing Positions[1]

Table ANTH-C82: Maximal Male Static Push Forces Exerted Horizontally on a Vertical Surface, while Standing [1]

Force-plate height*	Distances†	Number of subjects	Force (lb) Means	Force (lb) S.D.
 40 percent of acromial height‡	80	43	428	166
	90	43	364	121
	100	43	372	113
	110	43	434	127
	120	43	441	134
	130 percent of thumb-tip reach§	43	397	116
			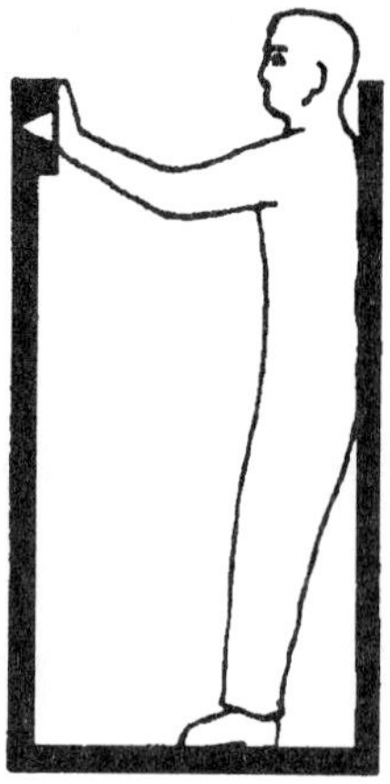 Both hands	
100 percent of acromial height§	50	39	131	32
	60	40	150	36
	70	39	221	61
	80	40	289	90
	90	40	220	68
	100	35	145	57
			Preferred hand	
	50	39	59	15
	60	40	67	16
	70	39	81	22
	80	40	117	32
	90	40	111	38
	100 percent of thumb-tip reach§	35	96	39
 100 percent of acromial height§	50	30	83	31
	60	41	78	28
	70	41	117	37
	80	42	159	43
	90 percent of span¶	37	73	30

Table ANTH-C82: Continued[1]

Force-plate height*	Distances†	Number of subjects	Force (lb) Means	S.D.
 50	80	41	149	40
50	100	42	174	48
50	120	37	175	37
70	80	41	161	36
70	100	42	164	52
70	120	37	184	31
90	80	42	141	33
90	100	42	152	44
90	120	37	194	32
percent of acromial height‡				
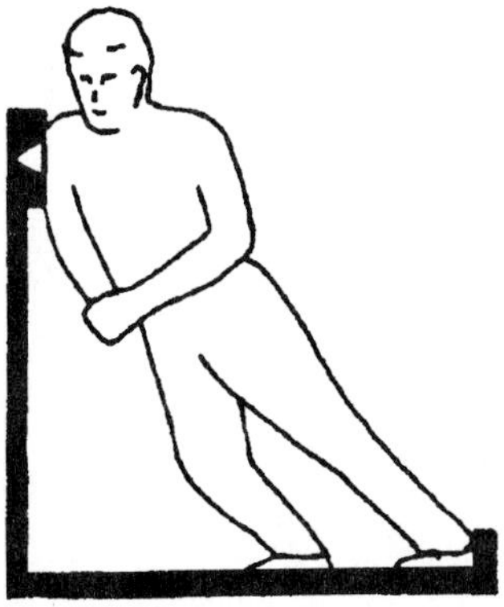 60	70	43	171	38
60	80	43	192	40
60	90	43	178	32
70	60	43	130	25
70	70	43	157	28
70	80	43	163	32
80	60	43	117	29
80	70	43	139	29
80	80	43	143	30
percent of acromial height‡				
 70	70	41	140	33
70	80	41	155	35
70	90	41	132	30
80	70	41	123	28
80	80	41	122	28
80	90	41	120	18
90	70	41	97	21
90	80	41	101	21
90	90	41	109	18
percent of acromial height‡				

*Height of the center of the force plate (8 in. high by 10 in. long) upon which force is applied.

†Horizontal distance between the vertical surface of the force plate and the opposing vertical surface (wall or footrest, respectively) against which the subjects braced themselves.

‡ See shoulder height.

§See thumb-tip reach.

¶The maximal distance between a person's fingertips as he extends his arms and hands to each side.

Table ANTH-C83: Maximal Static Hand Forces at Various Elbow Angles Exerted on a Verticle Hand Grip by Seated Males[1]

A. Left hand

Direction of force	Elbow angle (deg.)	Means	Percentiles (lb) 5th	Percentiles (lb) 95th	S.D.
Push, horizontal	60	80	22	164	31
	90	83	22	172	35
	120	99	26	180	42
	150	111	30	192	48
	180	126	42	196	47
Pull, horizontal	60	64	26	110	23
	90	80	32	122	28
	120	94	34	152	34
	150	112	42	168	37
	180	117	50	172	37
To the left, horizontal	60	32	12	62	17
	90	33	10	72	19
	120	30	10	68	18
	150	29	8	66	20
	180	30	8	64	20
To the right, horizontal	60	50	17	83	21
	90	48	16	87	22
	120	45	20	89	21
	150	47	15	113	27
	180	43	13	92	22
Up, vertical	60	44	15	82	18
	90	52	17	100	22
	120	54	17	102	25
	150	52	15	110	27
	180	41	9	83	23
Down, vertical	60	46	18	76	18
	90	49	21	92	20
	120	51	21	102	23
	150	41	18	74	16
	180	35	13	72	15

B. Right hand

Direction of force	Elbow angle (deg.)	Means	Percentiles (lb) 5th	Percentiles (lb) 95th	S.D.
Push, horizontal	60	92	34	150	38
	90	87	36	154	33
	120		36	172	43
	150	123	42	194	45
	180	138	50	210	49
Pull, horizontal	60	63	24	74	22
	90	88	37	135	30
	120	104	42	154	31
	150	122	56	189	36
	180	121	52	171	37
To the left, horizontal	60	52	20	87	19
	90	50	18	97	23
	120	53	22	100	26
	150	54	20	104	25
	180	50	20	104	26
To the right, horizontal	60	42	17	82	20
	90	37	16	68	18
	120	31	15	62	17
	150	33	15	64	18
	180	35	14	62	24
Up, vertical	60	49	20	82	18
	90	56	20	106	22
	120	60	24	124	24
	150	66	18	118	28
	180	43	14	88	22
Down, vertical	60	51	20	89	21
	90	54	26	88	20
	120	58	26	98	23
	150	47	20	80	18
	180	41	17	82	18

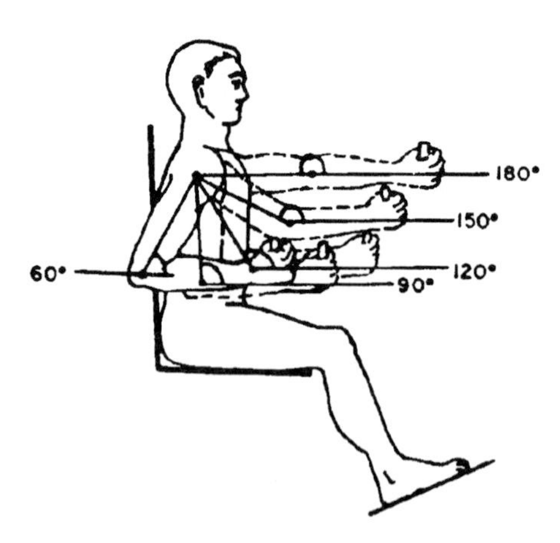

Figure ANTH-C63: Maximal Static Hand Forces at Various Elbow Angles Exerted on Verticle Hand Grip by Males

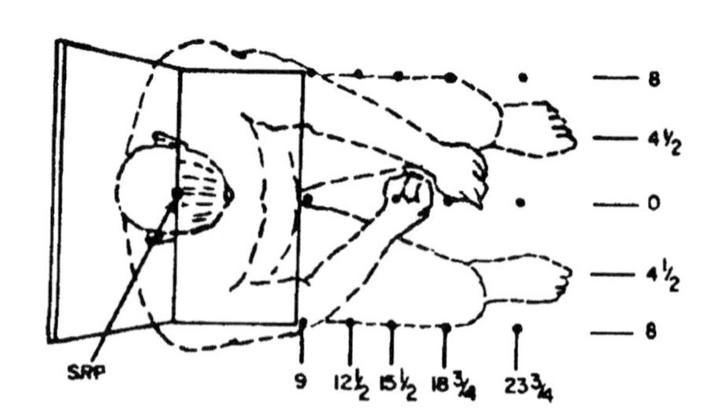

Figure ANTH-C64: Maximal One- and Two- Handed Static Forces Exerted on an Aircraft Joystick by SeatedSeated Males

Table ANTH-C84: Maximal One- and Two- Handed Static Forces Exerted on an Aircraft Joystick by Seated Males[1]

Handle distance from SRP (in.)	Handle distance from midsagittal plane of body (in.)	A. Push (Percentiles (lb)) Right hand† 5th	Right hand† 50th	Right hand† 95th	Both hands† 50th	B. Pull (Percentiles (lb)) Right hand† 5th	Right hand† 50th	Right hand† 95th	Both hands† 50th
9	0	26	46	67	99	34	57	86	106
	4½ (left)	18	33	54	88	28	45	66	106
	8 (left)	12	29	44	77	26	40	67	93
	4½ (right)	34	58	82	99	39	62	88	106
	8 (right)	37	65	95	99	39	58	86	106
12½	8 (left)	18	36	68	110	33	53	77	120
	8 (right)	43	74	102	110	49	80	108	120
15½	0	43	86	160	165	54	83	113	146
	8 (left)	23	60	118	121	39	64	98	133
	8 (right)	53	100	164	143	55	89	119	146
18¾	0	64	124	177	154	56	86	127	160
	8 (left)	36	72	114	121	45	74	108	146
	8 (right)	70	125	198	154	58	99	126	160
23¾	0	54	106	141	110	62	102	138	173
	8 (left)	29	64	104	88	51	90	129	173
	8 (right)	56	100	147	99	58	103	133	173

Control distance from SRP (in.)	Control distance from midsagittal plane of body (in.)	C. To the left (Percentiles (lb)) Right hand† 5th	Right hand† 50th	Right hand† 95th	Both hands† 50th	D. To the right (Percentiles (lb)) Right hand† 5th	Right hand† 50th	Right hand† 95th	Both hands† 50th
9	0	30	47	66		23	38	49	
	4½ (left)	31	49	67		31	48	64	
	8 (left)	24	44	65		34	55	74	
	4½ (right)	26	46	78		15	27	51	
	8 (right)	26	44	72		12	22	43	
12½	8 (left)	23	44	70		31	48	70	
	8 (right)	22	39	59		16	24	46	
15½	0	24	38	52		20	28	39	
	8 (left)	20	35	58		25	43	63	
	8 (right)	24	40	70		13	22	49	
18¾	0	8	32	53		15	25	35	
	8 (left)	16	30	56		22	36	61	
	8 (right)	22	39	70		14	24	50	
23¾	0	14	29	46		13	20	30	
	8 (left)	11	21	49		19	31	48	
	8 (right)	20	37	76		12	22	51	

Note: In the neutral position, the stick is grasped 13½ inches above SRP.

Unpublished data, Anthropology Branch, Aerospace Medical Research Laboratories.

† All subjects used an aircraft type of seat with standard lap belt and shoulder harness.

Table ANTH-C85: Maximal One- and Two- Handed Static Forces Exerted on an Aircraft Joystick by Seated Males[1]

Direction of force	Preferred hand		Both hands	
	Means lb	S.D. lb	Means lb	S.D. lb
Forward push	167	33.4	199	28.3
Backward pull	134	12.7	204	14.3
Lateral push*	88	25.0	101	22.4
Lateral pull†	63	20.1	88	19.9

Note: The stick is located 20 in. forward of Seat Reference Point (SRP); see

*Lateral push applied by the palm of either hand against the control handle.

†Lateral pull applied by the fingers of either hand against the control handle.

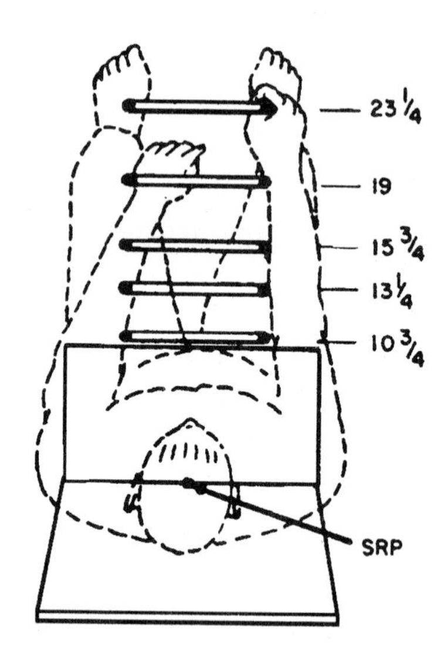

Figure ANTH-C65: Maximal One- and Two- Handed Static Forces Exerted on an Aircraft Control Wheel by Seated Males

Table ANTH-C86: Maximal One- and Two- Handed Static Forces Exerted on an Aircraft Control Wheel by Seated Males[1]

Control distance from SRP (in.)	Wheel rotation (deg.)	A. Push (Percentiles) Right hand† 5th	A. Push Right hand† 50th	A. Push Right hand† 95th	A. Push Both hands† 50th	B. Pull (Percentiles) Right hand† 5th	B. Pull Right hand† 50th	B. Pull Right hand† 95th	B. Pull Both hands† 50th
10¾	0	52	86	135	147	44	66	102	126
	45 (left)	48	84	149	147	40	67	111	126
	90 (left)	32	67	125	103	23	55	109	98
	45 (right)	40	67	128	147	39	67	97	126
	90 (right)	19	52	112	88	18	43	87	98
13¼	90 (left)	32	54	93	88	33	67	120	112
	90 (right)	25	51	83	88	31	60	102	112
15¾	0	61	90	155	177	66	94	145	154
	90 (left)	32	59	139	118	42	71	144	140
	90 (right)	32	53	102	132	49	80	130	140
19	0	64	121	235	265	73	106	169	196
	90 (left)	37	88	171	162	60	88	127	154
	90 (right)	33	67	140	162	61	94	149	168
23¼	0	105	171	242	265	77	125	182	234
	90 (left)	82	131	211	177	73	117	162	182
	90 (right)	49	117	197	191	74	110	186	196

Control distance from SRP (in.)	Control rotation (deg.)	C. To the left (Percentiles) Right hand† 5th	C. To the left Right hand† 50th	C. To the left Right hand† 95th	C. To the left Both hands† 50th	D. To the right (Percentiles) Right hand† 5th	D. To the right Right hand† 50th	D. To the right Right hand† 95th	D. To the right Both hands† 50th
10¾	0	26	46	88	92	20	48	96	91
	45 (left)	21	54	123	102	24	69	121	132
	90 (left)	23	47	91	102	27	59	101	101
	45 (right)	31	54	120	133	24	51	118	111
	90 (right)	21	42	104	122	15	54	112	121
13¼	90 (left)	26	44	86	102	21	52	98	111
	90 (right)	25	45	99	122	19	51	111	101
15¾	0	27	46	112	102	27	59	97	101
	90 (left)	27	43	82	82	19	53	96	101
	90 (right)	29	50	86	112	20	46	91	91
19	0	25	44	95	102	30	63	104	101
	90 (left)	22	43	76	82	27	46	94	101
	90 (right)	33	52	104	122	22	41	87	81
23¼	0	20	39	86	92	35	60	98	101
	90 (left)	21	38	73	71	26	42	82	91
	90 (right)	26	55	109	102	22	40	68	71

Note: Wheel grips are 15 in. apart. In the neutral position, they are 18 inches above SRP.

Unpublished data, Anthropology Branch, Aerospace Medical Research Laboratories.

† All subjects used on aircraft type of seat, with standard lap belt and shoulder harness.

Table ANTH-C87: Maximal Static Hand Forces Exerted, at Various Elbow Angles, on a Horizontal HandGrip by Seated Males[1]

A. Left hand

Direction of force	Elbow angle (deg.)	Means	Percentiles (lb) 5th	95th	S.D.
Push, horizontal	60	86	33	138	35
	90	60	27	93	28
	120	43	17	71	17
	150	37	15	69	18
	180	32	12	59	13
Pull, horizontal	60	39	20	64	18
	90	37	17	65	18
	120	30	12	56	14
	150	32	15	52	13
	180	34	16	61	15
To the right, horizontal	60	42	20	66	15
	90	38	17	60	12
	120	34	17	53	8
	150	31	17	54	11
	180	28	15	41	8
To the left, horizontal	60	36	18	51	15
	90	27	11	54	11
	120	22	10	39	10
	150	23	9	53	16
	180	20	10	49	13
Up, vertical	60	57	22	100	22
	90	77	37	123	24
	120	91	45	145	30
	150	100	58	159	32
	180	101	47	171	11
Down, vertical	60	74	18	139	35
	90	75	23	136	34
	120	75	29	148	40
	150	79	39	136	29
	180	76	34	138	31

B. Right hand

Direction of force	Elbow angle (deg.)	Means	Percentiles (lb) 5th	95th	S.D
Push, horizontal	60	94	40	156	36
	90	65	25	100	24
	120	46	23	70	15
	150	40	18	66	18
	180	32	17	59	12
Pull, horizontal	60	37	13	50	16
	90	32	14	54	13
	120	26	13	43	10
	150	29	12	48	10
	180	28	11	48	12
To the right, horizontal	60	41	19	72	19
	90°	31	12	64	15
	120	26	9	53	13
	150	21	9	39	11
	180	19	10	34	7
To the left, horizontal	60	48	16	73	18
	90	39	16	59	15
	120	34	15	47	11
	150	32	18	45	7
	180	31	16	57	13
Up, vertical	60	49	23	79	20
	90	69	28	112	29
	120	91	41	138	30
	150	99	43	165	38
	180	95	35	156	35
Down, vertical	60	81	23	158	35
	90	83	22	142	35
	120	92	37	161	35
	150	90	40	154	34
	180	87	41	143	31

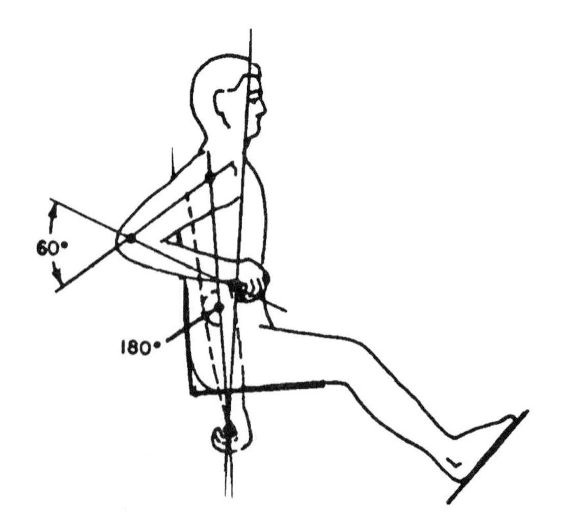

Figure ANTH-C66: Hand Forces Exerted at Various Elbow Angles

Table ANTH-C88: Maximal Static Hand Forces, at Various Elbow Angles, Exerted on a Horizontal HandGrip by Seated Males[1]

A. Left hand

Direction of force	Elbow angle (deg.)	Means	Percentiles (lb) 5th	Percentiles (lb) 95th	S.D.
Push, horizontal	60	89	35	176	42
	90	59	25	104	27
	120	40	15	80	18
	150	38	13	69	30
	180	30	14	47	10
Pull, horizontal	60	54	23	87	23
	90	42	13	68	21
	120	40	14	66	18
	150	40	16	62	15
	180	40	17	70	18
To the right, horizontal	60	38	16	64	12
	90	32	12	46	12
	120	31	14	55	13
	150	32	12	62	15
	180	29	12	43	9
To the left, horizontal	60	42	17	81	20
	90	33	16	52	12
	120	28	14	45	8
	150	26	12	43	10
	180	27	8	44	10
Up, vertical	60	49	20	89	22
	90	75	24	131	29
	120	94	38	152	33
	150	104	44	164	36
	180	111	45	173	40
Down, vertical	60	58	20	138	41
	90	80	23	160	43
	120	84	35	136	33
	150	84	43	136	29
	180	78	36	124	28

B. Right hand

Direction of force	Elbow angle (deg.)	Means	Percentiles (lb) 5th	Percentiles (lb) 95th	S.D.
Push, horizontal	60	96	34	172	39
	90	65	25	117	24
	120	43	20	71	17
	150	36	17	59	14
	180	32	12	58	15
Pull, horizontal	60	51	16	93	25
	90	43	13	74	19
	120	40	11	63	17
	150	37	11	66	17
	180	39	15	73	19
To the right, horizontal	60	44	18	73	19
	90	39	18	72	24
	120	34	17	64	15
	150	32	15	60	14
	180	29	14	48	12
To the left, horizontal	60	36	13	70	17
	90	31	13	48	12
	120	30	12	46	11
	150	31	12	52	14
	180	28	10	44	10
Up, vertical	60	45	17	78	22
	90	63	21	107	27
	120	88	41	143	33
	150	103	37	161	40
	180	113	51	165	34
Down, vertical	60	59	20	132	35
	90	80	17	143	37
	120	92	29	148	13
	150	93	37	150	35
	180	87	44	135	32

Table ANTH-C89: Maximal Static Hand Forces , at Various Elbow Angles, Exerted on a Verticle HandGrip by Seated Males[1]

A. Left hand

Direction of force	Elbow angle (deg.)	Means	Percentiles (lb) 5th	95th	S.D.
Push, horizontal	60	52	17	87	21
	90	54	18	91	22
	120	63	21	108	27
	150	65	24	111	26
	180	67	26	116	28
Pull, horizontal	60	57	17	97	24
	90	66	23	118	26
	120	75	22	126	30
	150	70	21	122	28
	180	61	18	111	26
To the left, horizontal	60	24	8	49	12
	90	22	6	45	10
	120	20	6	38	9
	150	20	5	56	15
	180	22	4	57	19
To the right horizontal	60	44	11	99	24
	90	40	13	92	22
	120	38	9	91	23
	150	34	8	79	23
	180	31	10	67	19
Up, vertical	60	35	13	71	17
	90	40	15	78	18
	120	40	11	81	21
	150	31	7	62	17
	180	18	5	44	12
Down, vertical	60	30	10	51	12
	90	31	12	57	12
	120	31	11	57	14
	150	28	10	48	11
	180	25	7	41	10

B. Right hand

Direction of force	Elbow angle (deg.)	Means	Percentiles (lb) 5th	95th	S.D.
Push, horizontal	60	66	24	119	26
	90	64	26	103	23
	120	73	29	128	28
	150	74	29	127	30
	180	69	31	123	26
Pull, horizontal	60	61	21	113	26
	90	73	24	121	30
	120	86	31	147	34
	150	81	29	133	33
	180	69	31	118	26
To the left, horizontal	60	49	16	91	22
	90	46	16	87	21
	120	48	15	97	25
	150	45	15	93	26
	180	37	12	71	17
To the right, horizontal	60	30	12	57	12
	90	28	13	51	11
	120	28	11	58	12
	150	28	12	60	14
	180	25	9	61	14
Up, vertical	60	44	13	85	21
	90	52	15	94	22
	120	50	13	91	21
	150	41	13	83	23
	180	23	8	47	12
Down, vertical	60	34	13	61	13
	90	36	16	60	13
	120	35	15	61	15
	150	34	15	60	13
	180	29	13	47	10

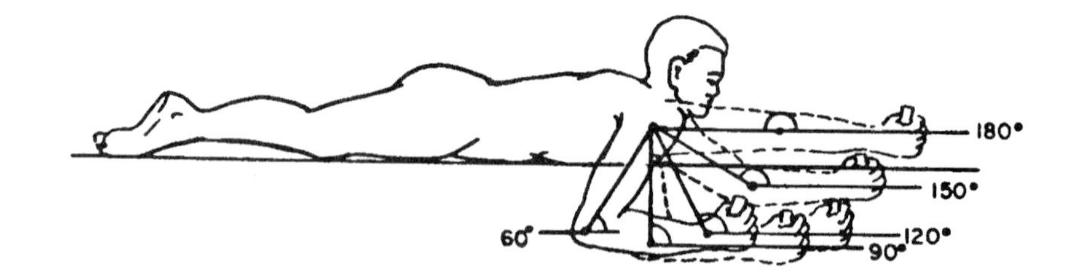

Figure ANTH-C67: Hand Forces Exerted at Various Elbow Angles by Prone Males

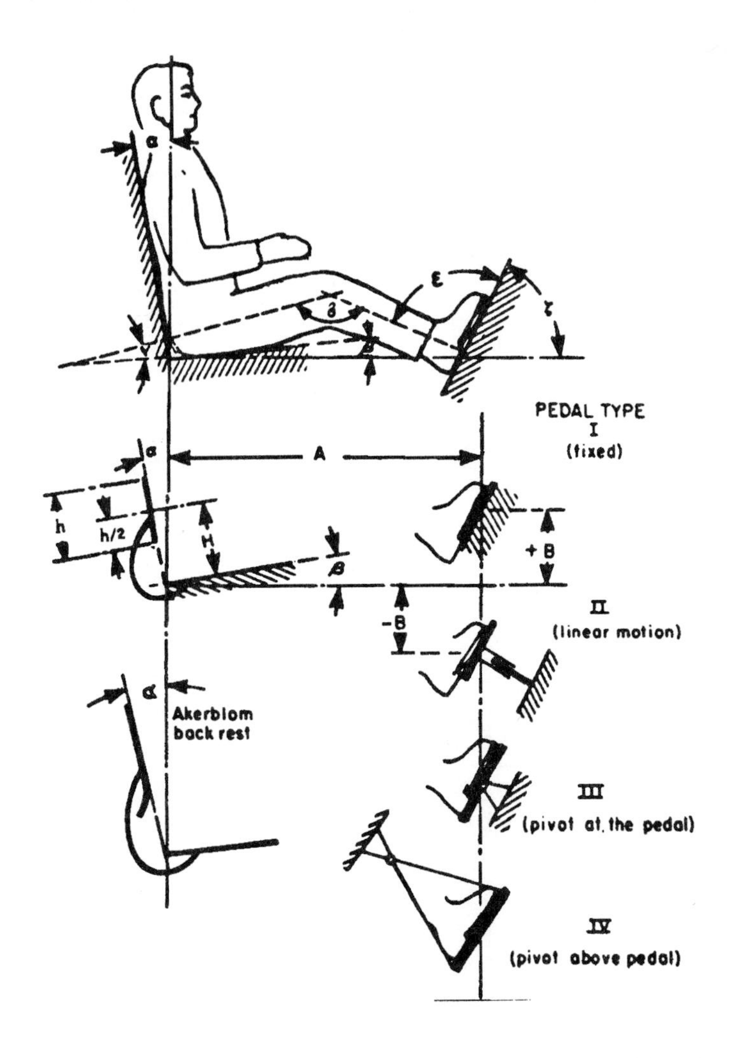

Figure ANTH-C68: Dimensional and Angular Factors Affecting Pedal Force Applicability

Table ANTH-C90: Maximal Static Leg Thrust Exerted on a Fixed Pedal by Seated Males[1]

Test conditions									Force			
Type of pedal	Type of seat	α (deg).	β (deg).	γ (deg).	δ (deg.)	ε (deg.)	A % of total leg reach	B (in.)	Means (lb)	S.D. (lb)	Direction	Subjects N
	H = 12 to 14 in. h = 5 in.	(†)	(†)	15	160			5	383	‡68	horizon-tal.	20 men
				0	160			−4½	319	‡76		
Type I; fixed pedal pushed with ball of foot.	Åkerblom backrest 20 in. high; most anti-rior part of the lumbar pad 7 in. above the seat pan.	20, above the lumbar pad.	3			90	95	12	305	70	Approx. in the line from hip joint to the ankle.	60 men.
							85		231	65		
							95	4	304	65		
							85		270	67		
							95	−2	315	69		
							85		283	65		
							95	−8	298	63		
							85		254	70		
							95	−14	235	52		
							85		235	67		

† Probably 0°, but not explicitly stated in the original publication.
‡ Calculated from $SD = \sqrt{N} \cdot SE$.

Table ANTH-C91: Maximal Static Leg Thrust Exerted by Seated Subjects on a Pedal Pivoted at the Instep[1]

Test conditions							Force		
Type of pedal	Type of seat	α (deg.)	β (deg.)	γ (deg.)	δ (deg.)	ϵ (deg.)	Means (lb)	Direction	Subjects N
	H = 8 in. h = 5 in.	(†)	(†)	10–20 - - - -	130–150 - - - -	90	appr. 440 appr. 340	5° to 15° below horizontal.	1 man.‡ 2 women.‡
Type III; pivoted near the instep; large enough to accommodate the entire foot.	Back rest supports pelvis and back.		(†)	−6 −15 −10 −10 −9	94 149 162 165 167	90	73 227 385 346 250	Approx. in the line from the hip joint to the center of the pedal.	6 "powerfully built men".§
				8 10 5	93 136 164	90	87 270 559		
				19 16 15 15 15 15	67 129 160 160 160 169	90	89 319 845 ¶(691) **(689) 530		
				36 33 34	88 106 125	90	135 184 443		
				48 49	72 81	90	133 130		

† Probably 0°, but not explicitly stated in the original publication.

¶ 32 drivers of the Royal Armoured Corps.
**12 school boys, aged 14–18.

Table ANTH-C92: Maximal Static Upward Pull Forces Exerted With Two Hands on a Horizontal Bar[1]

Population	N	Percentiles (lb) 5th	50th or Mean	95th	S.D.
British women:					
College students	460	(160)	216	(272)	34.4
Factory workers:					
Employed	3076	(119)	183	(247)	38.8
Unemployed	413	(101)	165	(229)	39.2
British men:					
College students	1704	(271)	367	(463)	58.9
Factory workers:					
Employed	10344	(251)	363	(474)	67.7
Unemployed	1250	(214)	315	(415)	60.8
U.S. Air Force:					
Cadets (men)	914	375	520	665	90
"Leglift"	914	1010	1480	1950	290

Table ANTH-C93: Maximal Static Grip Force Exerted by Males[1]

Population	Percentiles (lb) 5th	50th or Mean	95th	S.D.
Air Force personnel, general:				
Right hand	*(59)	104	*(148)	27.3
Left hand	*(56)	94	*(134)	23.7
Air Force personnel, aircrewmen:				
Right hand	105	134	164	18.0
Left hand	96	124	154	16.0
Air Force rated officers:				
Preferred hand	98	124	154	16.8
Army personnel:				
Right hand	106	137	172	----
Left hand	99	132	168	----
Navy personnel:				
Mean of both hands	95	119	143	14.4
Industrial workers:				
Preferred hand	92	117	143	15.4
Truck and bus drivers:				
Right hand	91	121	151	18.1
Left hand	86	113	140	16.4
Rubber industry workers:				
Right hand	*(89)	124	*(159)	21.2
Left hand	*(86)	122	*(159)	22.2
University men:				
Right hand	*(74)	108	*(142)	21.0
Left hand	*(65)	95	*(124)	18.0
Same subjects, force exerted over one minute:				
Right hand	*(42)	62	* (82)	12
Left hand	*(39)	55	*(71)	10

* Percentiles in parentheses were computed from the 50th percentile using the S.D.

Table ANTH-C94: Maximal Static Force Exerted by 100 Males in Attempted Flexion of the Extended Fingers of the Right Hand[1]

Finger	Mean	S.D.
Thumb	16	3.8
Index	13	2.8
Middle	14	4.3
Ring	11	3.8
Little	7	2.5

Note: For measurements on the thumb, the ulnar side of the hand was placed on a flat surface; for measurements on the other fingers, the back of the hand touched the surface. The experimenter held down the subject's wrist. Force was measured perpendicular to the extended finger.

Table ANTH-C95: Maximal Static Force Exerted Between Thumb and Fingers by Males Civilians[1]

	Force (lb)	
Type of prehension	Avg.	S.D.
Palmar	21.5	5.4
Tip	21.0	4.8
Lateral	23.2	4.8

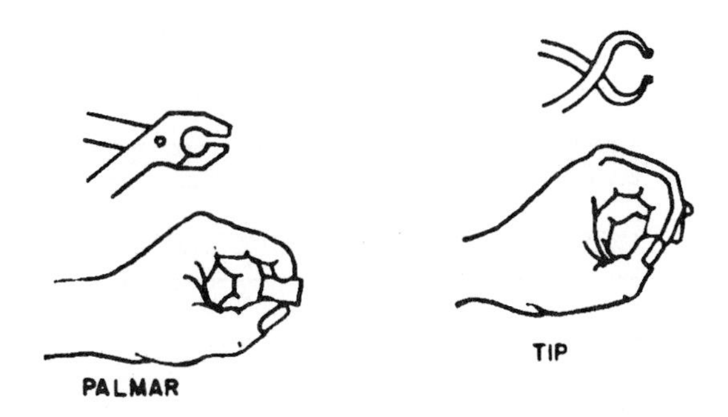

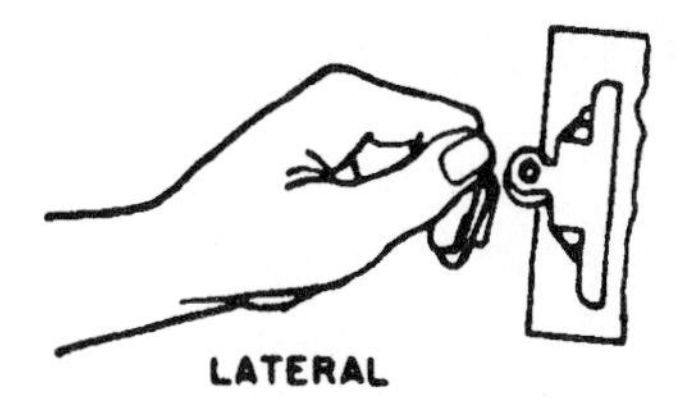

Figure ANTH-C69: Maximal Static Forces Exerted Between Thumb and Fingers

Table ANTH-C96: Maximal Static Grip Force Exerted Females[1]

Population	Percentiles (lb) 5th	50th	95th	S.D.
Navy personnel				
Mean of both hands	58	73	87	8.8
Industrial workers:				
Preferred hand	57	74	91	10.3

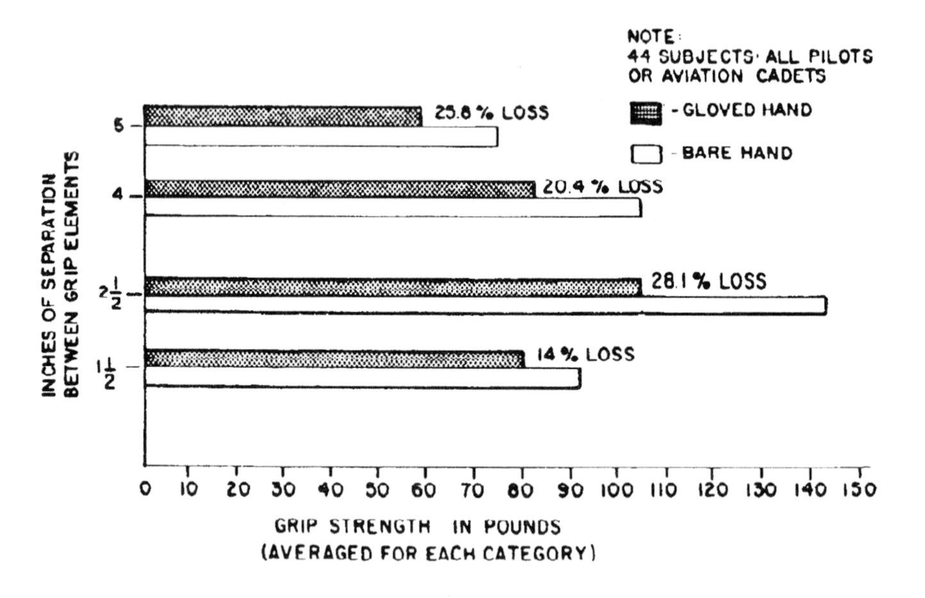

Figure ANTH-C70: Grip Strength of Pilots

REFERENCES

1. VanCott, H.P., and Kinkade, R.G. 1972. *Human Engineering Guide to Equipment Design. (Revised)* Washington,DC: US Government Printing Office.

CHAPTER 1: ANTHROPOMETRY

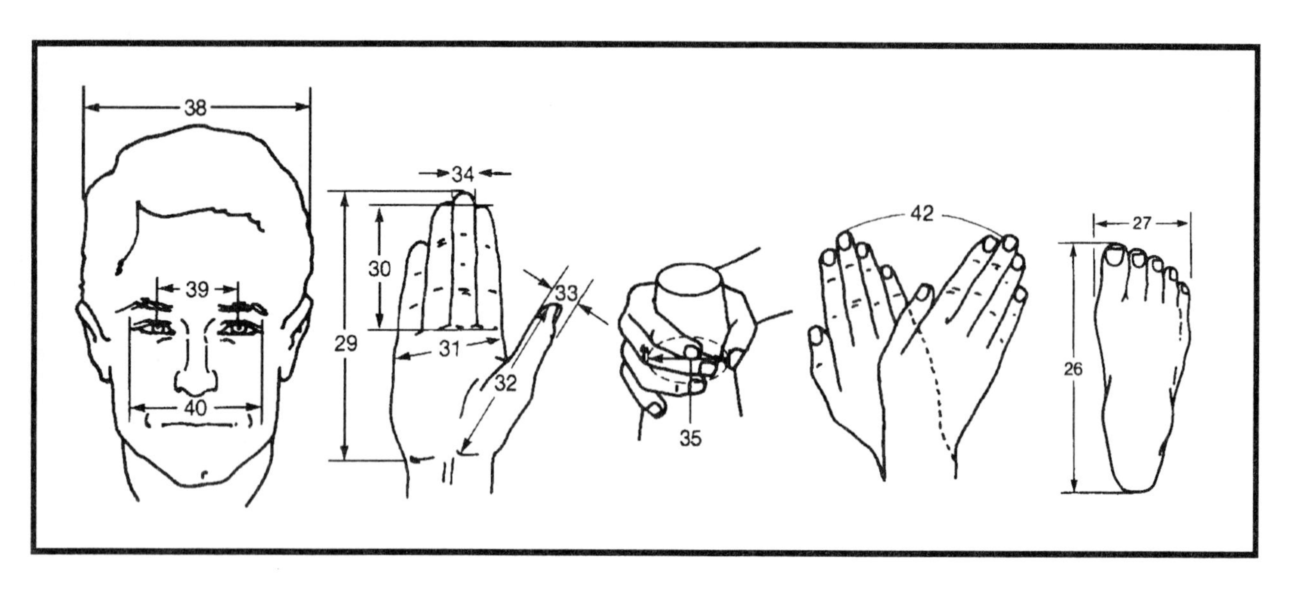

Section D: International Populations

Only partial anthropometric data was available. Nevertheless, some data is better than no data. However, some caution should be used with data in Tables D1-D5, as the data is at least 20 years old. It is included since it is still being referenced and there is little cross-cultural data available.

Table ANTH-D1: Weight, Stature, and Seated Height of Nude International Males[1]

Population	Group	N	Weight Kg	Weight SD	Weight Lb	Weight SD	Stature CM	Stature SD	Stature IN	Stature SD	Sitting height CM	Sitting height SD	Sitting height IN	Sitting height SD
Africa	Dinka Nilotes	279	58.4	5.92	128.3	13.0	181.3	6.12	71.5	2.41	----	----	----	----
Australia	Mixed Army rec.	3580	72.6	----	159.7	----	174.0	5.67	68.5	2.23	90.5	----	35.6	----
Belgium	Recruits 20-yr.	17018	64.1	7.8	141.1	17.7	171.5	6.10	67.5	2.40	----	----	----	----
Belgium	Flying personnel	2450	66.5	6.9	146.2	15.2	173.9	5.79	68.4	2.28	----	----	----	----
Bulgaria	Civilians aged 26	114	67.7	8.13	148.9	17.9	169.8	5.89	66.9	2.32	89.4	3.68	35.2	1.45
Canada	RCAF pilots	314	76.6	9.91	168.5	21.8	176.5	5.89	69.5	2.32	91.5	3.35	36.0	1.30
Canada	Civilians 30–40	400	76.4	----	168.0	----	172.9	----	68.1	----	----	----	----	----
Ceylon	Sinhalese	635	----	----	----	----	160.4	6.02	63.1	2.37	81.5	3.45	32.1	1.36
China (Taiwan)	Young men	31	52.6	4.44	115.7	9.77	167.6	5.22	65.9	2.06	90.4	2.92	35.6	1.15
China (Taiwan)	Military 20–30	1049	55.7	----	122.3	----	164.5	----	64.8	----	----	----	----	----
France	Pilots, cadets	7084	----	----	----	----	171.3	5.81	67.4	2.29	----	----	----	----
France	Pilots, cadets	1000	65.8	7.03	143.0	15.5	----	----	----	----	----	----	----	----
France	Recruits, 20-yr.	234	65.4	7.17	143.9	15.8	169.5	6.05	66.7	2.38	----	----	----	----
East Germany	Civilians 30–39	1651	72.0	----	158.4	----	171.0	----	67.4	----	----	----	----	----
East Germany	Civilians 30–39	12200	----	----	----	----	170.9	7.19	68.3	1.83	----	----	----	----
West Germany	Recruits 20–yr.	316202	68.5	----	150.5	----	175.2	6.6	69.0	1.68	----	----	----	----
West Germany	Recruits 20–yr.	378133	68.4	8.9	150.4	19.6	----	----	----	----	----	----	----	----
Greece	Mixed military	1084	67.0	7.62	147.4	16.8	170.5	5.88	67.1	2.31	90.3	3.01	35.6	1.19
Great Britain	RAF bomb personnel	11772	66.1	7.06	145.1	15.5	173.2	6.2	68.2	2.44	----	----	----	----
Great Britain	RAF Pilots	4357	72.1	9.27	158.6	20.4	177.3	6.15	69.8	2.42	93.3	3.3	36.7	1.30
India	Civilians 19–60	499	50.9	8.00	123.0	17.6	163.0	6.10	64.3	2.40	85.1	3.04	33.5	1.20
India	Civilians mixed	3774	----	----	----	----	163.9	5.57	64.5	2.19	----	----	----	----
Italy	Mixed military	1358	70.3	8.43	154.7	18.6	170.6	6.23	67.2	2.45	89.7	3.20	35.3	1.26
Israel	Civilians 40–44	3339	72.2	11.0	158.8	24.2	168.4	6.50	66.2	2.56	89.2	3.7	35.1	1.46
Japan	JASDF pilots	239	61.1	5.86	133.2	12.9	166.9	4.80	65.7	1.89	90.8	2.62	35.8	1.03
South Korea	ROKAF pilots	264	62.9	6.53	138.2	14.4	168.7	4.61	6.64	1.81	90.8	2.81	35.8	1.11
Mexico	Rural	5217	----	----	----	----	160.2	----	6.31	----	83.8	----	32.3	----
Mexico	Mixed rural	1502	54.1	----	119.0	----	----	----	----	----	----	----	----	----
Norway	Recruits 20-yr.	5765	70.1	7.46	154.2	16.4	177.5	6.03	69.9	2.37	92.7	3.47	36.5	1.37
Thailand	Mixed military	2950	56.3	10.3	123.8	22.7	163.4	5.3	64.4	2.09	86.4	3.1	34.0	1.22
Turkey	Mixed military	915	64.6	8.23	142.1	18.1	169.3	5.76	66.7	2.27	89.7	3.15	35.3	1.24

Table ANTH-D1: Weight, Stature, and Seated Height of Nude International Males[1]

Population	Group	N	Weight Kg	Weight SD	Weight Lb	Weight SD	Stature CM	Stature SD	Stature IN	Stature SD	Sitting height CM	Sitting height SD	Sitting height IN	Sitting height SD
USA:														
Air Force—														
1950	Flyers	4063	74.4	9.27	163.7	20.9	175.5	6.19	69.1	2.44	91.3	3.27	35.9	1.29
Army	Separatees	24449	70.3	9.33	154.8	20.5	174.0	6.38	68.5	2.51	90.9	3.4	35.8	1.34
Army	Pilots	500	75.4	8.59	165.8	18.9	176.5	5.72	69.5	2.25	90.4	3.22	35.6	1.27
Navy—														
1964	Pilots	1549	77.7	8.66	171.4	19.09	177.6	5.91	69.9	2.33	92.1	3.16	36.3	1.25
Nat'l. Health														
Survey	Total males 17–79	3091	76.4	----	168.0	----	173.2	----	68.2	----	90.4	----	35.6	----
FAA	Tower trainees 21–50	678	73.4	10.0	161.8	22.1	176.7	6.35	69.6	2.50	92.0	3.27	36.2	1.29
Vietnam	Mixed military	2129	51.1	6.0	112.4	13.2	160.5	5.5	63.2	2.16	85.0	3.3	33.5	1.30

*Computed from published means and standard deviations.
†Taken by USAF technique.

Table ANTH-D2: Stature of Nude International Males[1]

Population	Group	N	Percentiles 1 cm	1 in.	5 cm	5 in.	Means cm	Means in.	95 cm	95 in.	99 cm	99 in.	SD cm	SD in.
MILITARY														
Australia	Mixed army	3580	160.0	63.0	164.1	64.7	174.0	68.5	184.1	72.5	188.2	74.2	5.67	2.76
Belgium	Recruits, 20-yr.	17018	156.5	61.6	160.8	63.3	171.5	67.5	182.2	71.8	186.5	73.5	6.1	2.4
Belgium	Flyers, 17–50	2450	159.3	62.7	163.6	64.4	173.9	68.5	184.2	72.6	188.5	74.2	5.79	2.28
Canada	RCAF pilots	314	164.2	64.7	167.8	66.1	177.4	69.8	188.0	74.0	190.2	74.9	6.1	2.4
France	Pilots, Cadets	7084	157.2	61.8	161.1	63.5	171.0	67.4	181.1	71.4	185.6	73.1	5.81	2.3
France	Recruits 20-yr.	234	154.5	60.8	159.0	62.6	169.5	66.7	180.2	71.0	184.7	72.7	6.05	2.38
West Germany	Recruits 20-yr.	316202	158.8	62.5	163.6	64.5	175.2	69.0	187.0	73.6	191.8	75.5	6.6	2.6
Greece	Mixed military	1084	157.5	62.0	160.9	63.4	170.5	67.2	180.3	71.1	184.8	72.7	5.88	2.32
Great Britain	Bomber personnel	11772	157.7	62.1	162.2	63.8	173.2	68.3	184.0	72.4	188.5	74.2	6.2	2.44
Great Britain	RAF pilots	4357	161.8	63.7	166.3	65.5	177.3	69.8	188.1	74.0	192.6	75.8	6.15	2.42
Italy	Mixed military	1358	157.0	61.8	160.2	63.1	170.7	67.2	180.7	71.2	185.6	73.0	6.23	2.45
Japan	JASDF pilots	239	157.3	61.9	159.4	62.8	166.7	65.7	175.0	68.8	180.4	71.1	4.8	1.89
South Korea	ROKAF pilots	264	157.6	62.1	159.6	62.9	168.7	66.4	173.7	68.4	177.9	70.0	4.6	1.81
Norway	Recruits 20-yr.	5765	162.3	64.0	166.7	65.7	177.5	70.0	188.1	74.1	192.5	76.8	6.03	2.38
Thailand	Mixed military	2950	151.5	59.7	155.0	61.1	163.5	64.4	172.0	67.8	176.0	69.3	5.3	2.08
Turkey	Mixed military	915	157.8	62.2	160.6	63.2	169.0	65.5	179.2	70.6	182.4	71.6	5.73	2.26
USA:														
Air Force—'50	USAF flyers	4063	161.3	63.5	165.5	65.2	175.6	69.1	185.8	73.1	190.3	74.9	6.19	2.44
Army	Separatees	24449	157.8	62.2	162.4	64.0	173.8	68.4	185.0	72.8	189.7	74.7	6.3	2.48
Army	Pilots	500	163.4	64.4	167.0	65.8	176.1	69.4	186.1	73.3	190.0	74.8	5.7	2.25
Navy—'64	Pilots	1549	165.3	65.1	168.2	66.2	177.5	69.9	187.7	73.9	191.3	75.3	5.91	2.33
Vietnam	Mixed military	2129	148.1	58.3	151.6	59.7	160.4	63.2	169.6	66.8	173.0	68.2	5.5	2.16
CIVILIAN														
Africa	Dinka Nilotes	279	166.3	65.5	170.2	67.0	181.7	71.5	191.7	75.5	195.5	77.0	6.1	2.41
Bulgaria	Civilians 26-yr.	114	155.0	61.1	159.3	62.8	169.8	66.8	180.2	71.0	184.5	72.7	5.89	2.32
Ceylon	Sinhalese	635	145.6	57.3	150.0	59.1	160.4	63.2	171.1	67.4	175.5	69.2	6.02	2.37
China (Taiwan)	Young men	31	154.4	60.8	158.2	62.3	167.6	66.0	177.0	69.7	180.8	71.2	5.22	2.05
East Germany	Civilians 30–39	12200	153.2	60.3	158.4	62.4	170.9	67.3	183.9	72.4	189.1	74.5	7.19	2.83
India	Civilians 19–60	499	147.8	58.2	152.2	60.0	163.0	64.2	173.9	68.5	178.3	70.3	6.10	2.41
India	Mixed	3774	150.1	59.5	154.1	60.7	163.9	64.6	174.0	68.5	178.0	70.1	5.57	2.79
Israel	Aged 40–44	3339	152.1	59.9	156.9	61.8	168.4	66.3	180.0	70.9	184.8	72.7	6.5	2.56
USA: FAA	Tower trainees (21–50)	678	160.8	63.3	166.4	65.5	176.6	69.5	187.0	73.6	192.6	78.5	6.35	2.5
National Health Survey	Total males (18–79)	3091	156.7	61.7	161.5	63.6	173.5	68.3	185.0	72.8	189.4	74.6	----	----

Table ANTH-D3 Stature of Nude International Males[1]

Population	Group	N	1 cm	1 in.	5 cm	5 in.	Means cm	Means in.	95 cm	95 in.	99 cm	99 in.	SD cm	SD in.
MILITARY														
Canada	RCAF pilots	314	84.8	33.3	85.8	33.8	91.3	36.0	96.3	37.9	99.2	39.1	3.68	1.30
Greece	Mixed military	1084	83.3	32.8	85.4	33.6	90.3	35.6	95.2	37.5	97.0	38.2	3.01	1.19
Great Britain	RAF pilots	4357	32.8	33.4	87.1	34.3	92.7	36.5	98.0	38.6	100.2	39.5	3.3	1.30
Italy	Mixed military	1358	82.0	32.3	84.2	33.2	89.7	35.3	94.8	37.3	97.1	38.2	3.2	1.26
Japan	JASDF pilots	239	84.5	33.2	86.2	33.9	90.7	35.7	95.3	37.5	97.5	38.4	2.62	1.03
South Korea	ROKAF pilots	264	83.3	32.8	86.1	33.9	90.7	35.7	95.3	37.5	98.2	38.6	2.81	1.11
Norway	Recruits 20-year	5765	84.0	33.2	86.5	34.2	92.7	36.5	98.8	38.9	101.3	39.9	3.47	1.37
Thailand	Mixed military	2950	79.5	31.3	81.5	32.1	86.5	34.1	91.5	36.0	93.5	36.8	3.1	1.22
Turkey	Mixed military	915	83.1	32.7	84.8	33.4	89.7	35.3	95.1	37.4	97.3	38.3	3.15	1.24
USA:														
Air Force—														
'50	Flyers	4063	83.5	32.9	85.8	33.8	91.4	36.0	96.6	38.0	98.9	38.9	3.27	1.29
Army	Separates	24449	82.5	32.5	85.1	33.5	90.9	35.8	97.0	38.2	97.0	39.2	3.4	1.34
Army	Pilots	500	82.5	32.5	85.1	33.5	90.4	35.6	95.7	37.7	98.3	38.7	3.22	1.27
Navy—														
'64	Pilots	1549	85.0	33.4	87.0	34.2	92.1	36.3	97.4	38.4	100.0	39.4	3.16	1.25
Vietnam	Mixed military	2129	77.5	30.5	79.6	31.3	85.0	33.5	90.5	35.6	92.5	36.4	3.3	1.30
CIVILIAN														
Bulgaria	26-year	114	80.5	31.7	83.1	32.7	89.4	35.2	96.0	37.8	98.5	38.8	3.68	1.45
Ceylon	Sinhalese	635	73.2	28.8	75.7	29.8	81.5	32.1	87.6	34.5	90.2	35.5	3.45	1.36
China (Taiwan)	Young men	31	83.7	32.8	85.3	33.6	90.4	35.6	95.5	37.6	97.7	38.5	2.92	1.15
India	Ages 19–60	499	77.6	30.6	80.0	31.5	85.1	33.5	90.4	35.6	92.7	36.5	3.04	1.19
Israel	Ages 40–44	3339	80.5	31.7	84.3	33.2	89.2	35.1	94.3	37.1	96.3	37.9	3.7	1.46
USA														
FAA	Tower trainees (21–50)	678	83.8	33.0	86.2	33.9	92.0	36.2	97.5	38.4	100.1	39.4	3.27	1.29
Drivers	Truck, bus	360	83.5	32.8	86.1	33.9	92.7	36.5	99.5	39.2	101.5	40.2	3.99	1.57
Nat'l. Health Survey	Total males (18–79)	3091	81.0	31.9	84.3	33.2	90.6	35.7	96.4	38.0	98.8	38.9	----	----

Table ANTH-D4: Weight of Nude International Males[1]

Population	Group	N	1 kg	1 lb	5 kg	5 lb	Mean kg	Mean lb	95 kg	95 lb	99 kg	99 lb	SD kg	SD lb
MILITARY														
Belgium	Flying personnel	2450	49.1	108.0	54.2	119.1	66.5	146.2	78.5	172.7	83.4	183.5	6.9	15.2
Canada	RCAF pilots	314	56.2	123.4	59.7	131.4	76.6	167.8	76.3	202.9	92.1	233.5	9.91	21.8
France	Pilots, cadets	1000	48.2	106.0	53.3	117.2	65.8	144.8	78.3	172.0	83.4	183.3	7.03	15.5
France	Recruits 20-year	234	47.2	103.9	52.4	115.1	65.4	143.8	77.8	171.3	83.0	182.5	7.17	15.8
West Germany	Recruits 20-year	378133	46.1	101.3	52.5	115.5	68.4	150.4	84.1	185.0	90.5	199.0	8.9	19.6
Greece	Mixed military	1084	52.2	114.8	55.8	122.9	67.0	147.4	80.5	177.0	86.7	190.8	7.62	16.3
Great Britain	Bomber personnel	11772	48.6	106.9	53.7	118.1	66.1	145.1	78.8	173.3	83.9	184.5	7.06	15.5
Great Britain	RAF pilots	4357	49.0	107.7	55.8	122.8	72.1	158.6	88.6	195.0	95.4	209.9	9.27	20.4
Italy	Mixed military	1358	53.9	118.7	57.6	126.8	70.3	154.7	85.0	187.0	93.1	204.8	8.43	18.6
Japan	JASDF pilots	239	50.3	110.7	52.5	115.5	61.1	133.2	71.8	158.0	79.9	175.8	5.86	12.9
South Korea	ROKAF pilots	264	50.2	110.3	53.1	116.8	62.9	138.2	76.5	168.5	82.5	181.7	6.53	14.4
Norway	Recruits 20-year	5765	51.5	113.2	57.0	125.3	70.1	154.2	83.3	183.2	88.8	195.6	7.46	16.4
Thailand	Mixed military	2950	45.5	100.1	48.0	105.5	56.3	123.8	67.0	147.3	73.0	160.6	10.3	22.7
Turkey	Mixed military	915	48.2	106.0	52.5	115.5	64.6	142.1	79.2	174.1	87.6	192.7	8.23	18.1
USA:														
Air Force—'50	Flyers	4063	56.0	123.1	60.3	132.5	74.4	163.7	94.5	200.8	98.1	215.9	9.27	20.9
Army	Separatees	24449	47.1	103.5	54.0	118.8	70.3	154.8	87.0	191.3	93.8	203.2	9.33	20.5
Army	Pilots	500	56.2	123.6	61.8	135.9	75.4	165.8	89.8	199.7	96.6	212.5	8.59	18.9
Navy—'64	Pilots	1549	58.7	129.3	63.7	140.3	77.7	171.4	92.3	203.6	100.2	220.9	8.66	19.1
Vietnam	Mixed military	2129	39.8	87.5	42.4	93.1	51.1	112.4	61.5	135.2	70.0	154.0	6.0	13.2
CIVILIAN														
Africa	Dinka Nilotes	279	42.5	93.5	47.7	105.0	58.3	128.3	68.8	151.3	73.2	161.0	5.92	13.0
Bulgaria	Civilians age 26	114	47.3	104.1	53.3	117.3	67.7	148.9	82.0	180.4	88.0	193.5	8.13	17.9
China (Taiwan)	Young men	31	42.0	92.4	45.1	97.0	52.6	115.7	60.7	133.5	63.8	140.5	4.44	9.77
India	19–60	499	30.7	67.5	36.7	80.7	50.9	123.0	65.0	143.0	71.0	156.1	8.00	17.6
Israel	40 years	3339	44.9	98.8	52.9	116.3	72.2	158.8	91.8	202.0	99.8	208.3	11.0	24.2
USA: FAA	Tower trainees (21–50)	678	48.5	106.6	58.0	127.9	73.4	161.8	90.3	199.0	98.7	217.0	10.0	22.1
National Health Survey	Total males (18–79)	3091	50.9	112.0	57.3	126.0	76.3	168.0	98.6	217.0	109.5	241.0	----	----

Table ANTH-D5: National Groups That Do Not Exceed the 75th Percentile (3rd Quartile) For USAF Flying Personnel[2]

	Italian AF	Greek AF	Turkish AF	South Korean AF	Japanese AF
Stature	93	93	96	99	99
Weight	88	95	97	98	99
Sitting height	89	86	88	81	87
Buttock-knee length	91	96	97	100	----
Gluteal furrow height	93	96	97	100	----
Foot length	82	85	87	100	----
Foot breadth	32	39	39	73	----
Hand length	78	84	83	97	95
Hand breadth at metacarpale	69	79	86	98	83
Head circumference	89	95	98	94	91

Table ANTH-D6: Canadian Stature Variation by Age[2]

Age	Percentile 5th	Percentile 50th	Percentile 95th	Change from previous value (50th percentile)
18–19	162.8	173.2	183.1	
20–24	162	173.3	184.2	+0.1
25–29	159.8	173.5	187.9	+0.2
30–34	161	172.9	184.9	−0.6
35–44	159.2	171.7	184.4	−1.2
45–54	156.9	169.7	182.9	−2.0
55–64	153.9	167.6	181.1	−2.1
>64	153.9	165.4	177.3	−2.2

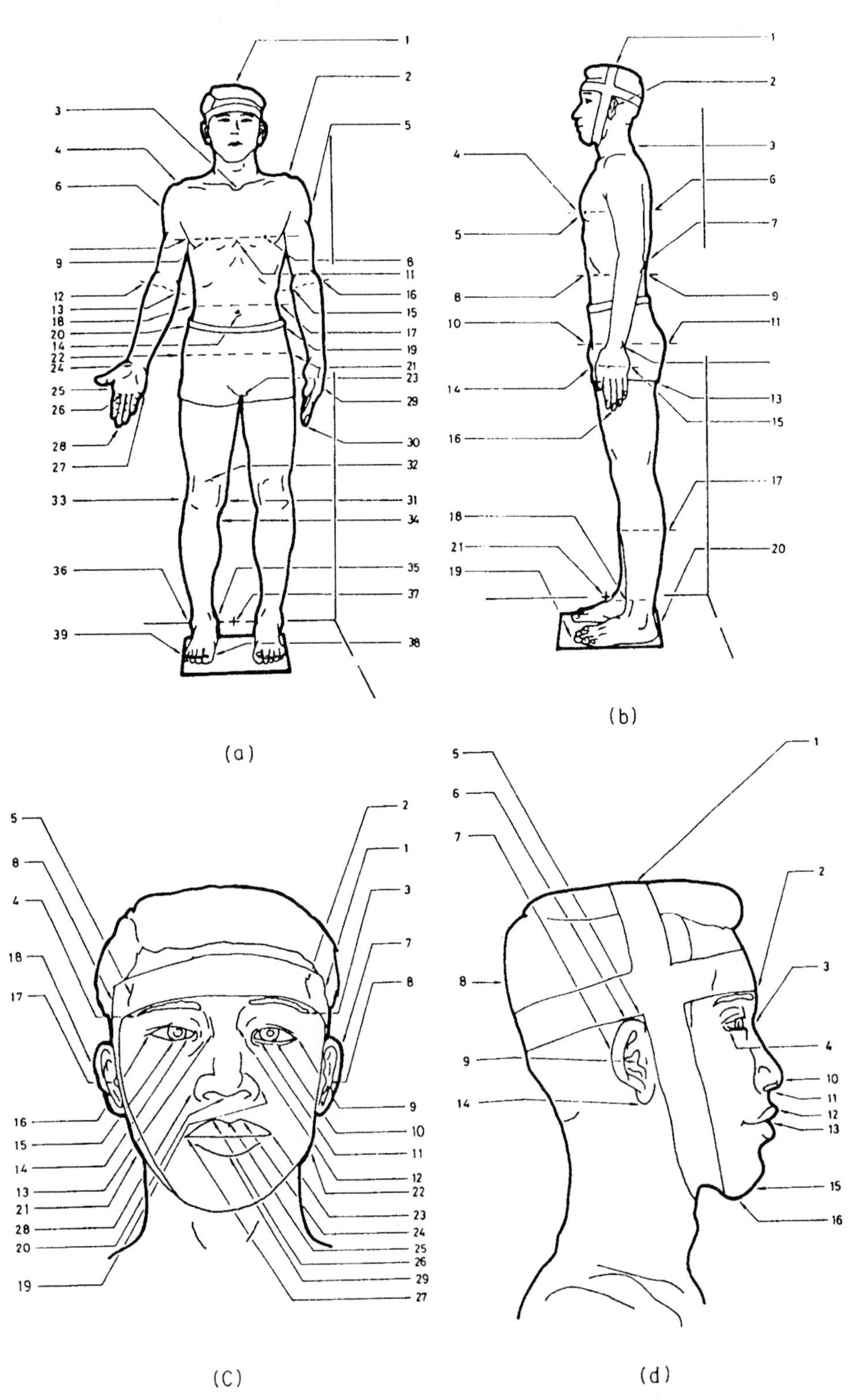

Figure ANTH-D1: Selected Reference Points on the Body

Table ANTH-D6: Measurements Used With Chinese Civilians[3] (Numbers DO NOT Match Figure on Previous Page)

1. Acromion-humerus length
2. Shoulder (acromiale) height, standing, right, vertical
3. Shoulder (acromiale) height, standing, left, vertical
4. Shoulder (acromiale) height, standing, average, vertical
5. Suprasternale height, vertical
6. Shoulder-fingertip length (acromion-dactylion)
7. Substernale height, vertical
8. Waist height (omphalion), vertical
9. Waist height, right, vertical
10. Waist height, left, vertical
11. Waist height, average, vertical
12. Hip (trochanteric) height, right, vertical
13. Hip (trochanteric) height, right, vertical
14. Hip (trochanteric) height, average, vertical
15. Crotch height, vertical
16. Palm length
17. Hand length
18. Knuckle (metacarpale III) height, vertical
19. Kneecap (patella) height, vertical
20. Tibiale height, vertical
21. Sphyrion height, medial malleolus height, vertical
22. Lateral malleolus height, vertical
23. Biacromial breadth
24. Shoulder (bideltoid) breadth, horizontal
25. Chest breadth
26. Bustpoint-bustpoint breadth
27. Elbow (humeral) breadth—palm facing camera.
28. Elbow (humeral) breadth—palm facing body
29. Waist breadth
30. Bi-iliocristale breadth
31. Hip breadth, standing
32. Hand breadth
33. Knee (femoral) breadth
34. Bimalleolar breadth
35. Stature, vertical
36. Tragion height, standing, vertical
37. Cervicale height, standing, vertical
38. Ulnar (olecranon-styloid process) length
39. Elbow (ulna) height, vertical
40. Buttock height, vertical
41. Wrist (stylion) height
42. Gluteal furrow height, vertical
43. Fingertip (dactylion) height, vertical
44. Calf height, vertical
45. Ankle height, vertical
46. Chest/bustpoint depth
47. Waist depth
48. Buttock depth
49. Hand breadth at thumb
50. Foot length
51. Philtrum length, vertical
52. Lip-to-lip length, vertical
53. Hand breadth
54. Minimum frontal breadth
55. Maximum frontal breadth
56. Biauricular breadth
57. Bitragion breadth
58. Biocular breadth
59. Interpupillary breadth
60. Interocular breadth
61. Nasal root breadth
62. Nose breadth
63. Bigonial breadth
64. Mouth breadth
65. Glabella to top of head, vertical
66. Sellion to top of head, vertical
67. Ectocanthus to top of head, vertical
68. Head height (tragion to top of head)
69. Pronasale to top of head, vertical
70. Subnasale to top of head, vertical
71. Stomion to top of head, vertical
72. Menton to top of head, vertical
73. Nose length (subnasale–sellion)
74. Ear length above tragion, vertical
75. Ear length, vertical
76. Head, diagonal (occiput to menton)
77. Nose protrusion
78. Menton-subnasale, length, vertical
79. Head length (glabella to back of head), horizontal
80. Head length (glabella to back of head), protuberant
81. Sellion to back of head, horizontal
82. Ectocanthus to back of head, horizontal
83. Ear breadth, horizontal
84. Tragion to back of head, horizontal
85. Pronasale to back of head, protuberant
86. Pronasale to back of head, horizontal
87. Subnasale to back of head, horizontal
88. Lip protrusion to back of head, horizontal
89. Menton to back of head, horizontal
90. Bitragion-coronal arc (directly measured)
91. Head circumference (directly measured)
92. Interscye, back (directly measured)
93. Bimalleolar breadth (directly measured)
94. Foot breadth (directly measured)
95. Foot breadth

Table ANTH-D7: Demographics of Sampled Chinese Males[3]

Age (in years)	Clerical Staff	Heavy Physical Workers	Light Physical Workers	Students	Housekeepers	Total
15–19	5	15	24	58	0	102
20–24	13	14	62	13	0	102
25–29	34	25	32	2	0	93
30–34	32	20	24	0	0	76
35–39	19	12	16	0	0	47
40–44	17	11	19	0	0	47
45–49	14	8	20	0	0	42
50–54	16	9	23	0	0	48
55–59	15	6	19	0	2	42
60–64	11	2	15	0	2	30
Total	176	122	254	73	4	629

Table ANTH-D8: Demographics of Sampled Chinese Females[3]

Age (in years)	Clerical Staff	Heavy Physical Workers	Light Physical Workers	Students	Housekeepers	Total
15–19	10	2	26	50	9	97
20–24	27	2	21	8	40	98
25–29	20	1	9	1	58	89
30–34	13	1	8	0	50	72
35–39	6	1	7	0	31	45
40–44	5	1	7	0	32	45
45–49	4	1	6	0	29	40
50–54	2	0	5	0	28	35
55–59	1	0	2	0	25	28
60–64	0	0	0	0	22	22
Total	88	9	91	59	324	571

Table ANTH-D9: Statistics for Stature Females (All Occupations)[3]

Age (in years)	N	Mean (cm)	S.D. (cm)	Var. (%)	Stature Ratio	Minimum (cm)	Maximum (cm)
15–19	65	155.99	4.97	3.188	1.0000	145.63	176.41
20–24	109	157.29	5.33	3.390	1.0000	143.05	167.95
25–29	71	156.11	5.02	3.219	1.0000	143.29	171.97
30–34	76	157.00	5.58	3.551	1.0000	142.80	171.63
35–39	43	156.35	4.98	3.186	1.0000	147.29	167.84
40–44	42	154.04	5.11	3.317	1.0000	141.85	164.41
45–49	33	153.89	4.35	2.829	1.0000	147.96	162.33
50–54	20	153.33	4.56	2.976	1.0000	141.72	162.81
55–59	19	153.93	4.49	2.915	1.0000	143.46	162.24
50–64	4	153.12	5.20	3.394	1.0000	148.95	160.67
•15–64•	482	155.96	5.23	3.351	1.0000	141.72	176.41

Percentiles (cm)

1st	2nd	5th	10th	25th	50th	75th	90th	95th	98th	99th
143.05	145.75	147.96	149.37	152.04	156.27	159.40	162.47	164.14	166.79	167.84

Table ANTH-D10: Comparison of Males Stature in 1986 and 1972[3]

Age (in years)	Mean Difference (cm)	SD (cm)	Z Value
15–19	4.88	6.72	5.26*
20–24	3.24	6.36	3.38*
25–29	2.13	6.53	2.74*
30–34	2.17	6.52	2.92*
35–39	1.86	6.48	2.07*
40–44	1.56	6.35	1.56
45–49	1.02	6.10	1.01
50–54	2.96	5.95	2.75*
55–59	0.12	5.81	0.11
60–64	2.45	5.67	1.54

* $P < 0.05$.

Table ANTH-D11: Comparison of Female Stature in 1986 and 1972[3]

Age (in years)	Mean Difference (cm)	S.D. (cm)	Z Value
15–19	2.06	5.83	2.83*
20–24	2.49	5.49	4.67*
25–29	2.42	5.81	3.47*
30–34	3.31	5.47	5.22*
35–39	3.06	6.25	3.19*
40–44	0.68	6.01	0.72
45–49	1.16	6.16	1.07
50–54	0.50	5.97	0.37
55–59	2.09	5.81	1.56
60–64	2.13	5.64	0.75

* $P < 0.05$.

Table ANTH-D12: Comparison of Body Dimensions Among Three Occupational Groups (in cm)[3]

	Clerical Staff		Light Physical Workers		Heavy Physical Workers		
Measurement	Mean	SD	Mean	SD	Mean	SD	t value
Males: #24	45.17	2.20	44.65	2.08	45.78	2.37	4.36*
#25	31.74	2.32	30.72	2.10	31.63	2.25	3.57*
#46	23.84	2.12	22.02	1.93	24.14	2.04	10.70*
#29	28.96	2.04	27.93	2.18	29.11	2.42	4.39*
#47	22.95	2.98	21.49	3.01	22.79	3.02	3.63*
#32	8.36	0.69	8.58	0.68	8.66	0.67	2.94**
#49	10.35	0.80	10.50	0.81	10.86	0.85	4.09*
#94	9.49	1.12	9.39	1.18	9.89	0.98	3.75*
Females: #24	40.65	1.96	41.45	2.33	43.14	3.50	2.80***
#25	27.44	1.48	27.80	1.90	28.72	2.43	1.93
#46	21.60	1.79	22.06	2.21	24.43	2.79	3.57*
#29	24.73	1.61	25.85	2.91	34.70	20.65	4.35*
#47	18.45	1.88	19.61	3.07	22.03	2.82	4.32*
#32	7.09	0.54	7.59	0.77	7.42	0.58	4.85*
#49	8.87	0.90	8.89	0.75	9.26	0.87	1.02
#94	8.70	0.43	8.90	0.52	9.16	0.62	2.44****

* $p < 0.001$; ** $p < 0.005$; *** $p < 0.01$; **** $p < 0.02$.

REFERENCES

1. Van Cott, H.P., and Kinkade, R.G., 1972. *Human Engineering Guide to Equipment Design. (Revised)* Washington, D.C.: US Government Printing Office. Library of Congress Number: 72-600054.

2. Pulat, B.M. 1992. *Fundamentals of Industrial Ergonomics.* NJ: Prentice-Hall. Reprinted by permission of Prentice-Hall.

3. Li, Chang-Chung, Hwang, Sheue-Ling, and Wang, Min-Yang. 1990. Static Anthropometry of Civilian Chinese in Taiwan Using Computer-Analyzed Photography. Reprinted with permission from *Human Factors* , Vol. 32, No. 3, 1990. Copyright 1990 by the Human Factors Society, Inc. All rights reserved

CHAPTER 2: WORKPLACE TABLES

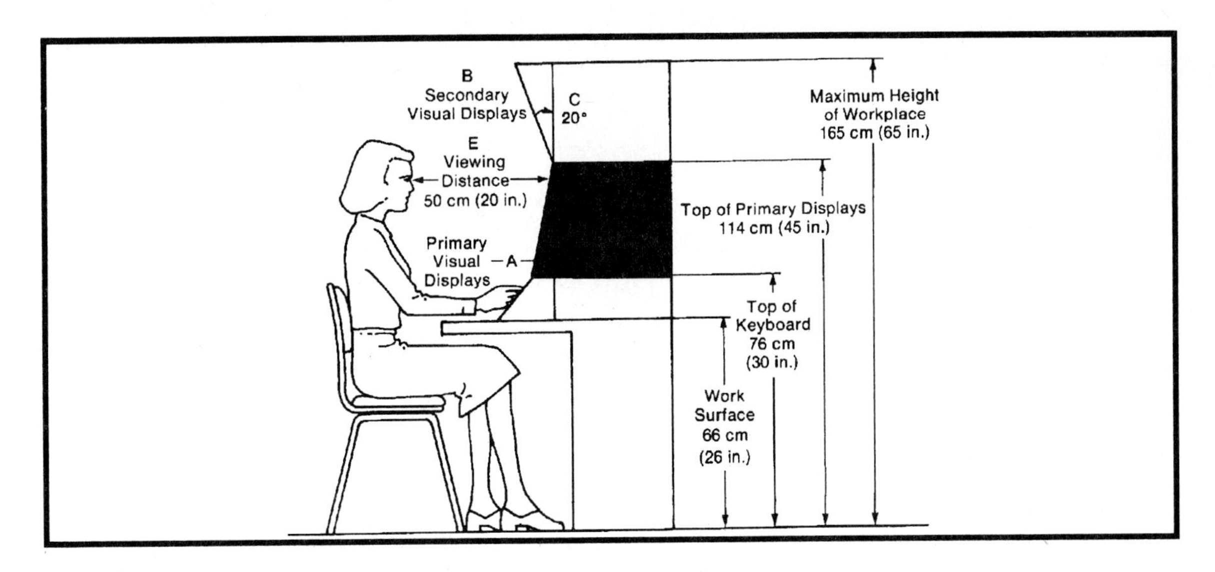

Section A: Seated and Standing Workplace Layout

This section presents information to be used when designing workstations that will be used for sitting or standing. Following the workplace layouts in this section will reduce fatigue and errors and increase worker productivity.

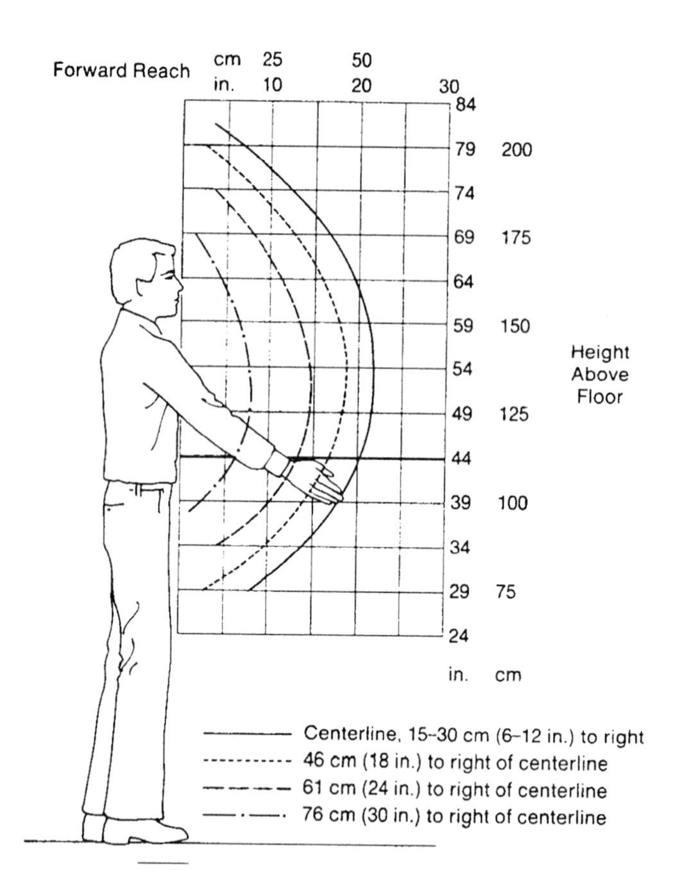

Figure WRKSTN-A1: Standing Reach Area, One Arm[1]

The four curves describe the forward reach from the front of the body at different heights of the hand above the floor for the 5th percentile person (See Anthropometry Section A). The outermost curve shows the forward reach capability within 30cm of the center of the body. Once the arm is positioned more than 30 cm to the right, there is a rapid reduction in forward reach capability at all heights above the floor. Maximum forward reach falls from 51 cm to 15 cm as the arm moves 76 cm to the right.

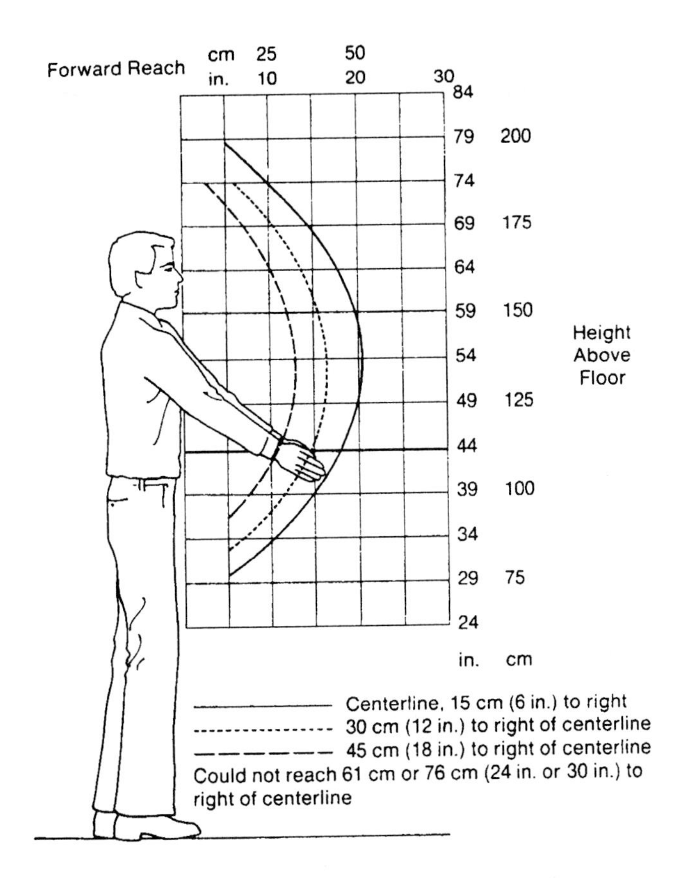

Figure WRKSTN-A2: Standing Reach Area, Two Arms[1]

The four curves describe the forward reach from the front of the body for both arms at different heights of the hands above the floor for the 5th percentile person (See Anthropometry Section A). Maximum forward reach falls from 51 cm to 36 cm with a 36 cm move to the right of the centerline. Forward reach is only marginally shorter two-handed tasks than it is for one-handed tasks within the 46 cm lateral limit, except at the lowest and highest points from the floor.

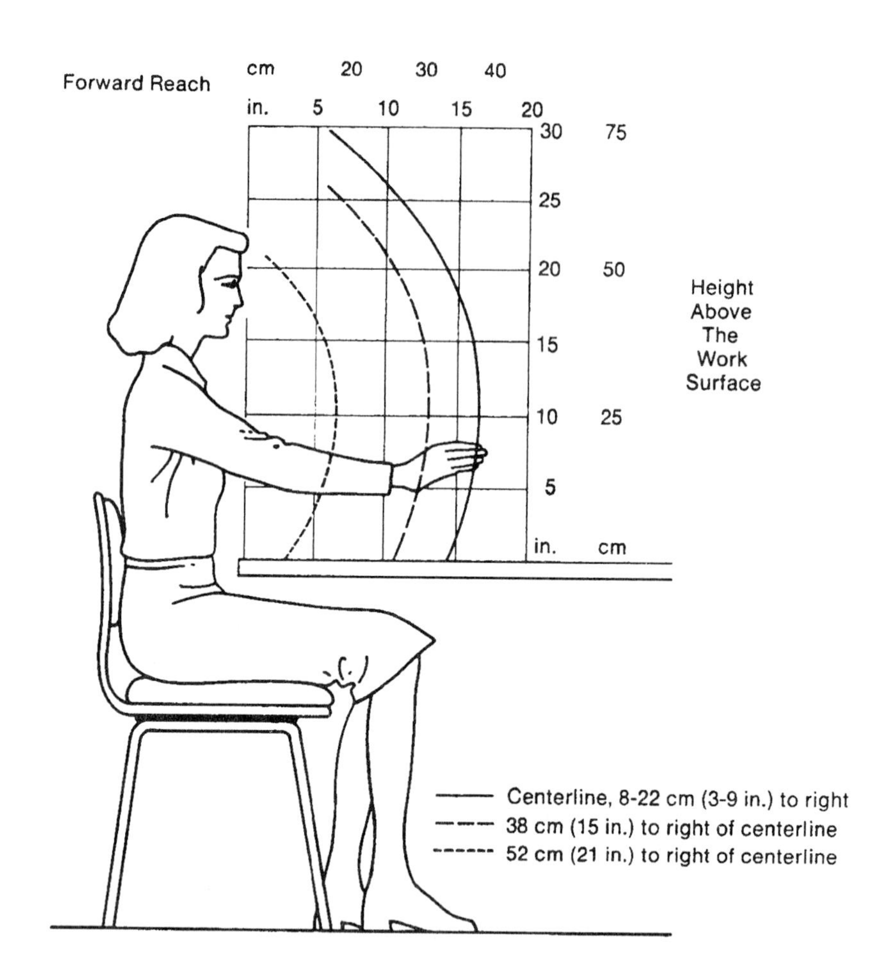

The three curves describe the seated reach workspace for a 5th percentile female's right hand. The forward reach capability is affected by the height of the hands above the work surface (vertical axis) and by the arm's distance to the right of the body's centerline.

Figure WRKSTN-A3: Forward Reach of a Small Operator, Seated[1]

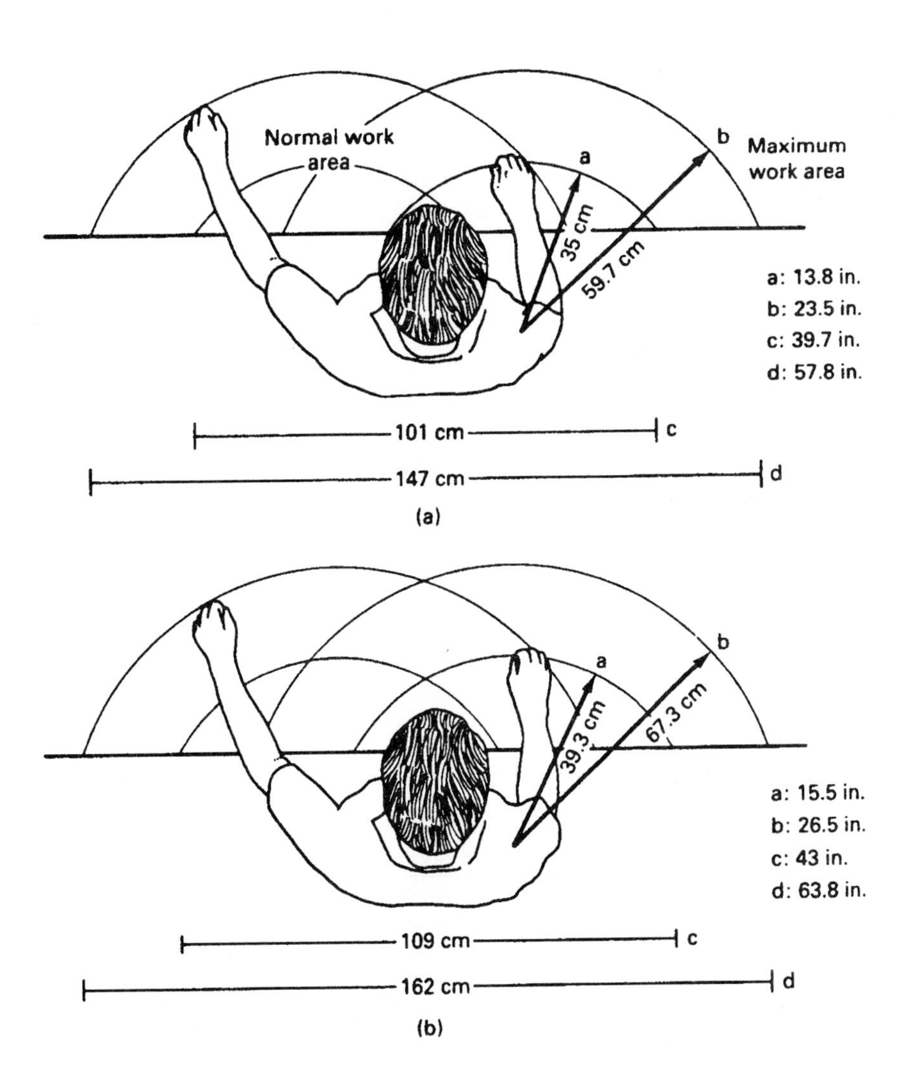

Figure WRKSTN-A4: Normal and Maximum Work Areas for (a) Women and (b) Men[2]

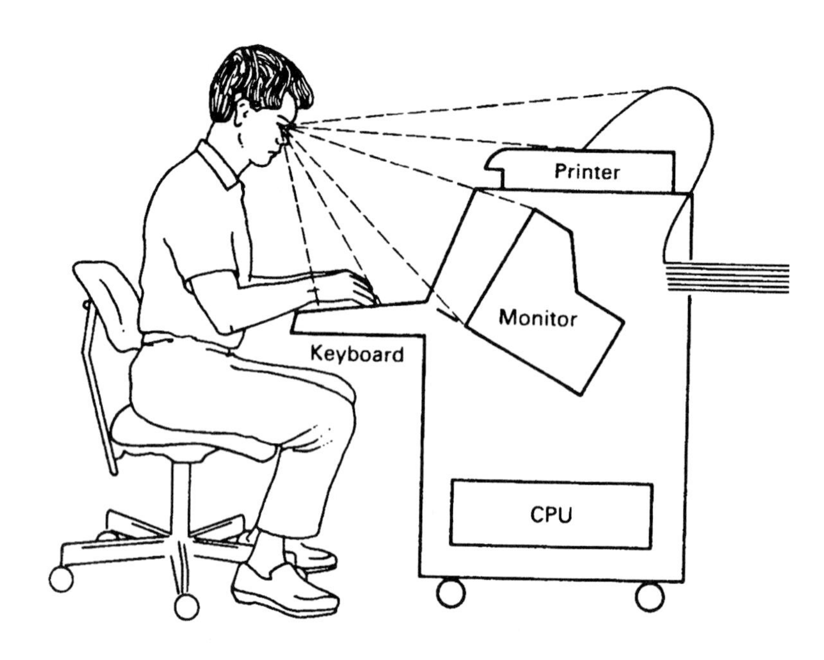

Figure WRKSTN-A5: Portable Work Area[2]

Table WRKSTN-A1: Recommended Work Surface Heights for Seated Operators [2,3]

Type of task	Work surface height [cm (in.)]	
	Male	Female
Fine work	99–105 (39–41)	89–95 (35–37)
Precision work	89–94 (35–37)	82–87 (32–34)
Writing, light assembly	74–78 (29–31)	70–75 (27–29)
Coarse or medium work	69–72 (27–28)	66–70 (26–27)

Table WRKSTN-A2: Recommended Work Surface Heights for Standing Operators [2,3]

Type of task	Work surface height [cm (in.)]	
	Male	Female
Precision work with elbows supported	109–119 (43–47)	103–113 (40–44)
Light assembly	99–109 (39–43)	87–98 (34–39)
Heavy work	85–101 (33–40)	78–94 (31–37)

Figure WRKSTN-A6: Reach Height Above Seat Surface[4]
(To be used with Table WRKSTN-A3)

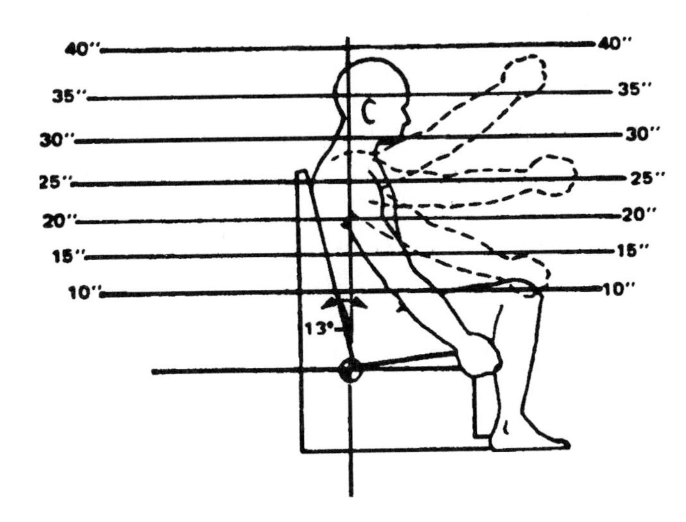

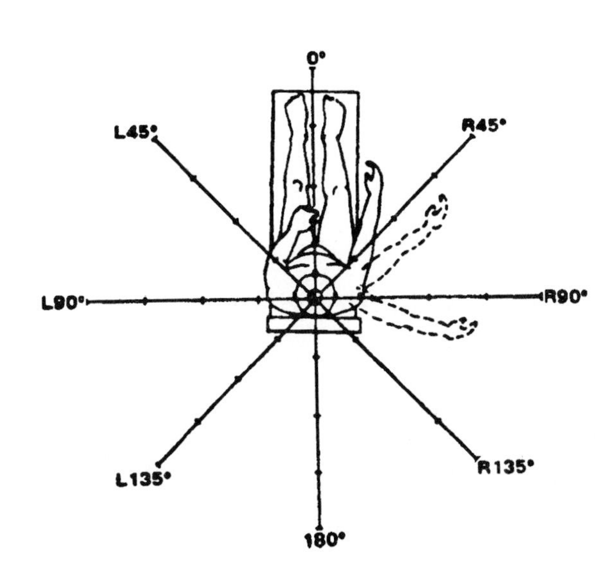

Figure WRKSTN-A7: Angle to Left or Right of Desk Surface[4]
(To be used with WRKSTN-A3)

Table WRKSTN-A3: Useful Dimensions Showing Portions of a Reach Envelope [4]

Desk top: angle to left or right	Height above seat surface using right arm/hand in inches (cm) 10	15	20	25	30	35	40
L 120						9.5 (25)	10.0 (26)
L 105						11.0 (28)	10.5 (27)
L 90						12.5 (31)	11.0 (28)
L 75						13.5 (34)	11.5 (29)
L 60			16.0 (40)	16.5 (42)	15.5 (39)	14.5 (37)	12.0 (30)
L 45		17.0 (43)	17.5 (45)	18.0 (46)	17.0 (43)	15.5 (39)	12.5 (32)
L 30		19.5 (50)	19.5 (49)	20.5 (51)	19.5 (49)	17.5 (44)	14.0 (35)
L 15		21.0 (53)	21.0 (54)	21.5 (55)	21.5 (54)	19.0 (48)	15.0 (39)
0		22.5 (57)	23.0 (58)	23.5 (60)	23.0 (58)	20.0 (51)	17.0 (43)
R 15		24.0 (61)	25.0 (64)	25.5 (65)	24.5 (62)	22.5 (57)	19.0 (48)
R 30	24.5 (62)	25.5 (65)	27.0 (69)	27.0 (69)	26.0 (66)	24.0 (61)	20.5 (52)
R 45	25.5 (65)	27.0 (69)	28.0 (71)	28.0 (71)	27.0 (69)	25.5 (65)	22.0 (57)
R 60	26.0 (66)	28.0 (71)	29.0 (73)	28.5 (72)	28.0 (71)	26.0 (66)	23.0 (58)
R 90	26.5 (67)	28.0 (71)	29.0 (74)	29.0 (74)	28.0 (71)	27.0 (68)	23.5 (60)
R 105	26.0 (66)	27.5 (70)	28.5 (73)	28.5 (72)	28.0 (71)	27.0 (68)	24.0 (61)
R 120	25.0 (63)	26.5 (67)	27.5 (70)	27.5 (70)	27.0 (69)	26.0 (66)	23.5 (60)
R 135	23.5 (60)						

Table WRKSTN-A4: Useful Body Dimensions for Clearances [4]

Dimension	95th Percentile	Example of use
Stature (A-2)	70.5 in. (5′11″) 181 cm	Eliminate overhead obstructions (without shoes or hat)
Head to seat height (C-1)	37.5 in. 96 cm	Privacy for taller users seated in a cubicle
Foot length (B-4)	10.5 in. 27 cm	Footroom under a typing table (without shoes)
Shoulder breadth (C-3)	19.5 in. 50 cm	Determining adequate room for three passengers in the backseat of a car
Hand width (D-2)	3.5 in. 9 cm	Ensuring a large enough space for reaching in a small area
Thigh clearance (E-6)	6.5 in. 17 cm	Ensuring sufficient room for placing knees under a visual display terminal (VDT)

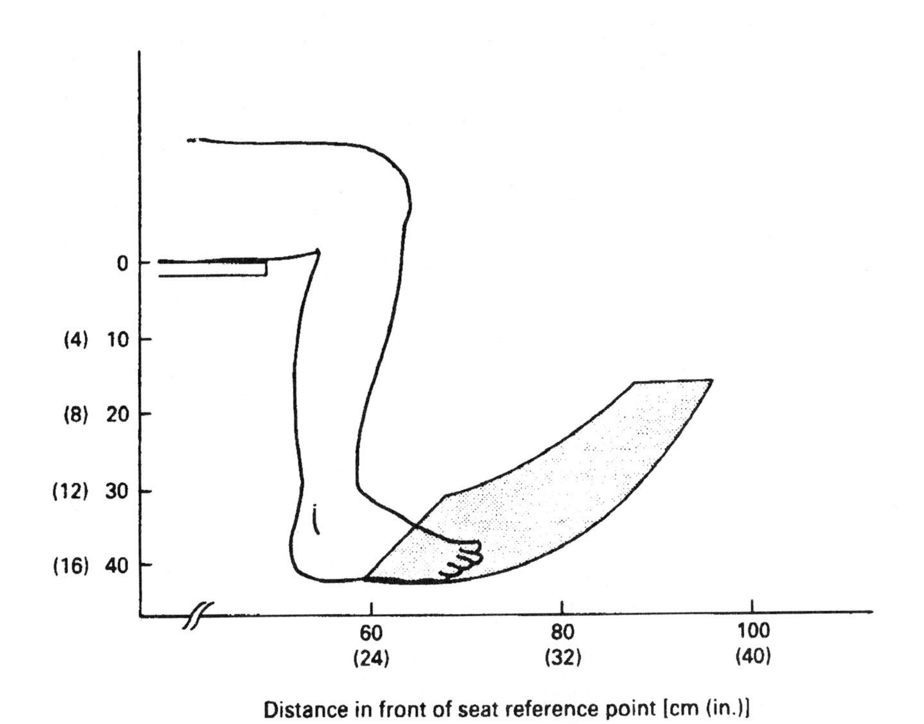

Figure WRKSTN-A8: Best Locations for Foot Controls[2]

Table WRKSTN-A5: Relationships Between Operator Physiological Systems, Environment, and Workplace Layout[5]

Considerations	Physiological systems affected		
	Musculo-skeletal	Cardio-vascular	Gastro-intestinal
Environments:			
Vibration	2	3	3
Oscillation	2	2	1
Acceleration	3	2	3
Impact	1	2	2
Noise	1	3	3
Workplace layout features:			
Improper postural support	1	2	2
Poor distribution of operator's body/limb weight	2	2	3
Awkward body or limb positions	1	3	3
Frequent requirement to use maximum reach or force	1	3	3

Rating criteria: 1 = Critical*, 2 = Important, 3 = Minor.

*Sound reasons can be presented for differential weighting of the above factors in terms of specific condition relationships; however, a good rule of thumb to follow is to recognize gastrointestinal and cardiovascular disturbance minimization should take precedence over musculoskeletal considerations, especially in systems requiring long-term operator exposure.

Table WRKSTN-A6: Choice of Workplace by Task Variables[5]

Parameters	Heavy Load and/or Forces	Intermittent Work	Extended Work Envelope	Variable Tasks	Variable Surface Height	Repetitive Movements	Visual Attention	Fine Manipulation	Duration > 4 Hours
Heavy Load and/or Forces		ST	ST	ST	ST	S/ST	S/ST	S/ST	ST/C
Intermittent Work			ST	ST	ST	S / S/ST	S / S/ST	S / S/ST	S / S/ST
Extended Work Envelope				ST	ST	S/ST	S/ST	S/ST	ST/C
Variable Tasks					ST	S/ST	S/ST	S/ST	ST/C
Variable Surface Height						S	S	S	S
Repetitive Movements							S	S	S
Visual Attention								S	S
Fine Manipulation									S
Duration > 4 Hours									

Note: S = sitting; ST = standing; S/ST = sit/stand (an alternative to standing all day; a standing workplace could be used but would not be the preferred choice); ST/C = standing, with chair available.

Job and workplace characteristics are looked at, two at a time, in relation to the preferred workplace choice: sitting, standing, sit/stand, or standing with a chair provided. More than one type of workplace may be acceptable for these task combinations; the most appropriate choice is indicated.

Table WRKSTN-A6: Choice of Workplace by Task Variables[5]

Table WRKSTN-A7: Levels of Workstation Adjustment[1]

Level	Characteristics	Examples
High	Instantaneous (<5 sec) Continuous Powered or mechanical assist	Pneumatic chair Air hoist Lowerators, levelators
Moderate	Takes 5 to 30 sec Incremental adjustment Manual effort	Chain hoist Pallet truck foot pedal Foldout steps Adjustable lighting fixtures Mechanically adjusted chairs
Low	Takes more than 30 sec Only 2 levels of adjustment Manual effort: pushing or lifting	Sliding, wing-nut chair or footrest adjustment Unpowered platform movement

Three levels of adjustment (column 1) are given. Their adjustment characteristics, including time to adjust, way of responding (continuous or incremental), and method of making the adjustment are identified in column 2. Examples of industrial equipment illustrating each level of adjustment are given in column 3. The higher the adjustability, the more likely it is that this feature will be used.

SEATED WORKSTATIONS

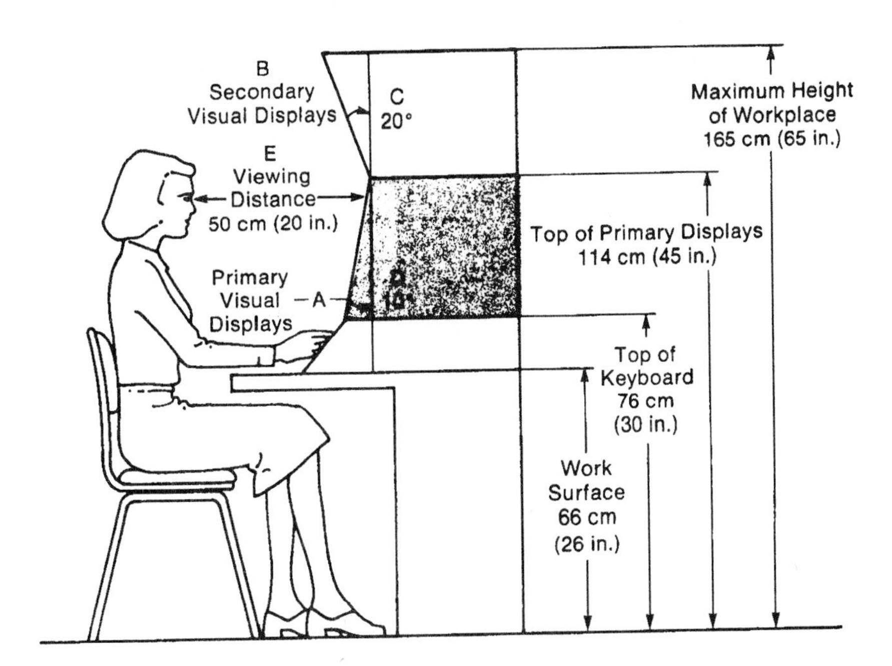

The preferred heights of primary (A) and secondary (B) visual displays for a seated console are shown. Primary displays are those most frequently monitored; secondary displays are less critical to the operation, but they give the operator information about the status of equipment, as needed. The console is at a 20° angle from the vertical plane at heights above 114 cm (45 in.) (C) and at a 10° angle below that height (D) in order to keep the display within 50 cm (20 in.) of the operator's eyes (E) in the resting posture. These suggested heights for primary displays will keep the visual angle within the ranges given in Figure IIA–21 for most people.

Figure WRKSTN:A9: Visual Dimensions for Seated Work[1]

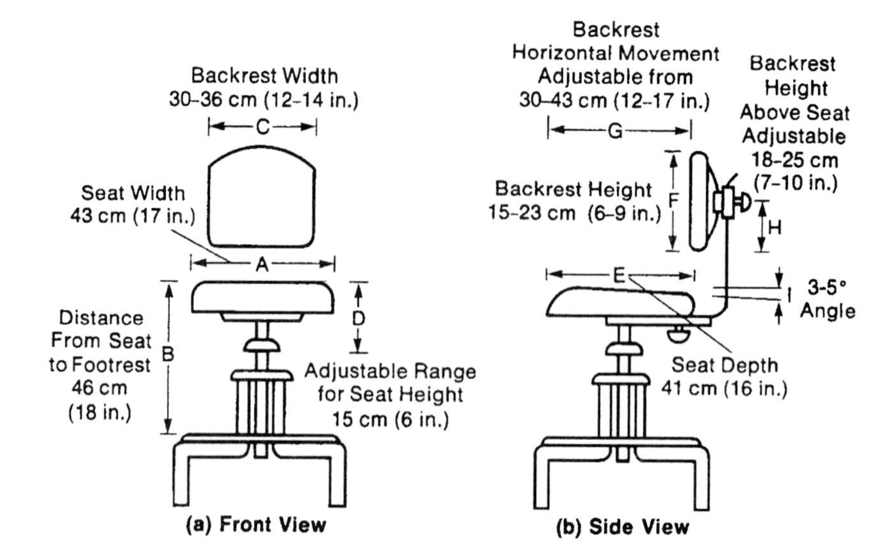

Dimensions for chair seat width (A), depth (E), vertical adjustability (D), and angle (I) and for backrest width (C), height (F), and vertical (H) and horizontal (G) adjustability relative to the chair seat are given, using both front (part a) and side (part b) views. The angle of the backrest should be adjustable horizontally from 30–43 cm (12–17 in.), by either a slide-adjust or a spring, and vertically from 18–25 cm (7–10 in.). This adjustability is needed to provide back support during different types of seated work. The seat should be adjustable within at least a 15-cm (6-in.) range. The height above the floor of the chair seat with this adjustment range will be determined by the type of workplace, with or without a footrest

Figure WRKSTN-A10: Recommended Chair Characteristics[1]

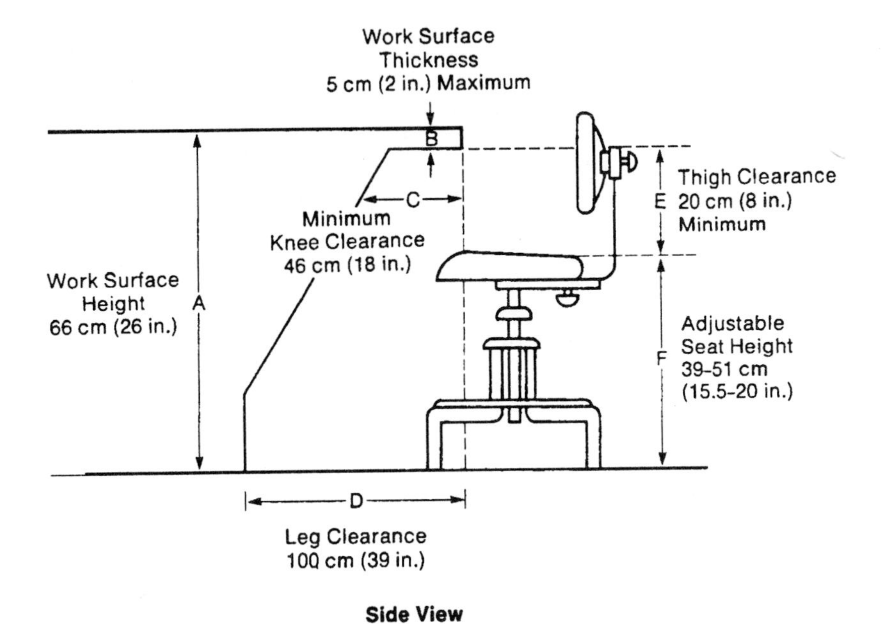

Side View

The heights, clearances, and work surface thickness of a seated workplace without a footrest are indicated. Because the feet must be able to rest on the floor, the chair has to adjust to lower levels (F), the forward leg clearance (D) must be greater, and the work surface height (A) has to be lower than in the case of the seated workplace with a footrest

Figure WRKSTN-A11: Recommended Dimensions for a Seated Workplace Without a Footrest[1]

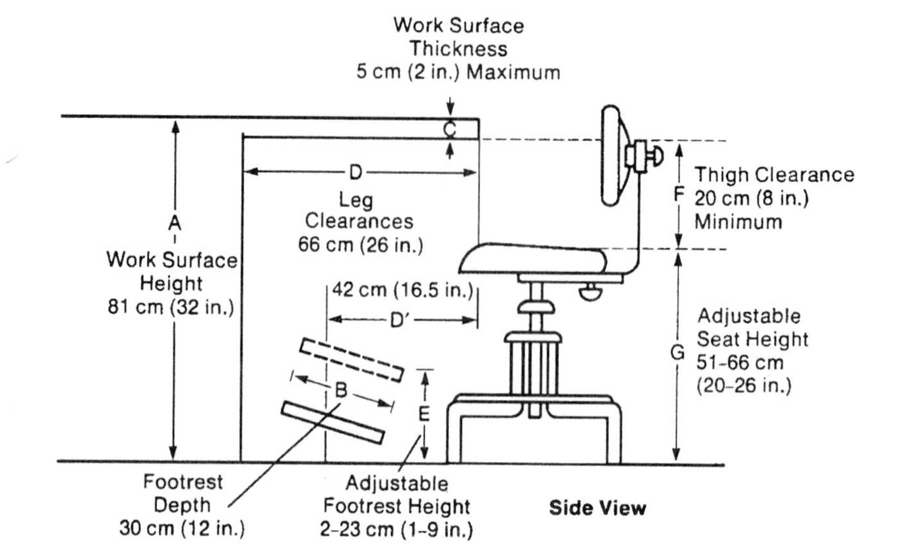

The heights, clearances, and work surface thickness of a seated workplace with a footrest and an adjustable chair are given. These design guidelines ensure that most people will be able to work comfortably at the workplace. The work surface height (A) given is the recommended level; chair (G) and footrest (E) heights should be adjustable to provide adequate thigh clearance (F) and leg comfort. When work surface height is 81 cm (32 in.), the table thickness (C) should not exceed 5 cm (2 in.). Two forward leg clearances are given: D is the recommended distance under the work surface, and D′ is the minimum distance from the edge of the work surface to about two-thirds of the depth (B) of the footrest. Seat and footrest adjustabilities are important in accommodating differences in size of people using a seated workplace.

Figure WRKSTN-A12: Recommended Dimensions for a Seated Workplace With a Footrest[1]

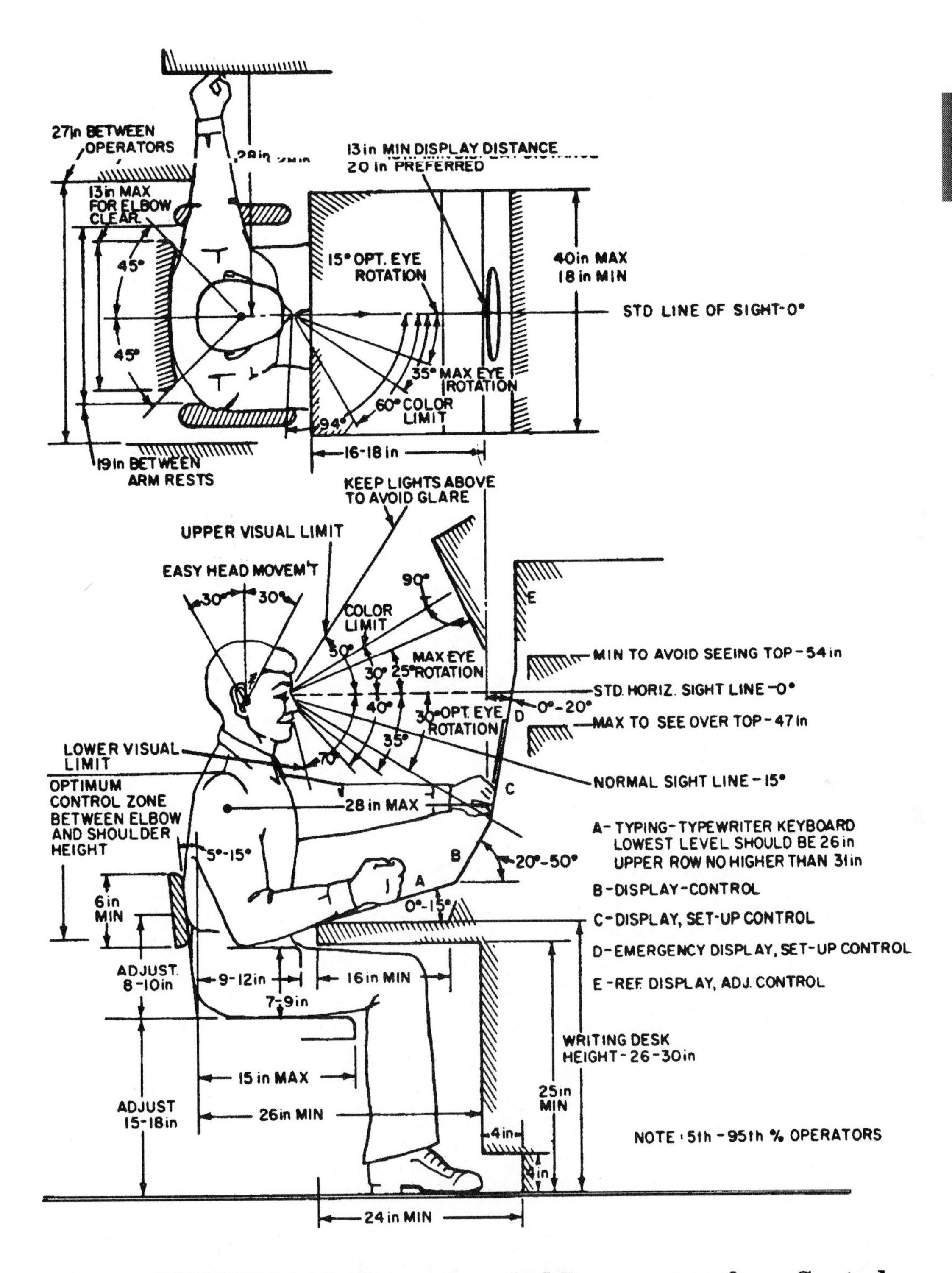

Figure WRKSTN-A13: Recommended Parameters for a Seated Operator Console[5]

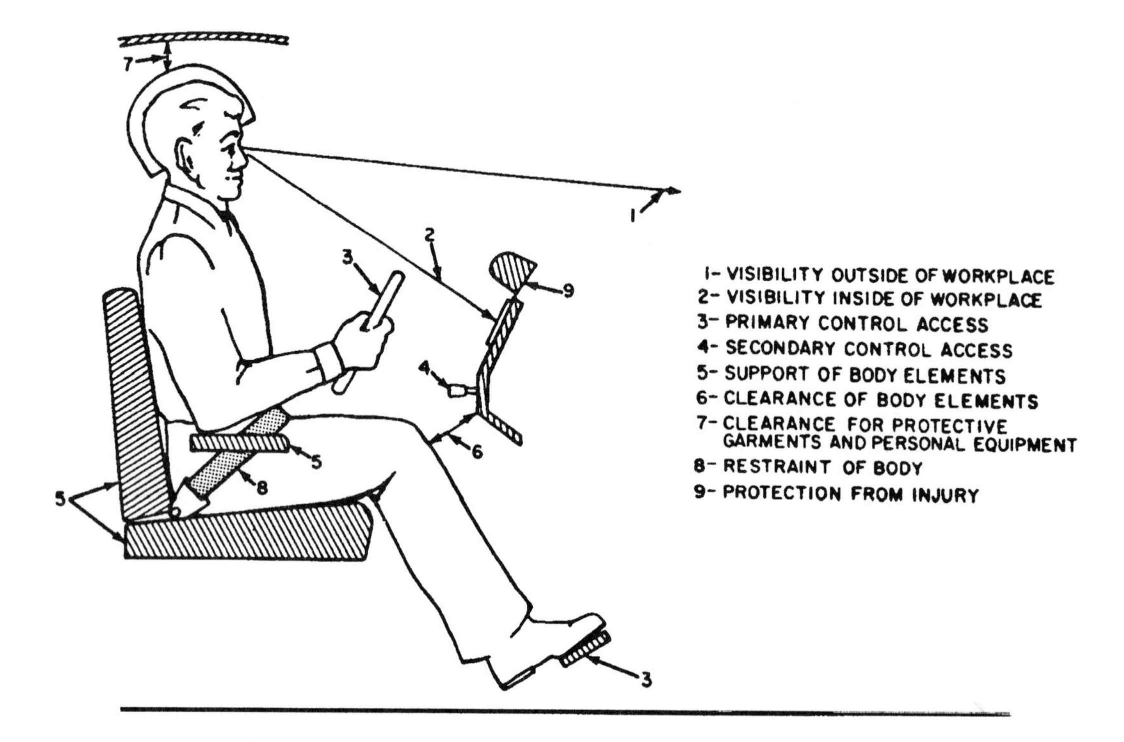

Figure WRKSTN-A14: A Checklist of Priorities for Considering Various Seated Workplace Function Requirements[5]

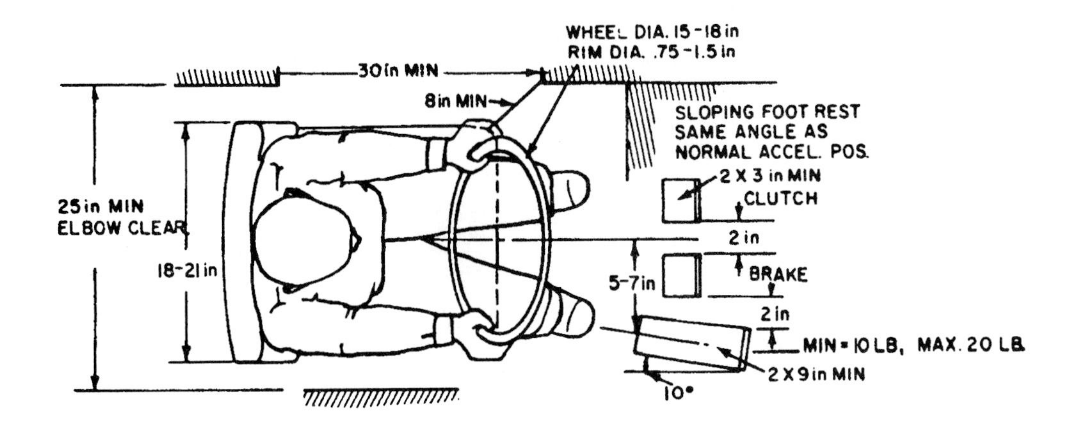

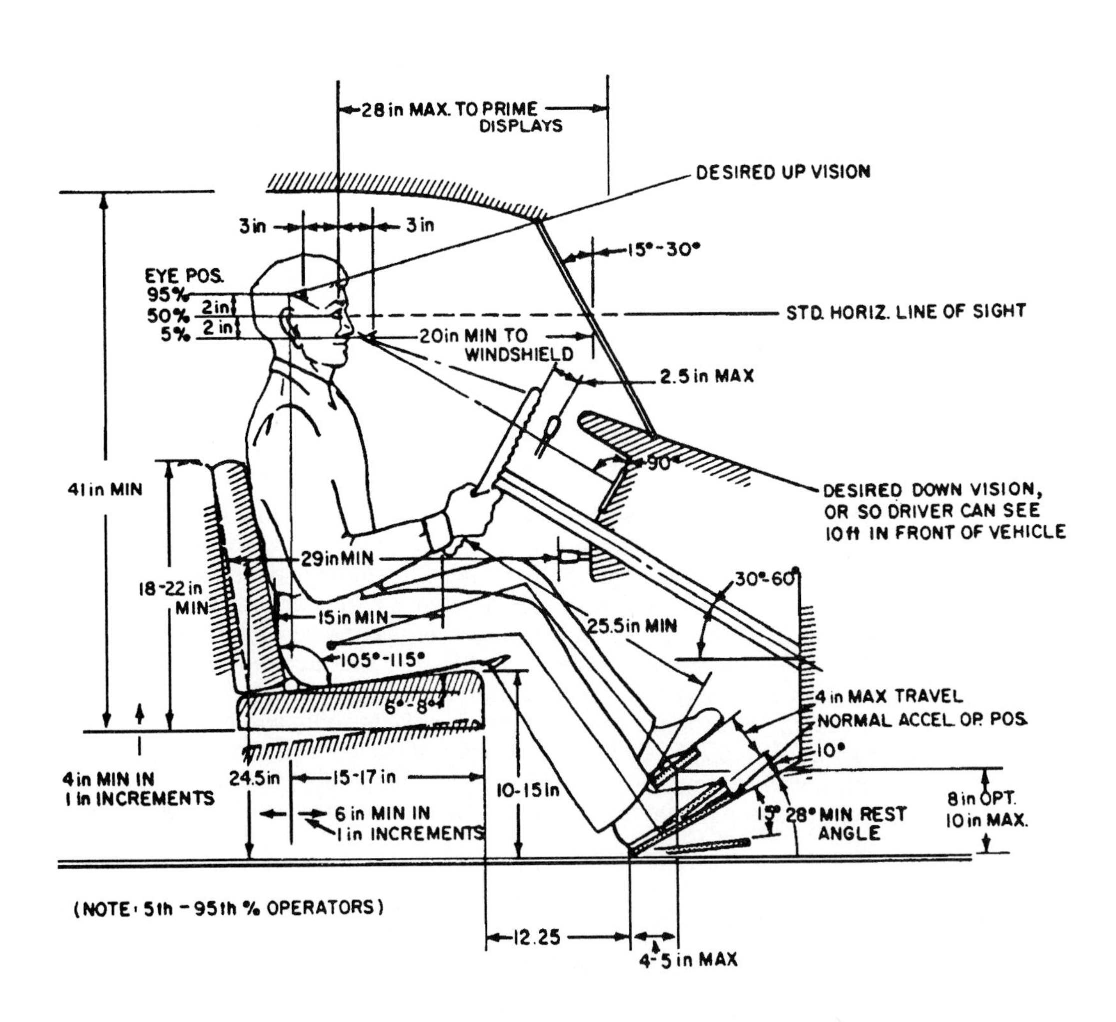

Figure WRKSTN-A15: Suggested Parameters for a Seated Vehicle Operator[5]

STANDING WORKSTATION LAYOUT

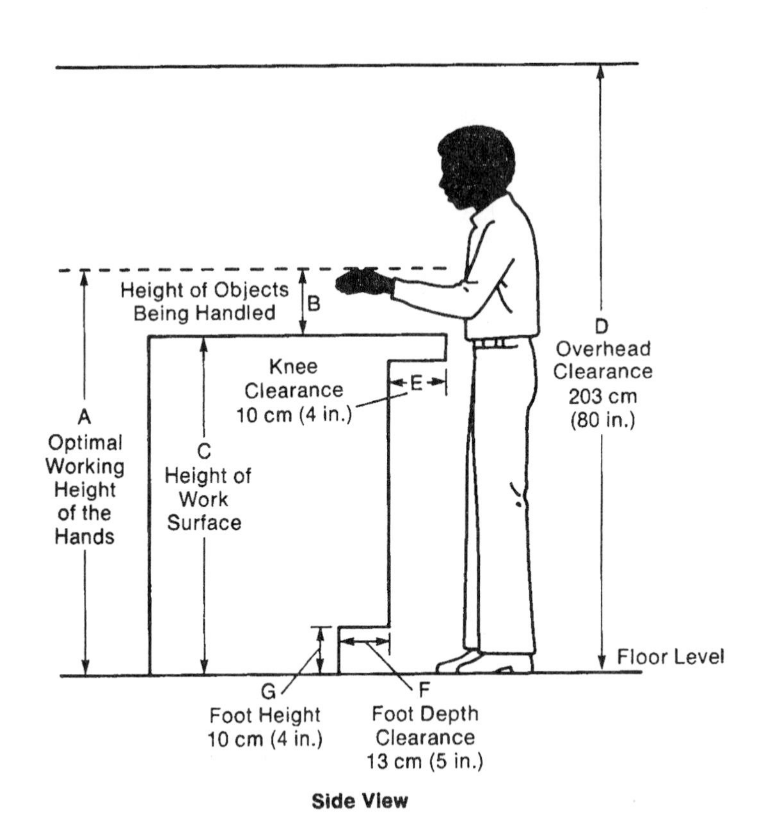

Side View

Workplace height (C) and overhead (D), knee (E), and foot (F and G) clearances are indicated for a typical standing workbench. The knee and foot clearances (E, F, and G) permit the operator to stand with knees bent and feet pointed straight ahead. Work height (C) varies according to the type of task being performed, the size of the objects worked on (B), and the location of the hands when doing it (A). Height A is the optimal working height of the hands, B is the typical height of the objects being assembled, packed, or repaired, and C is the height of the work surface without a product on it. Guidelines for determining the proper standing workplace height for a given task are further explained in the text.

Figure WRKSTN-A16: Standing Workstation Dimension[1]

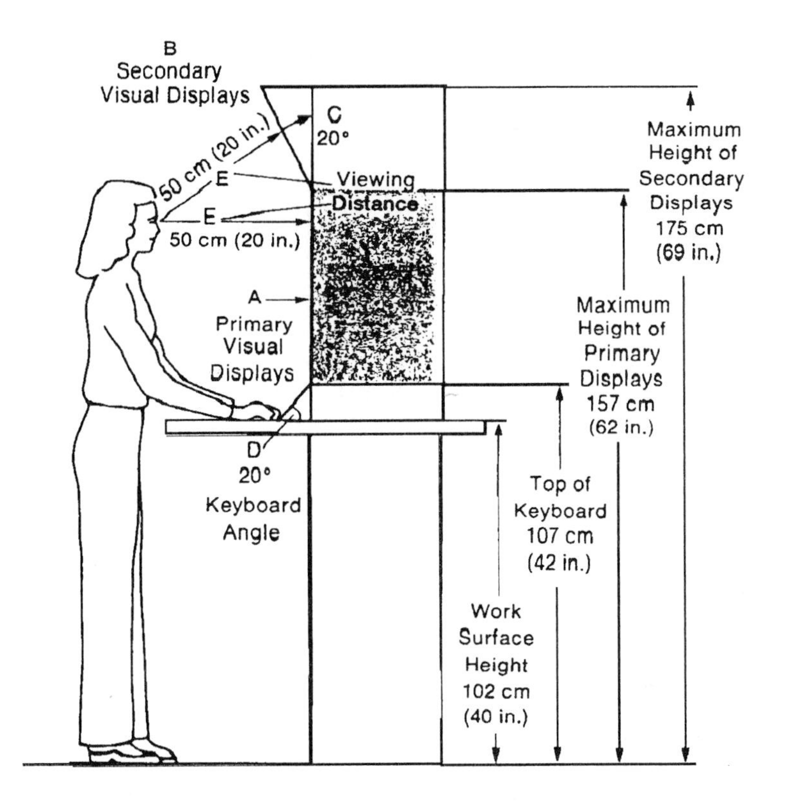

The recommended heights of primary (A) and secondary (B) visual displays for a standing console are given. Primary displays are those most critical to the operation. Secondary displays are consulted less frequently and usually give the operator information about the status of equipment or processes. The front surface of the console is angled at heights above 157 cm (62 in.) (C) to keep the distance to the operator's eyes at 50 cm (20 in.) (E). The keyboard is also at a 20° angle (D) from the work surface top. These guidelines for console design should reduce the potential for neck muscle fatigue

Figure WRKSTN-A17: Visual Dimensions for Standing Workstations[1]

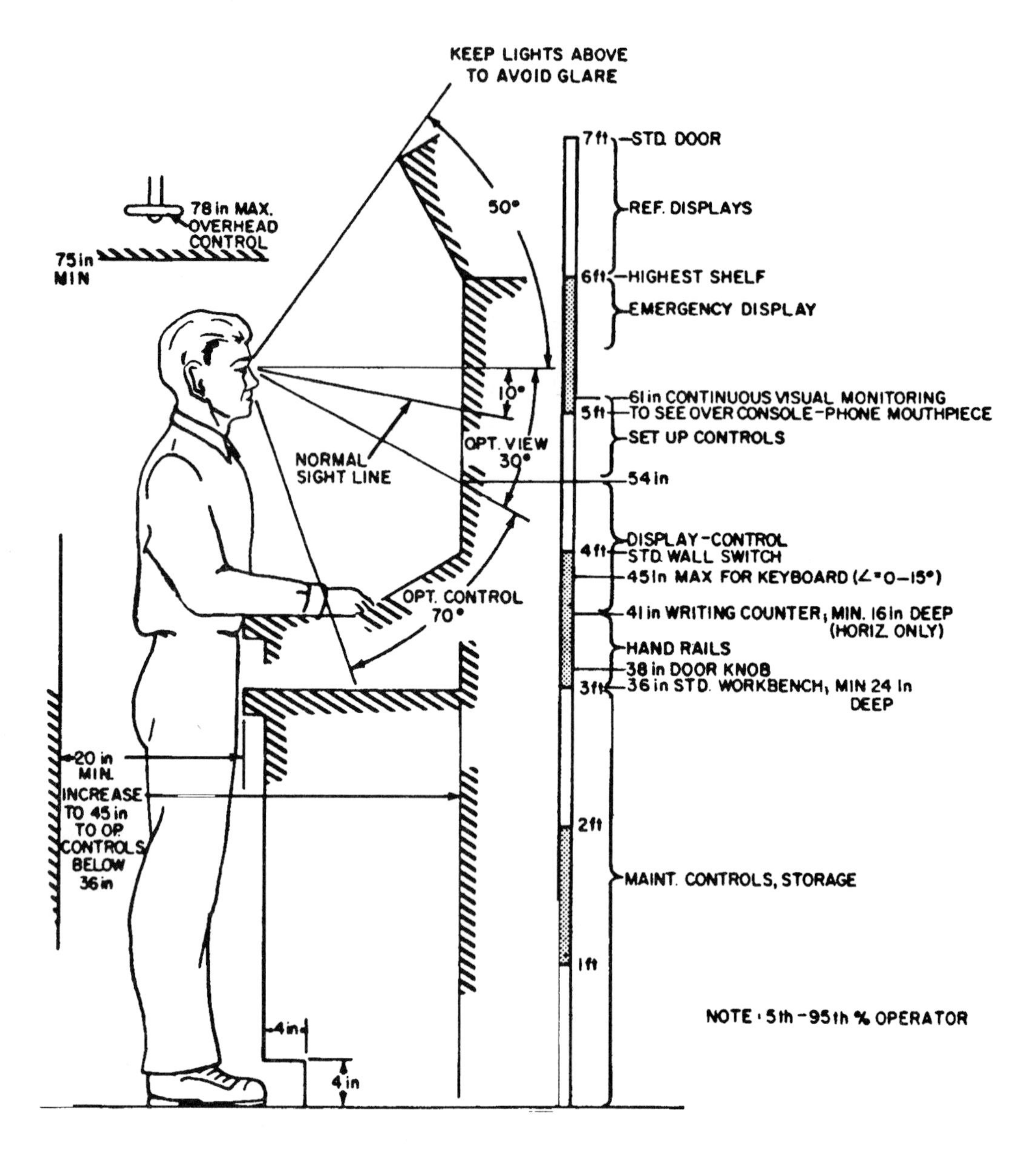

Figure WRKSTN-A18: Suggested Parameters for a Standing Workplace[5]

SHARED WORKSTATIONS

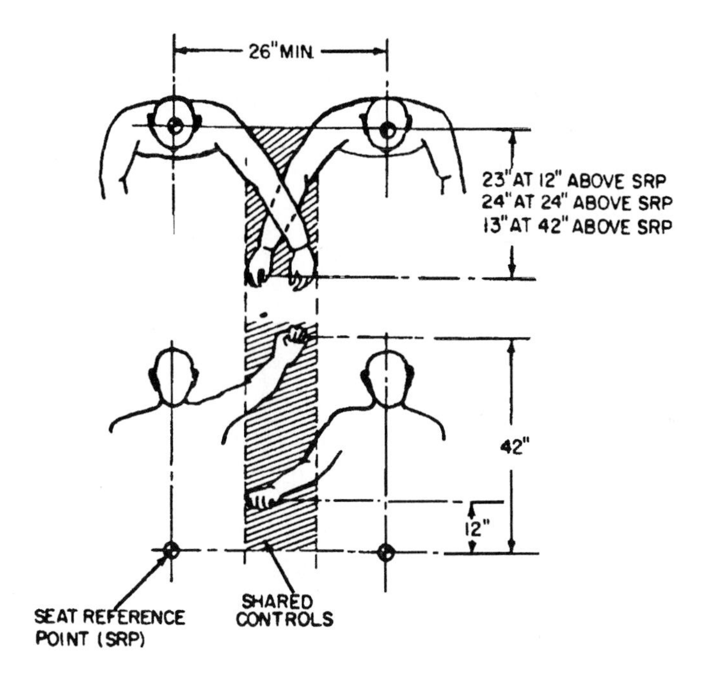

Figure WRKSTN-A19: Suggested Parameters for Side-by-Side Workplaces[5]

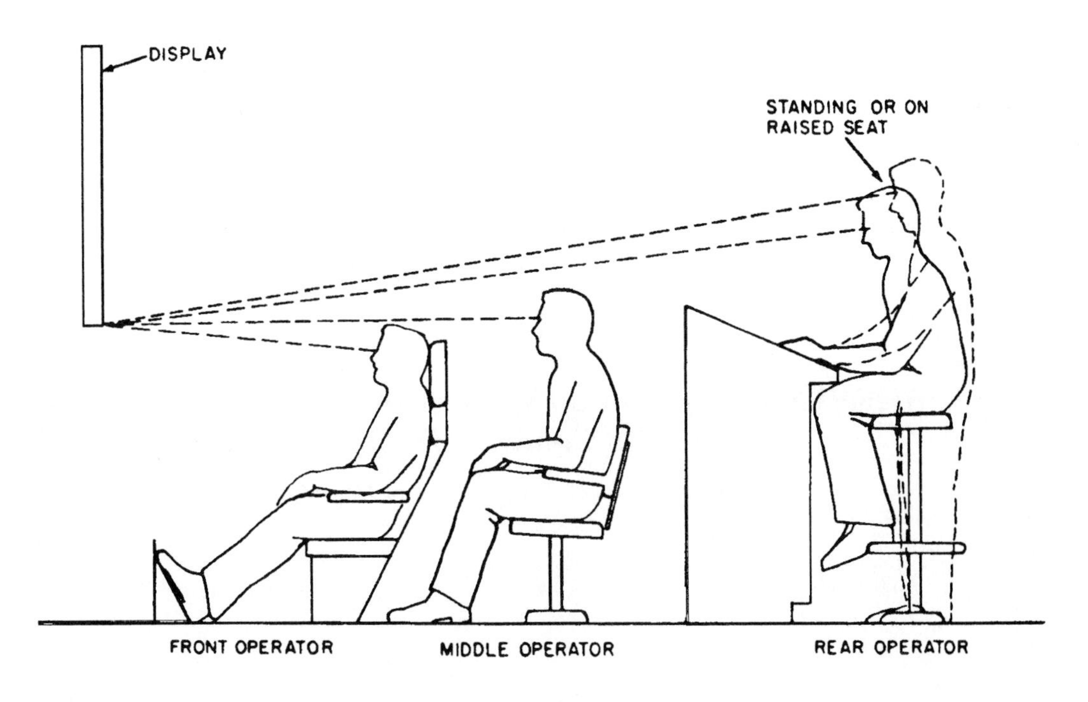

Figure WRKSTN-A20: Seating to Prevent Visual Obstruction of a Common Display[5]

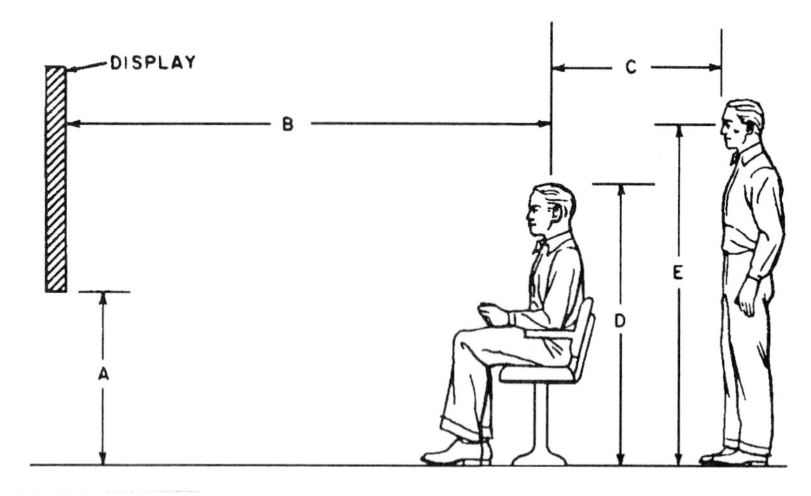

Figure WRKSTN-A20: Dimensions Used to Compute Positions of Standing and Seated Operators[5]

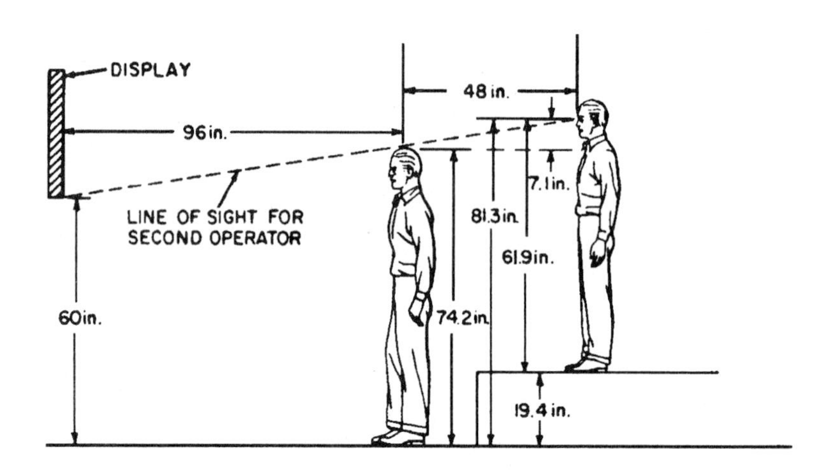

Figure WRKSTN-A21: Use of a Platform With a Standing Workplace With a Common Display[5]

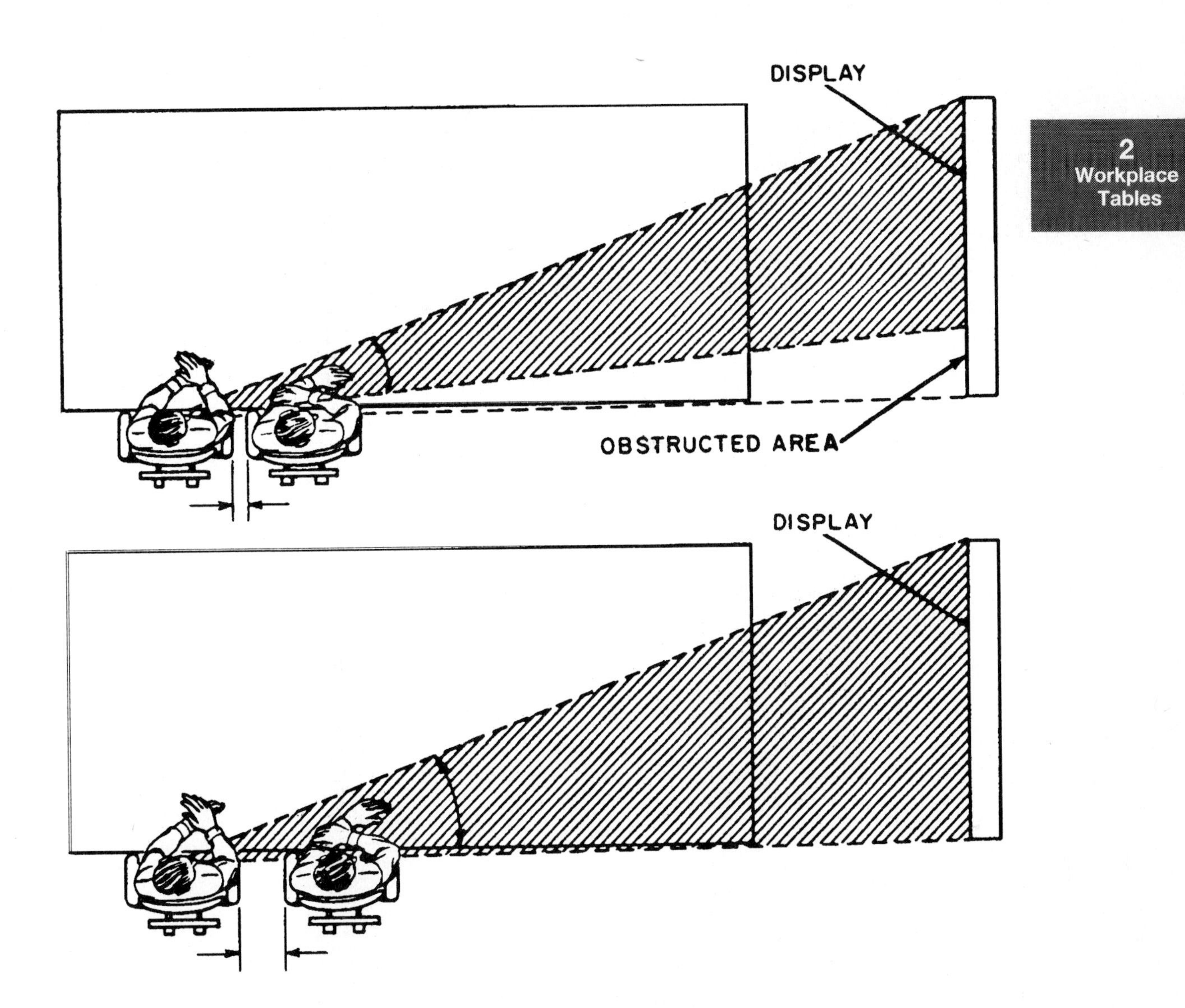

Figure WRKSTN-A22: Increasing Distance Between Viewers Increases Viewable Area of Shared Displays[5]

REFERENCES

1. Reprinted with permission from *Ergonomic Design for People at Work*, © Eastman Kodak Company, 1983, published by Van Nostrand Reinhold. Courtesy of Eastman Kodak Company.

2. Pulat, B.M. 1992. *Fundamentals of Industrial Ergonomics.* Englewood Cliffs, NJ: Prentice-Hall. Reprinted with Permission.

3. Ayoub, M.M. Work place design and posture. Reprinted with permission from *Human Factors*, Vol. 15, No. 3, 1973. Copyright 1973 by the Human Factors Society, Inc. All rights reserved.

4. Bailey, R.W. 1989. *Human Performance Engineering. (2nd Ed)* Englewood Cliffs, NJ: Prentice-Hall. Reprinted with Permission.

5. Van Cott, H.P., and Kinkade, R.G., 1972. *Human Engineering Guide to Equipment Design. (Revised)* Washington, D.C.: US Government Printing Office. Library of Congress Number: 72-600054.

CHAPTER 2: WORKPLACE TABLES

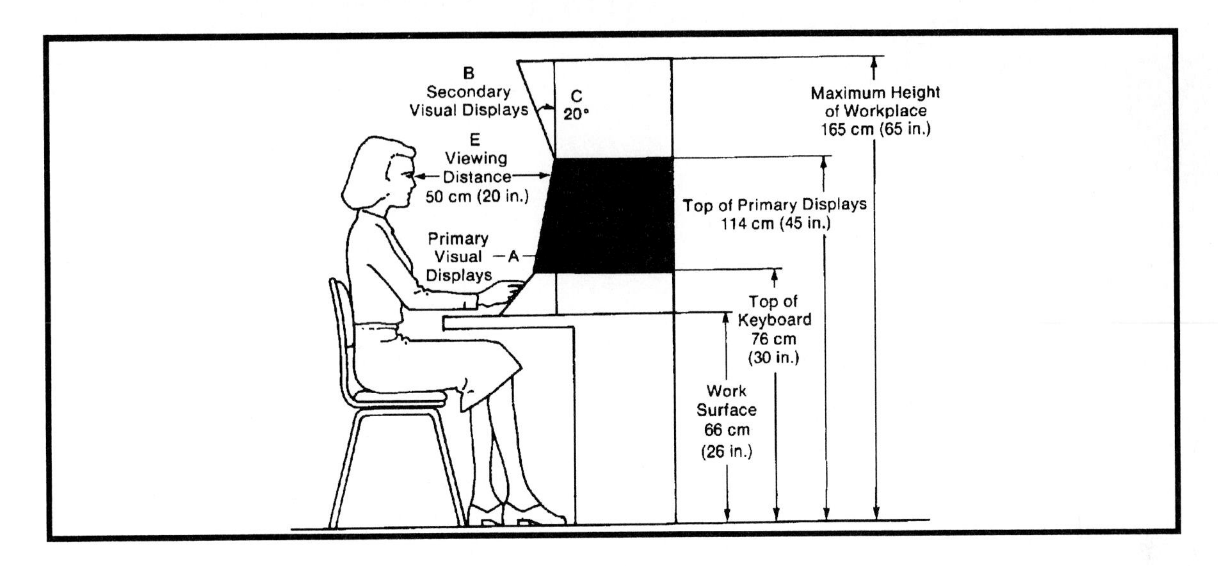

Section B: Ergonomic Layout of Auditoriums

This section presents information to be used when designing auditoriums. The layouts in this section attempt to ensure that the stage or display is equally visible to all.

AUDITORIUM LAYOUT

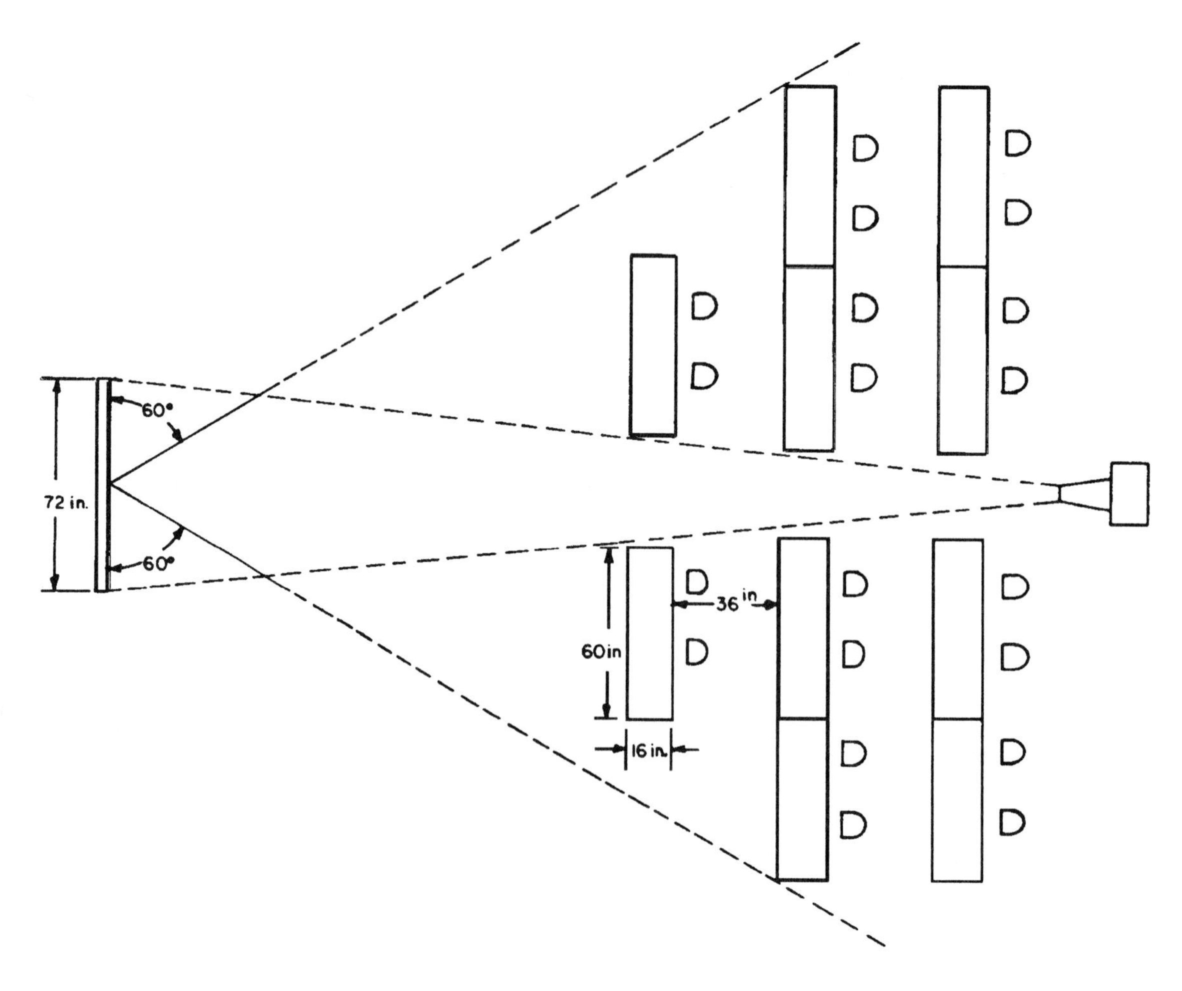

Using narrow tables in conference rooms permits natural head-on viewing by all participants

Figure WRKSTN-B1: Conference Room Arrangement Permitting Natural Head-on Viewing[1]

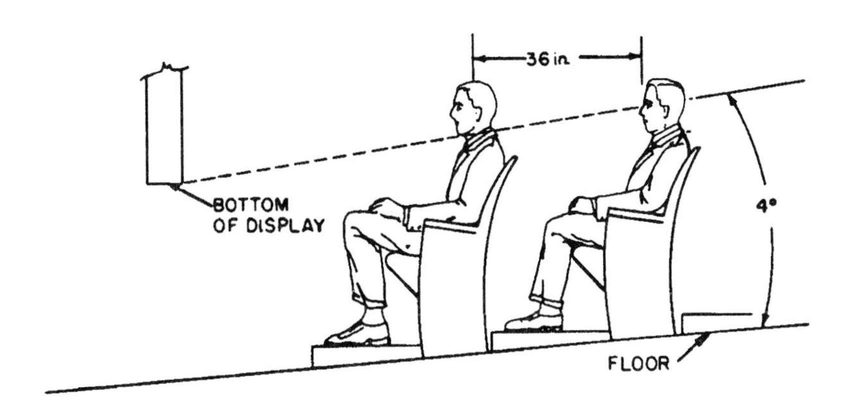

Figure WRKSTN-B2: Display Viewed From an Inclined Floor[1]

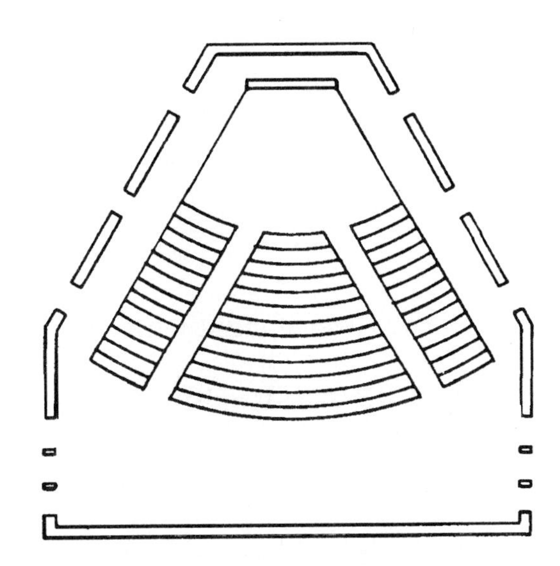

Figure WRKSTN-B3: Auditorium Seating Arrangement[1]

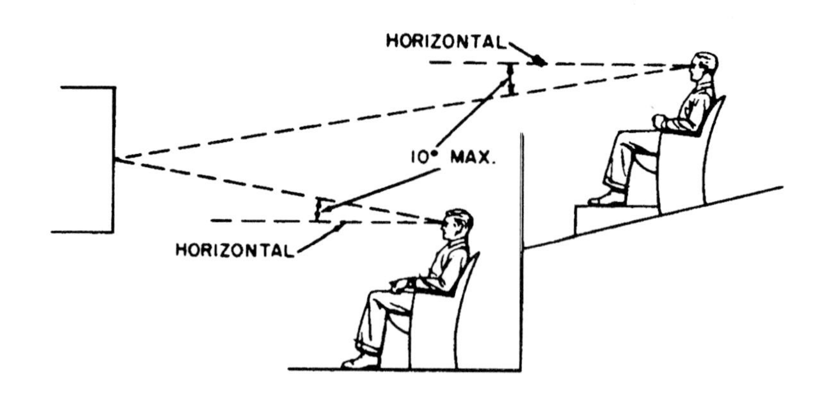

Figure WRKSTN-B4: Vertical Viewing Angle[1]

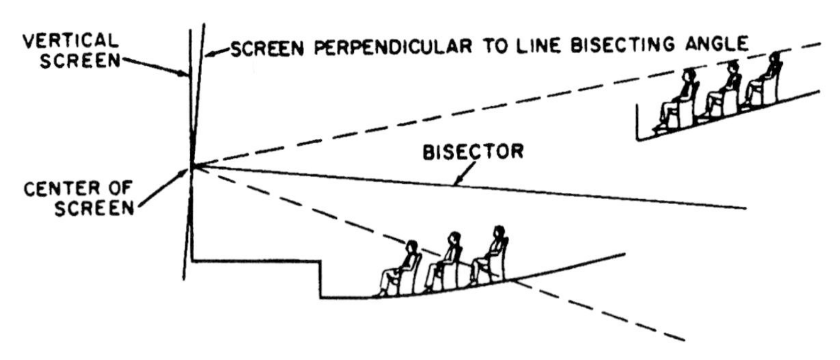

Position screens so that line of sight is normal for all viewers

Figure WRKSTN-B5: Auditorium Screen Positioning[1]

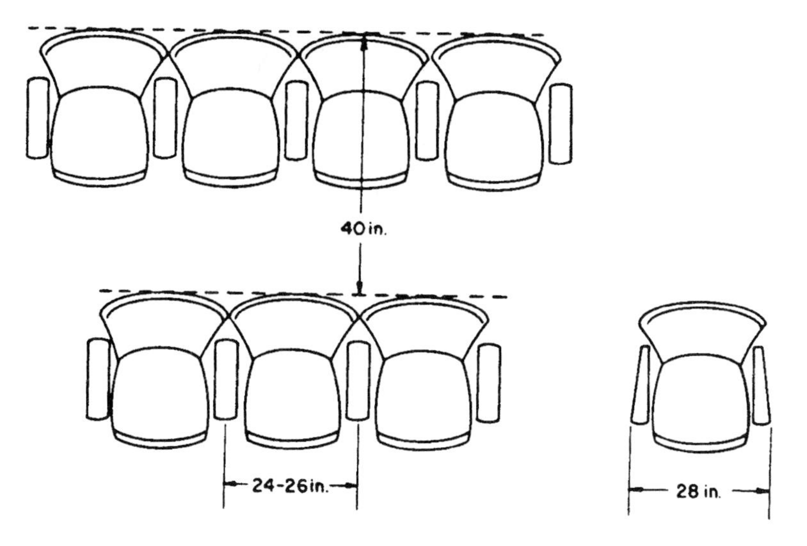

Figure WRKSTN-B6: Armrest Dimensions[1]

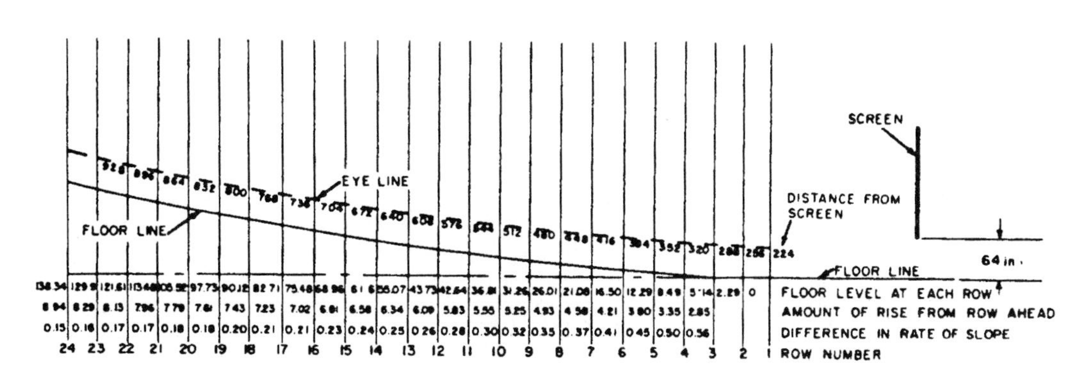

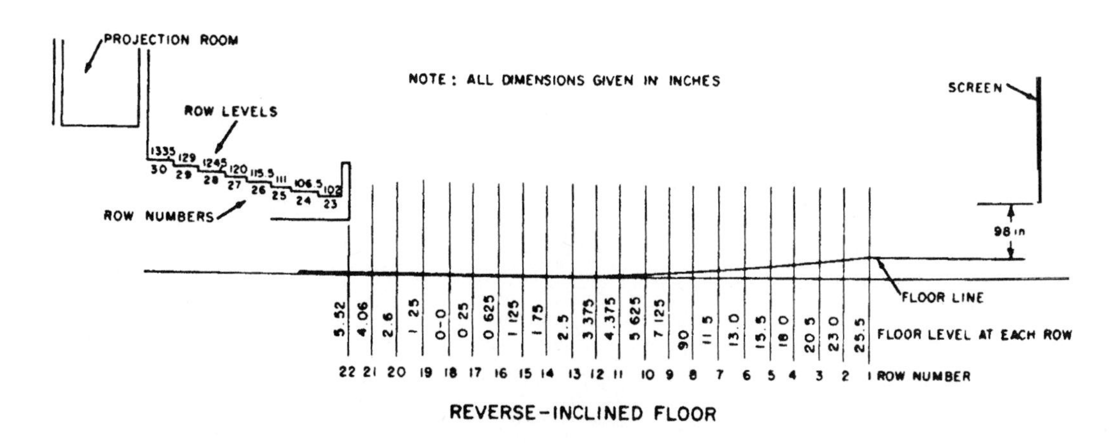

Figure WRKSTN-B7: Inclined and Reverse-Inclined Floor Auditorium Dimensions[1]

REFERENCES

1. Van Cott, H.P., and Kinkade, R.G., 1972. *Human Engineering Guide to Equipment Design. (Revised)* Washington, D.C.: US Government Printing Office. Library of Congress Number: 72-600054.

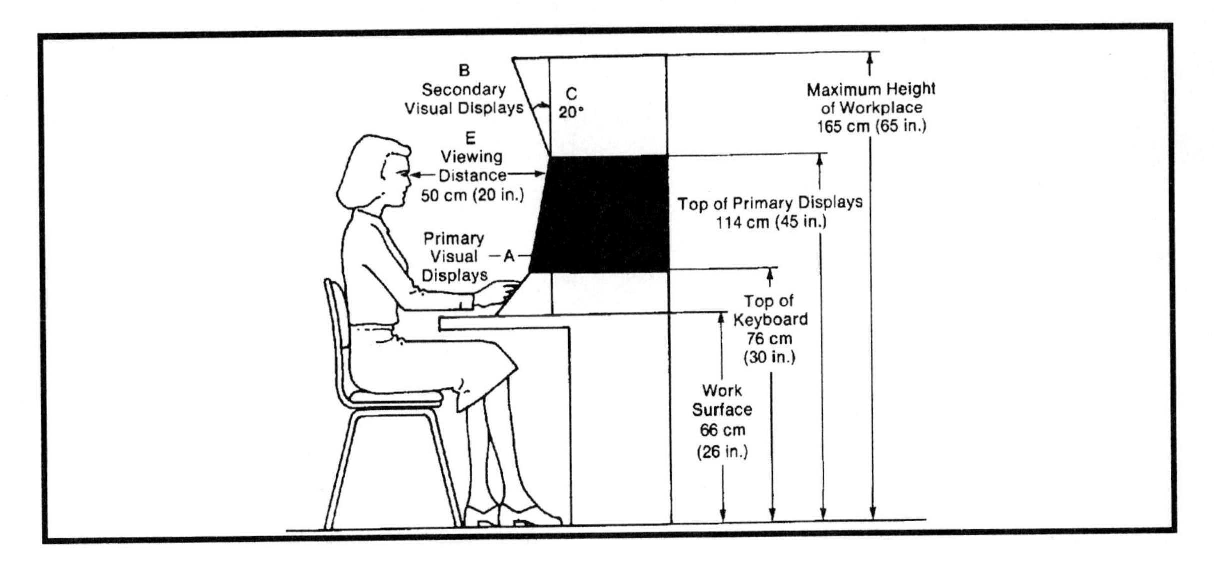

Section C: Workplace Clearances

This section presents information to be used when designing workplaces that require some type of clearance. Information in this section ranges from the design of hand ports to the design of corridors and ladders.

WORKPLACE CLEARANCES

Table WRKSTN-C1: Minimum Clearance for the Working Hand[1]

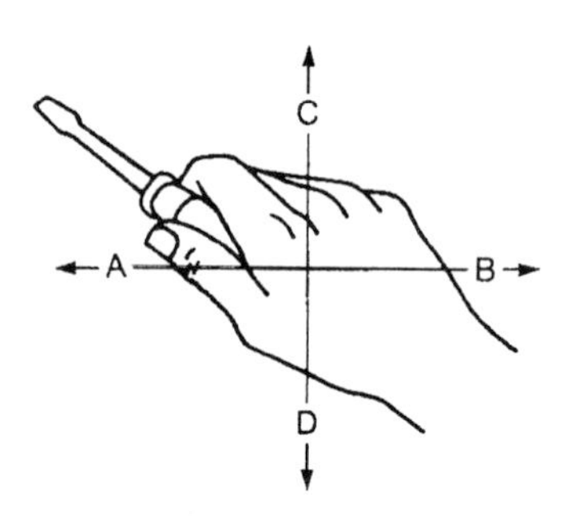

Hand Action	Minimum Dimensions For Hand Clearance			
	A, To Left	B, To Right	C, Up	D, Down
Turning Screwdriver [20-cm (8-in.) Length] or Spinate Wrench [15-cm (6-in.) Length]	38 mm (1.5 in.)	62 mm (2.4 in.)	62 mm (2.4 in.)	42 mm (1.7 in.)
Grasping, Turning, and Cutting with Needle-Nosed Pliers [14-cm (5-in.) Length] or Wire Cutters [13-cm (5-in.) Length]	53 mm (2.1 in.)	73 mm (2.9 in.)	46 mm (1.8 in.)	68 mm (2.7 in.)
Turning Socket Wrench [10-mm ($\frac{3}{8}$-in.) Base, 7-cm (3.2 in.) Shaft]	54 mm (2.1 in.)	83 mm (3.2 in.)	83 mm (3.2 in.)	73 mm (2.9 in.)
Turning Allen Wrench [5-cm (2-in.) Length]	36 mm (1.4 in.)	86 mm (3.4 in.)	94 mm (3.7 in.)	64 mm (2.5 in.)

The minimum space needed to work with common tools (identified in column 1) in maintenance or adjustment tasks is given. Measurements of tool length are given in column 1; the clearances assume that multiple rotations are needed, even where space is tight. The four directions of clearance are radial deviation (A), usually to the left; ulnar deviation (B), usually to the right; extension (C), or up; and flexion (D), or down. More space is desirable, if possible.

Table WRKSTN-C2: Selected Clearance for Arms and Hands[1]

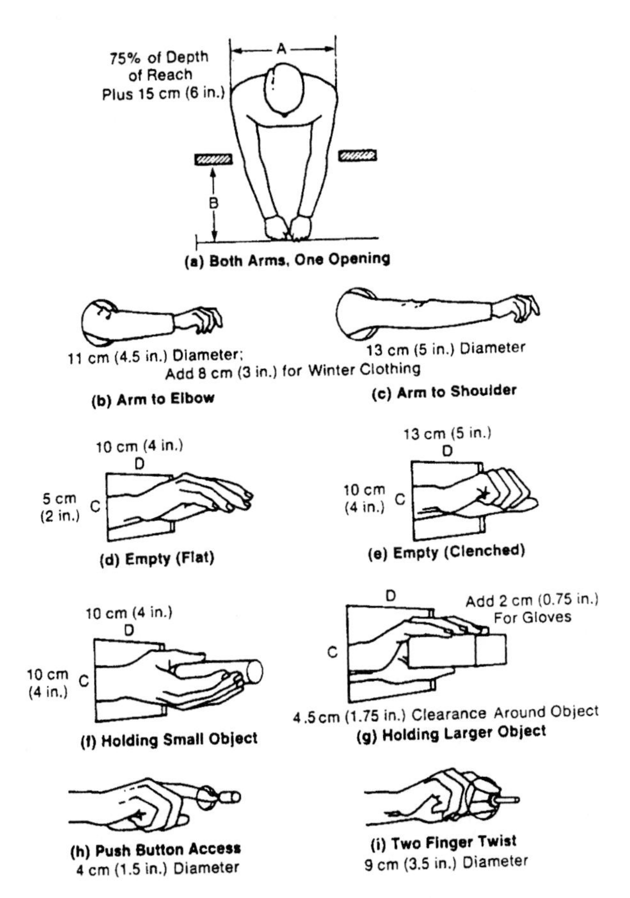

The dimensions for access ports in equipment that will permit the finger, hand, arm, or both arms to enter are given. If both arms must enter (part a), a minimum of 61 cm (24 in.) of horizontal clearance (A) is needed to provide a 61-cm (24-in.) forward reach (B). The port diameters for arm-to-elbow (part b) and arm-to-shoulder (part c) access must be increased if the operation is done under conditions where heavy clothing is worn. Height (C) and width (D) clearances for the hand when empty or holding an object are given in parts d, e, f, and g. These values should be increased by 2 cm (0.75 in.) if work gloves are worn. The access diameters shown in parts h and i are for one- or two-finger access. The size of the part being adjusted will determine the proper diameter of the two-finger access port; the larger the part, the larger is the opening that is needed to access it.

Table WRKSTN-C3: Upright, Kneeling and Prone Clearances[1]

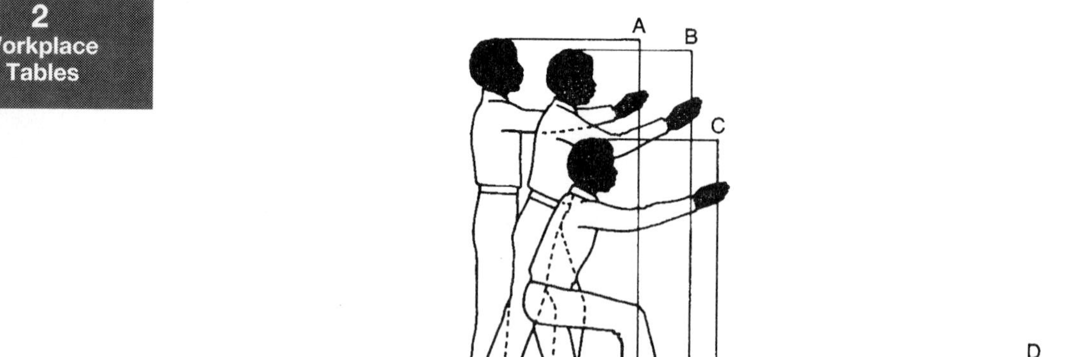

Position	Minimum Dimensions Vertical	Horizontal
1. Standing	203 cm (80 in.)	76 cm (30 in.)
2. Standing, Legs Braced	203 cm (80 in.)	102 cm (40 in.)
3. Kneeling	122 cm (48 in.)	117 cm (46 in.)
4. Prone Arm Reach	46 cm (18 in.)	243 cm (96 in.)

The vertical dimension (column 2 of the accompanying table) and horizontal dimension (column 3) required for a person doing tasks while standing erect (1), with a foot braced (2), kneeling (3), and lying prone with the arms outstretched (4) are given. The breadth should be at least 61 cm (24 in.).

Table WRKSTN-C4: Work Clearances, Horizontal[1]

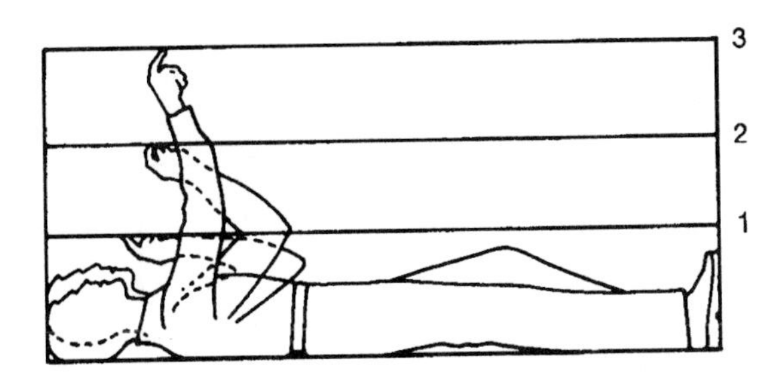

Position	Minimum Dimensions	
	Vertical	Horizontal
1. Lying for inspection	46 cm (18 in.)	193 cm (76 in.)
2. Restricted space for small tools and minor adjustments; power from elbow extension not possible	61 cm (24 in.)	193 cm (76 in.)
3. Space for reasonable arm extension; 152–203-mm (6–8-in.) length power tools could be used	81 cm (32 in.)	193 cm (76 in.)

The minimum horizontal space for a work area when a person has to lie in a supine position (on the back) is shown; in the third column of the accompanying table, this dimension is given as 193 cm (76 in.). This space will accommodate most people comfortably. The vertical clearance dimensions (1, 2, and 3) vary with the task to be performed; more space is needed if the arms have to exert force or use tools (2 and 3).

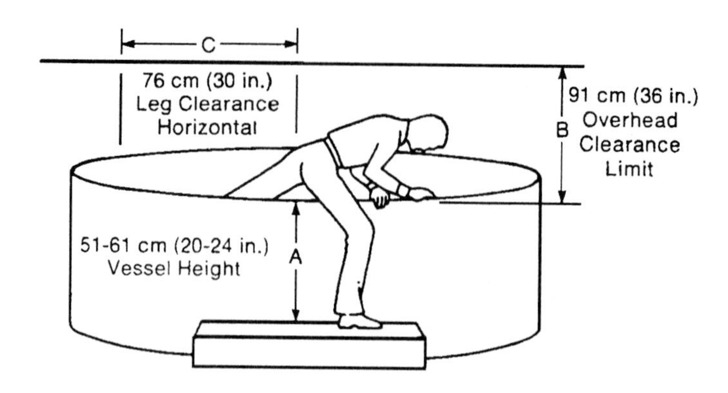

The dimensions of the work area around an open-top vessel, such as a chemical reactor or a tank, are shown. The distance between the walking surface (often a platform on the floor) and the top of the vessel (A) can be about 10 cm (4 in.) less and still accommodate most workers. The clearance above the vessel to any overhead obstruction (B), such as pipes or an overhead hoist, is needed to minimize the operator's risk of bumping his or her head and shoulders. The horizontal distance from the point where the operator enters the vessel to the nearest vertical barrier (C) should be at least 76 cm (30 in.) to permit leg extension.

Figure WRKSTN-C1: Clearances for Entering Open-Top Vessels[1]

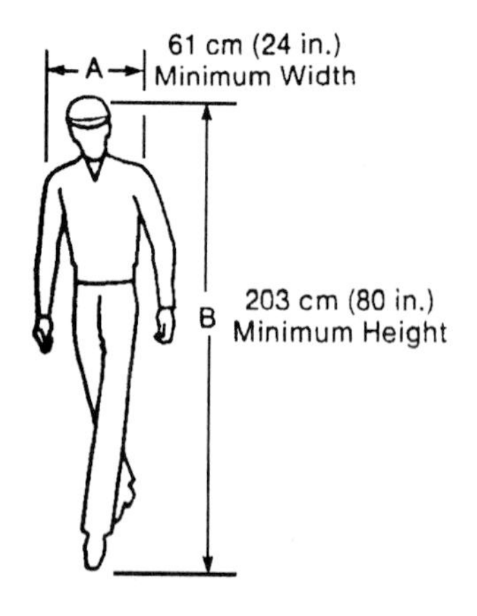

The minimum amount of space needed to permit a person to walk normally is shown. The minimum width (A) includes about 5 cm (2 in.) of clearance on either side of the shoulders of a very broad-shouldered person.

Figure WRKSTN-C2: Minimum Clearances for Walking[1]

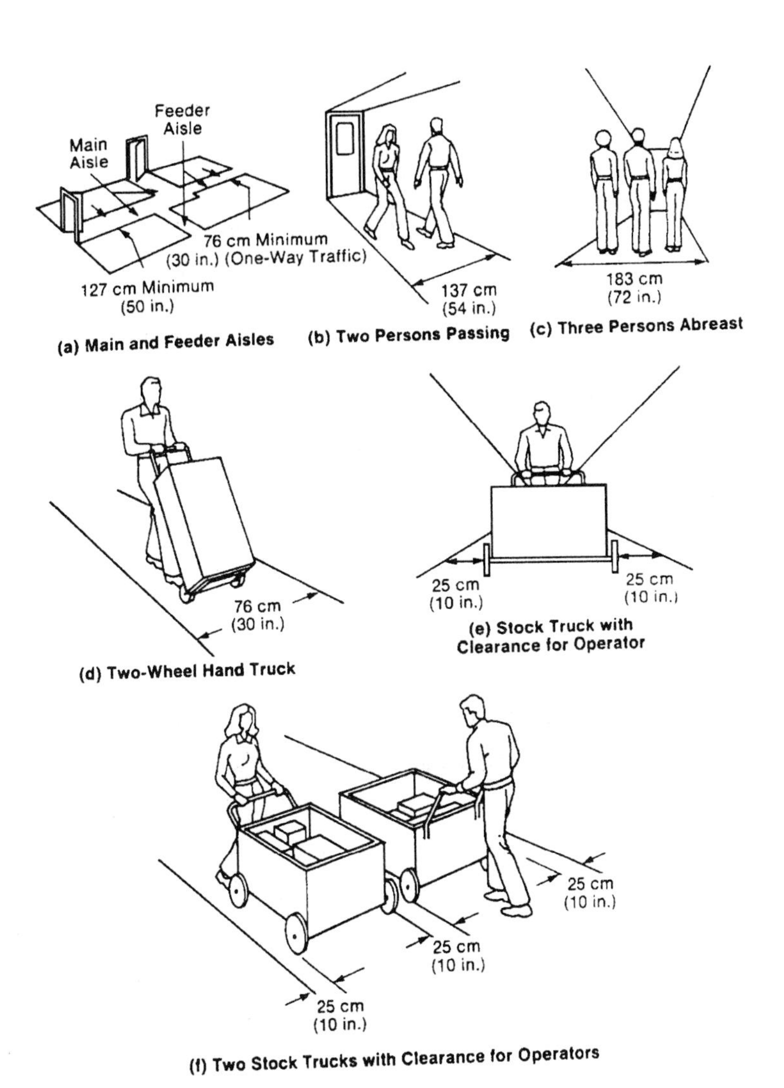

The aisle widths shown in these illustrations are the minimum values for traffic and for handling trucks in production areas. Main aisles should be wider than feeder aisles (see part a). Both widths should be determined by the traffic needs (see parts b, c, and d) but should not be less than the values given in part a. Where trucks and carts are used, there should be 25 cm (10 in.) of clearance on either side of them (see part e) and between them if there is two-way traffic in the corridor (see part f). Minimum aisle width will be set by truck width, its clearance needs, and the traffic pattern in the production area.

Figure WRKSTN-C3: Minimum Clearances for Aisles and Corridors[1]

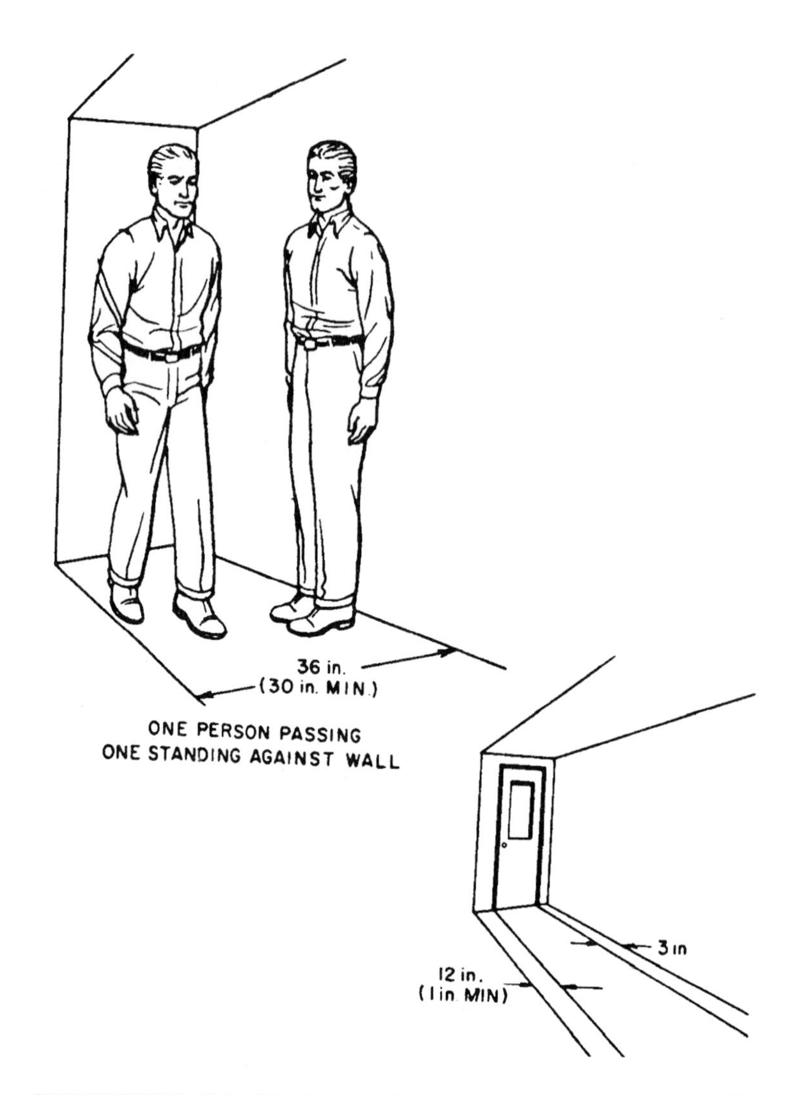

Figure WRKSTN-C4: Aisle and Door Arrangements[3]

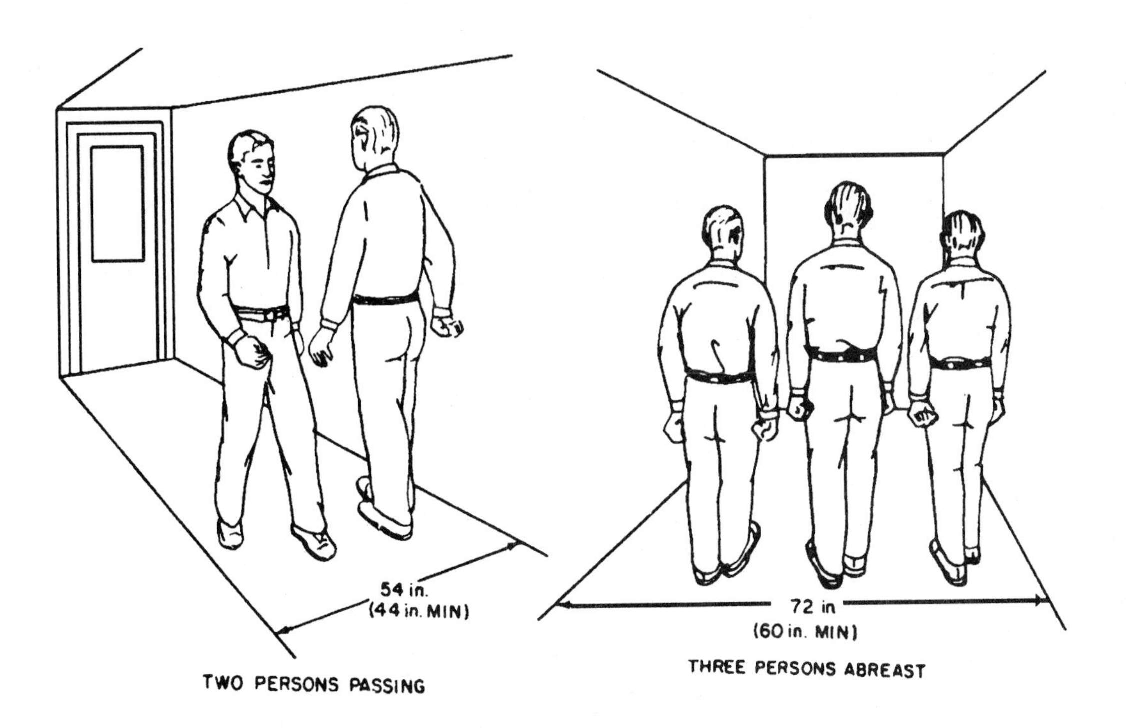

Figure WRKSTN-C5: Aisle Widths for Two- and Three- Person Flow[3]

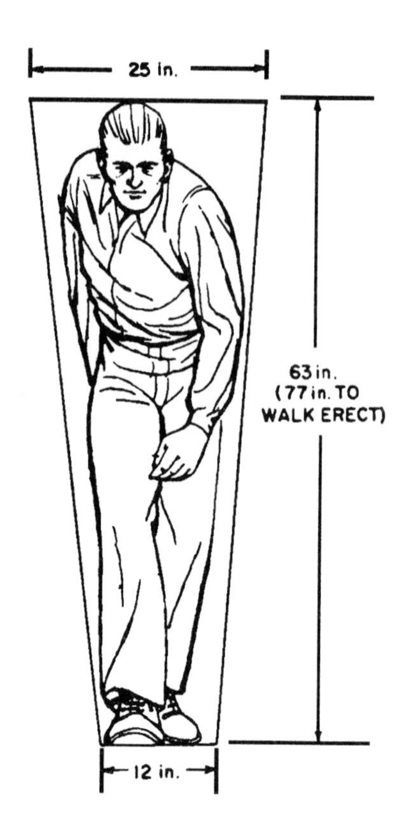

Figure WRKSTN-C6: Minimum Tunnel Dimension[3]

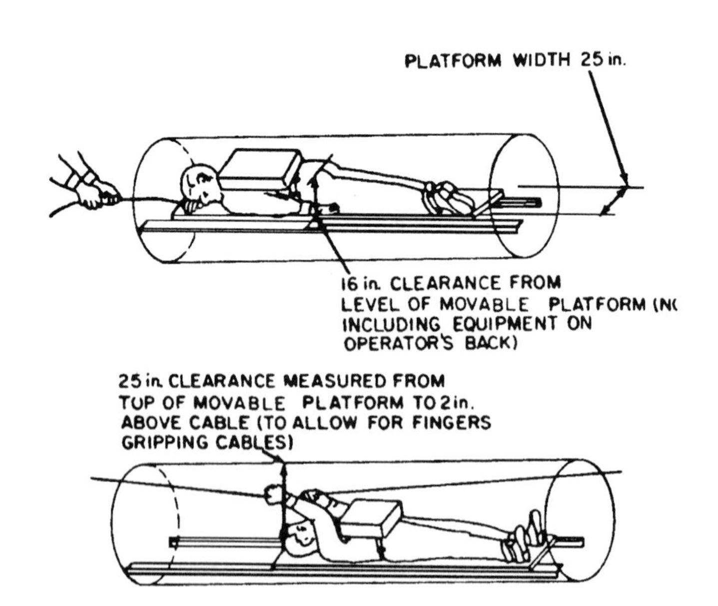

Figure WRKSTN-C7: Minimum Crawl Tunnel Dimensions[3]

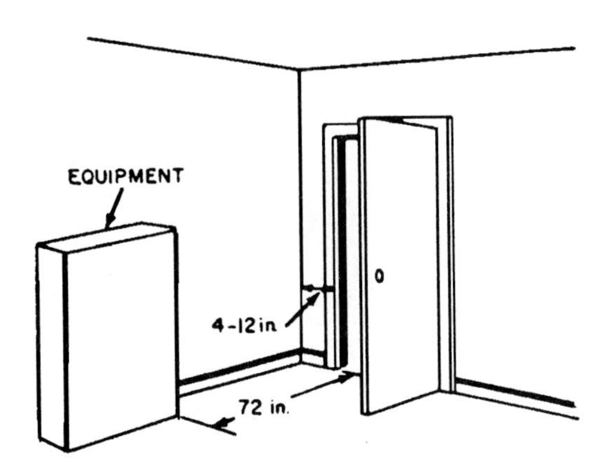

Figure WRKSTN-C8: Relation of Equipment to Door Movement[3]

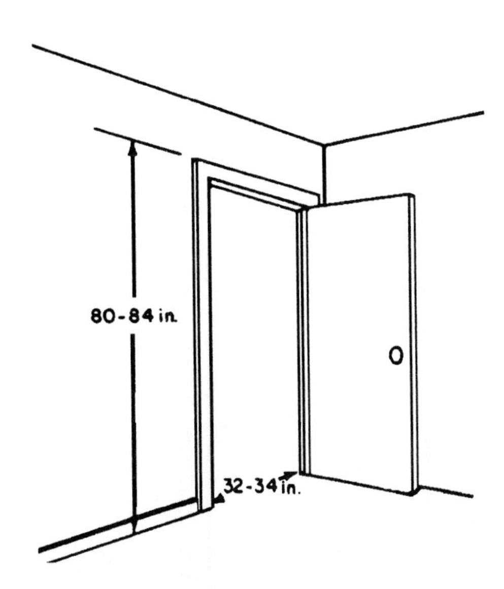

Figure WRKSTN-C9: Door Dimensions for One Person at a Time[3]

2
Workplace
Tables

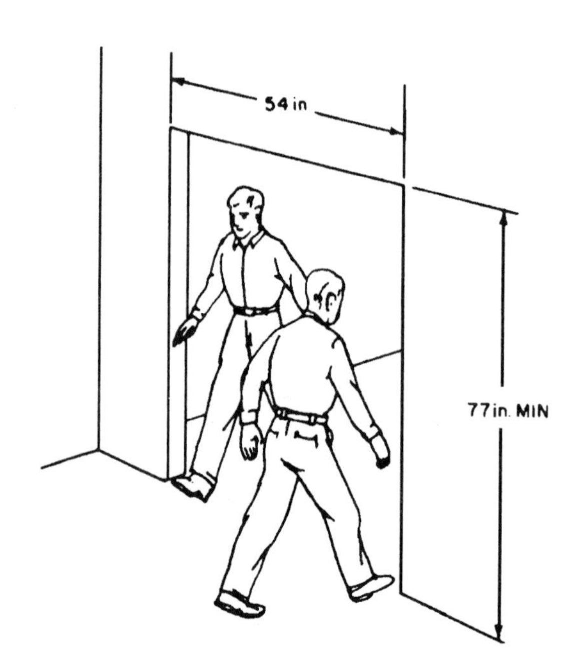

Figure WRKSTN-C10: Archway Dimension for Two-Person Flow[3]

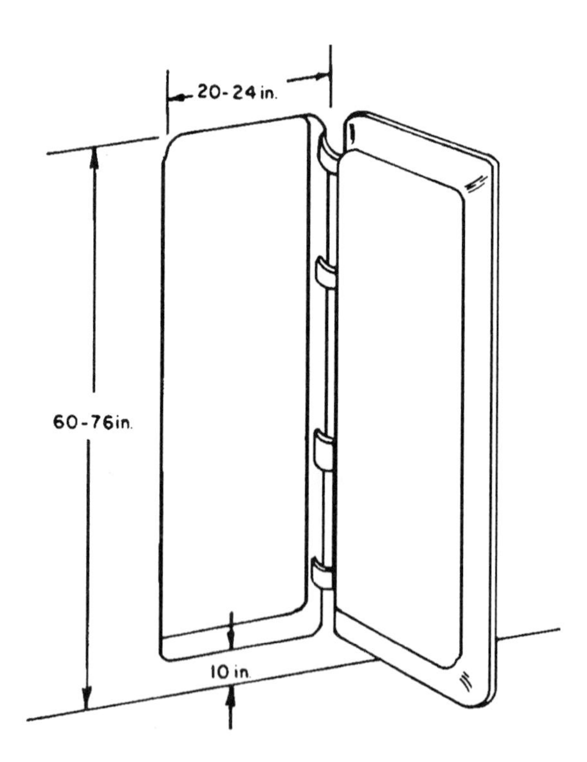

Figure WRKSTN-C11: Armored Door Dimensions[3]

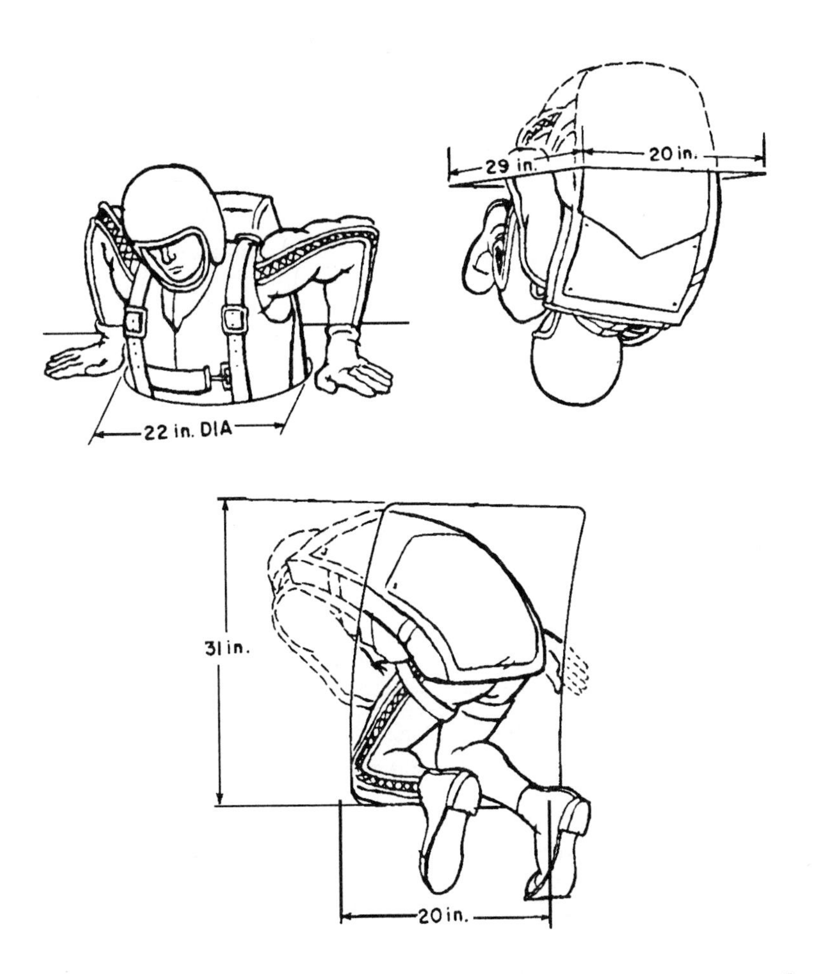

Figure WRKSTN-C12: Escape Hatch Dimensions[3]

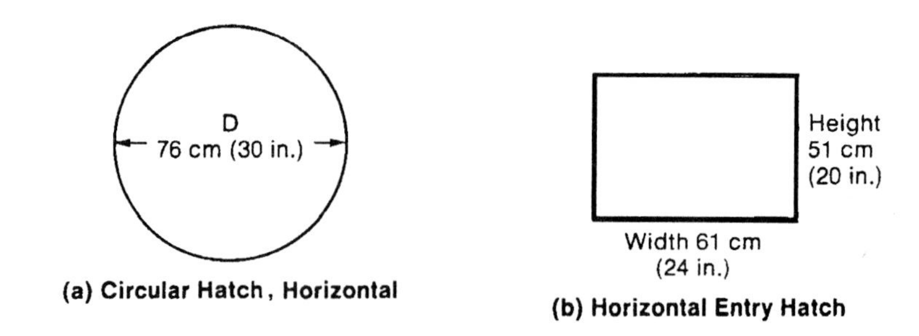

(a) Circular Hatch, Horizontal

(b) Horizontal Entry Hatch

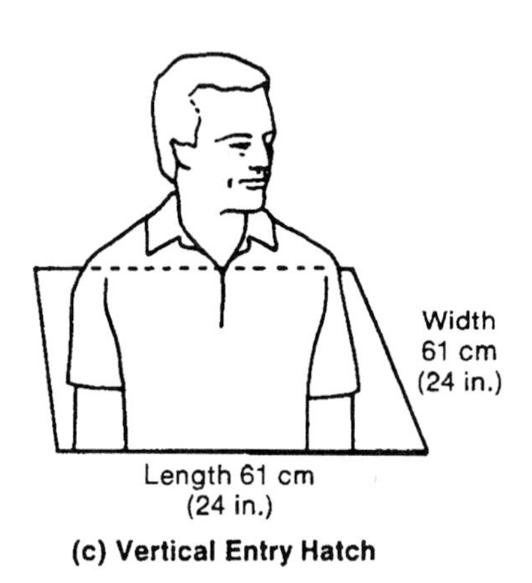

(c) Vertical Entry Hatch

Minimum dimensions for three full-body access ports are shown: a horizontal, circular hatch (part a), such as a pipe, with diameter D of 76 cm (30 in.); a rectangular, horizontal entry hatch (part b), with a height of 51 cm (20 in.) and a width of 61 cm (24 in.); and a 61-cm (24-in.) square, vertical port (part c). People wearing heavy clothing need 10–20 cm (4–8 in.) more clearance than shown here.

Figure WRKSTN-C13: Minimum Full-Body Access Port[1]

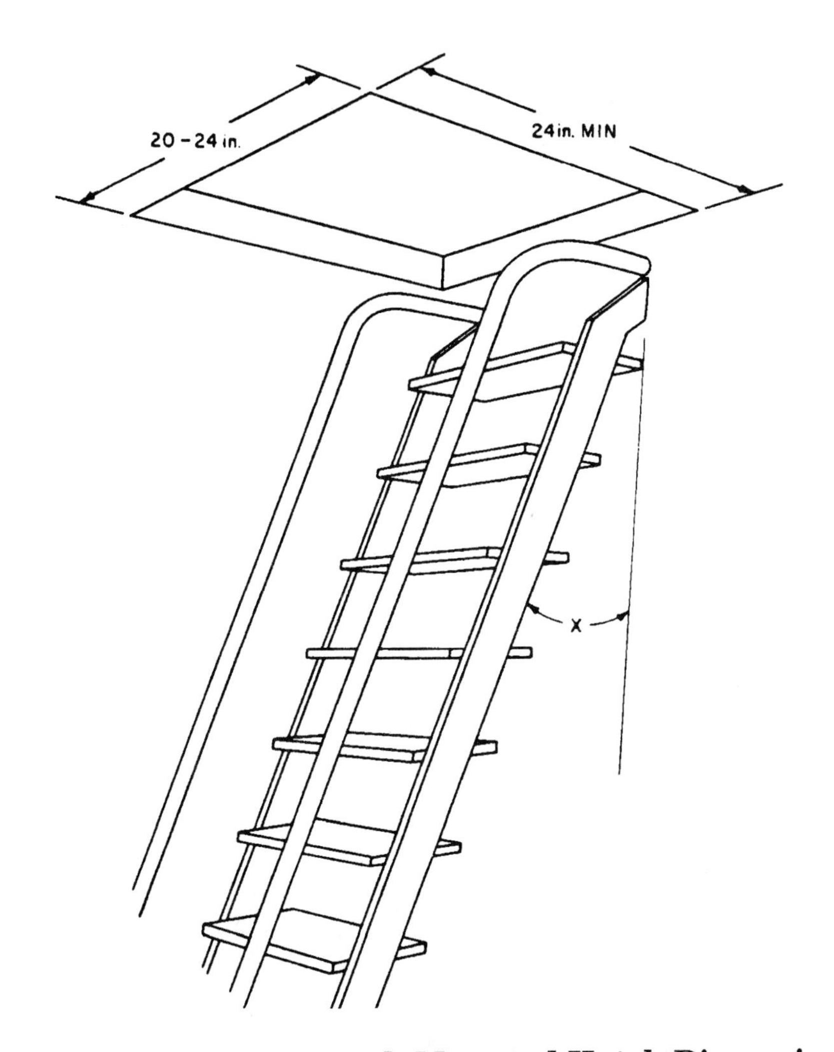

Figure WRKSTN-C14: Deck-Mounted Hatch Dimensions[3]
(as Angle X increases, Depth of Hatch Must Increase)

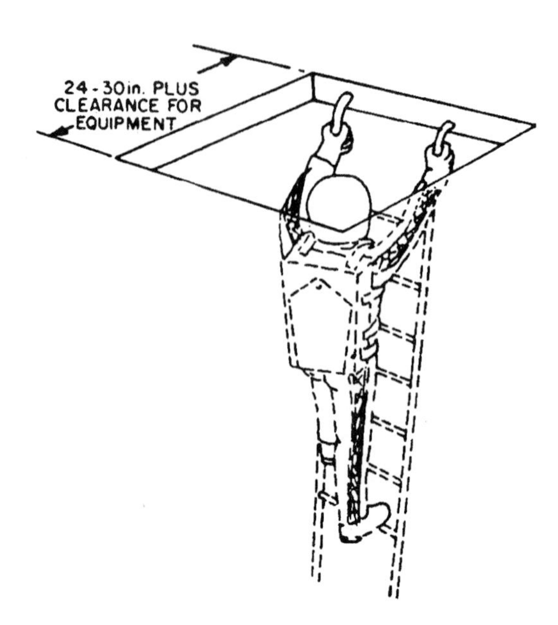

Figure WRKSTN-C15: Added Clearance in Width Required for Vertical Hatch Use by Workers Wearing Equipment[3]

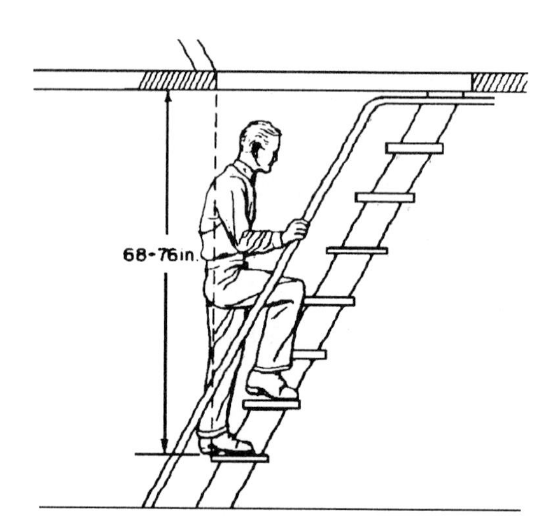

Figure WRKSTN-C16: Minimum Hatch Edge to Ladder Tread Distance[3]

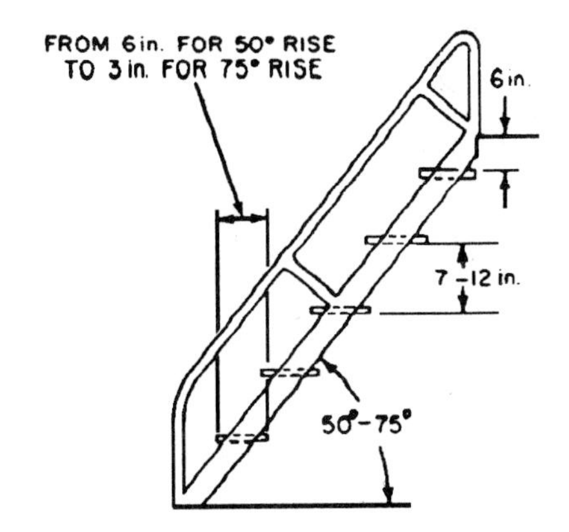

Figure WRKSTN-C17: Ship's Ladder Dimensions[3]

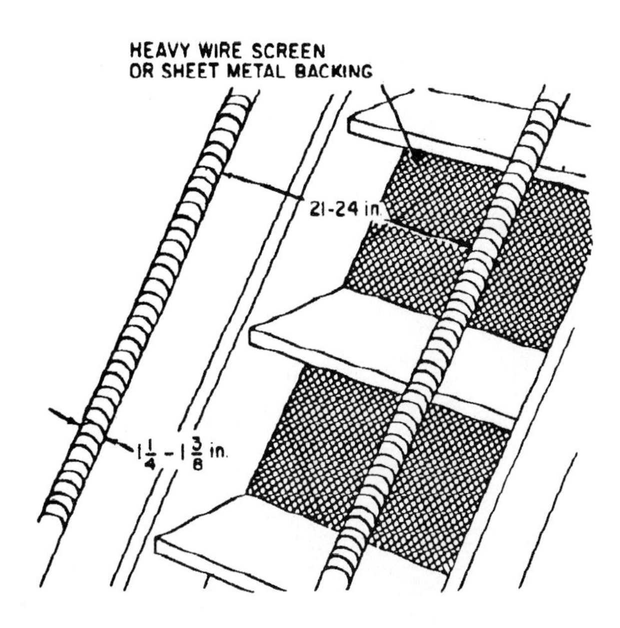

Figure WRKSTN-C18: Handrail Arrangements[3]

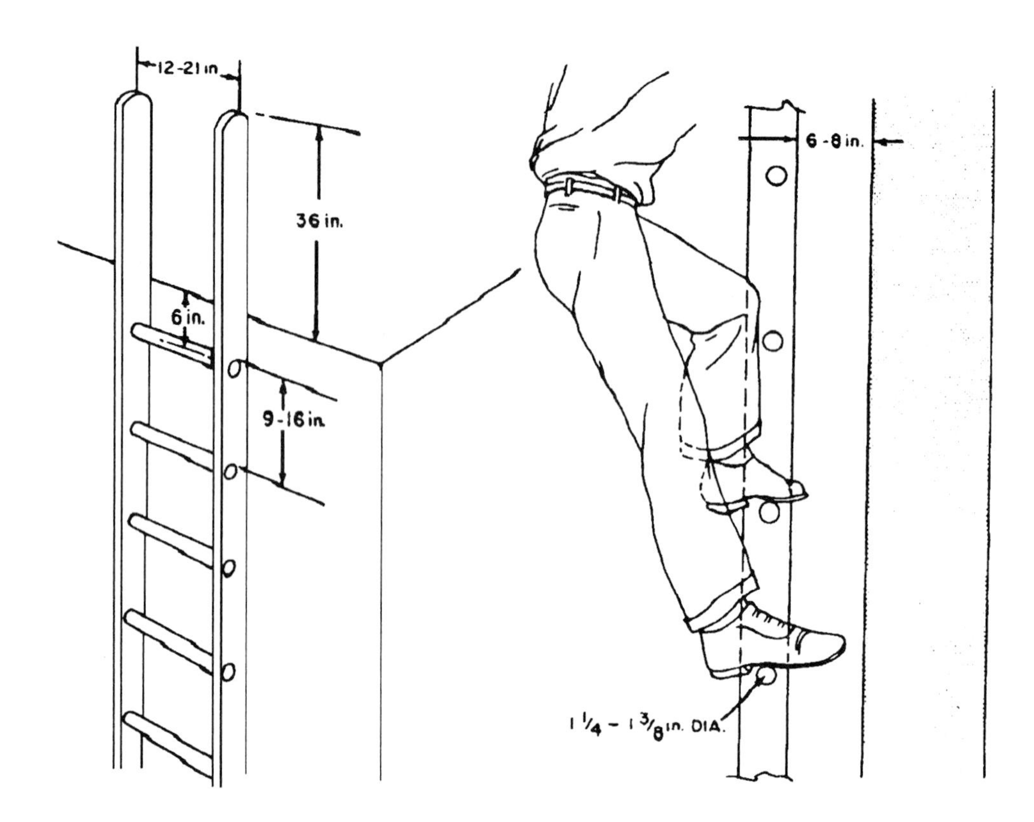

Figure WRKSTN-C19: Vertical Ladder Design[3]

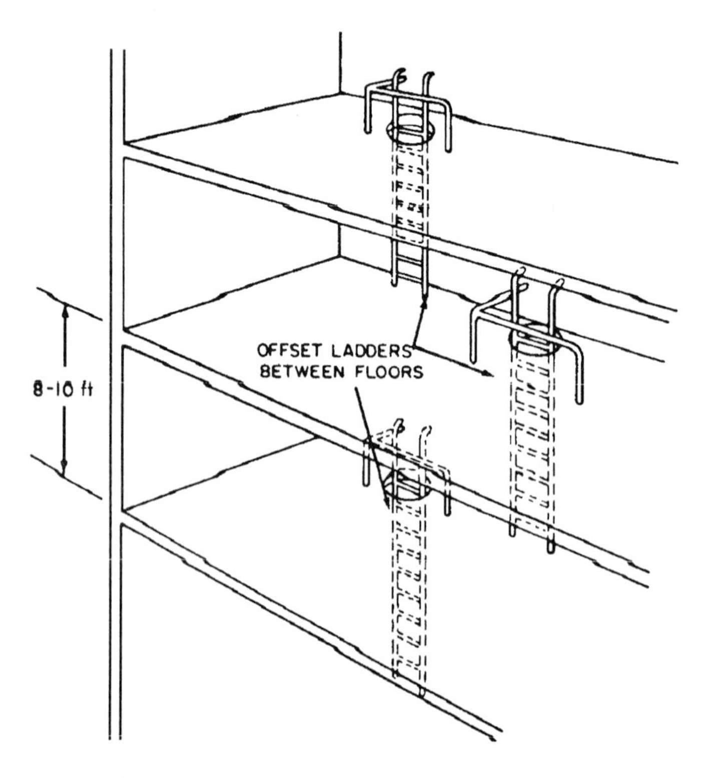

Figure WRKSTN-C20: Offset Ladders Between Floors[3]

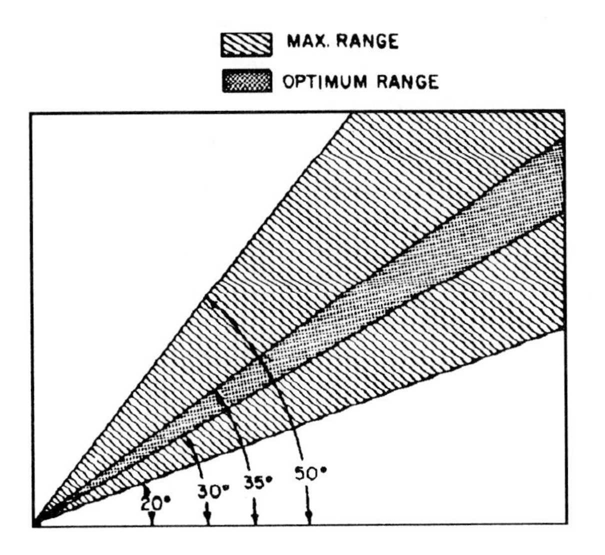

Figure WRKSTN-C21: Stair Rise Angles[3]

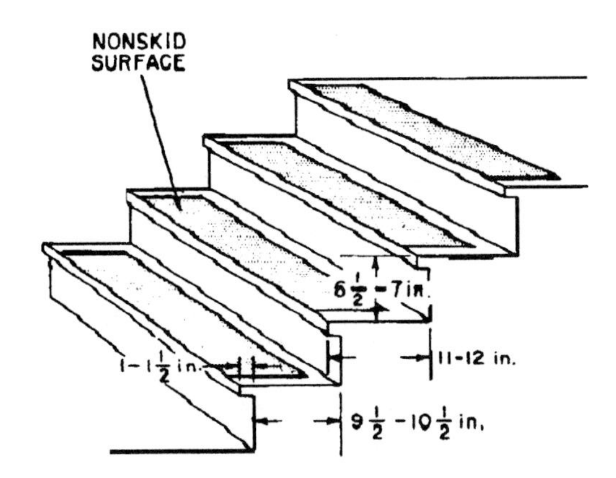

Figure WRKSTN-C22: Stair Tread Dimensions[3]

REFERENCES

1. Reprinted with permission from *Ergonomic Design for People at Work*, © Eastman Kodak Company, 1983, published by Van Nostrand Reinhold. Courtesy of Eastman Kodak Company.

2. Pulat, B.M. 1992. *Fundamentals of Industrial Ergonomics.* Englewood Cliffs, NJ: Prentice-Hall. Reprinted with Permission.

3. Van Cott, H.P., and Kinkade, R.G., 1972. *Human Engineering Guide to Equipment Design. (Revised)* Washington, D.C.: US Government Printing Office. Library of Congress Number: 72-600054.

CHAPTER 2: WORKPLACE TABLES

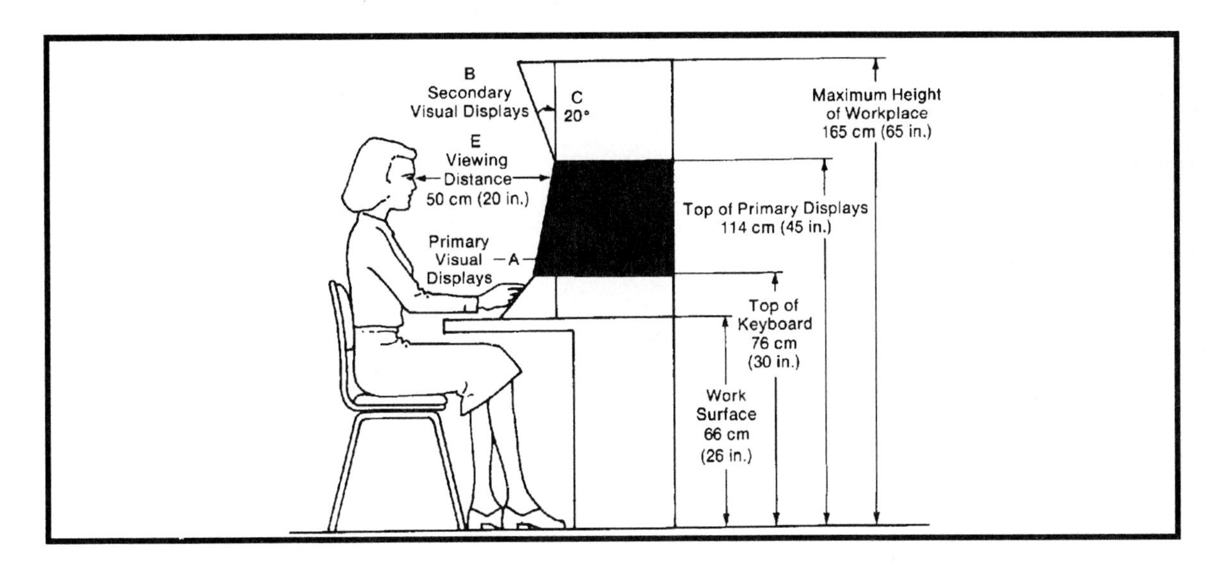

Section D: Workplace Illumination and Visual Angle

This section presents information to be used when illuminating workplaces.

WORKPLACE ILLUMINATION

Table WRKSTN-D1: Luminance Conversion Factors[1]

	Candela per meter squared (cd/m^2)	Footlambert (ftL)	Millilambert (mL)	Nit (nt)
1 Candela per meter squared =	0.2919	0.3142	1	
1 Footlambert =	3.426		1.076	3.426
1 Millilambert =	3.183	0.9290		3.183
1 Nit =	1	0.2919	0.3142	

Table WRKSTN-D2: Various Luminance Levels in Millilamberts[1]

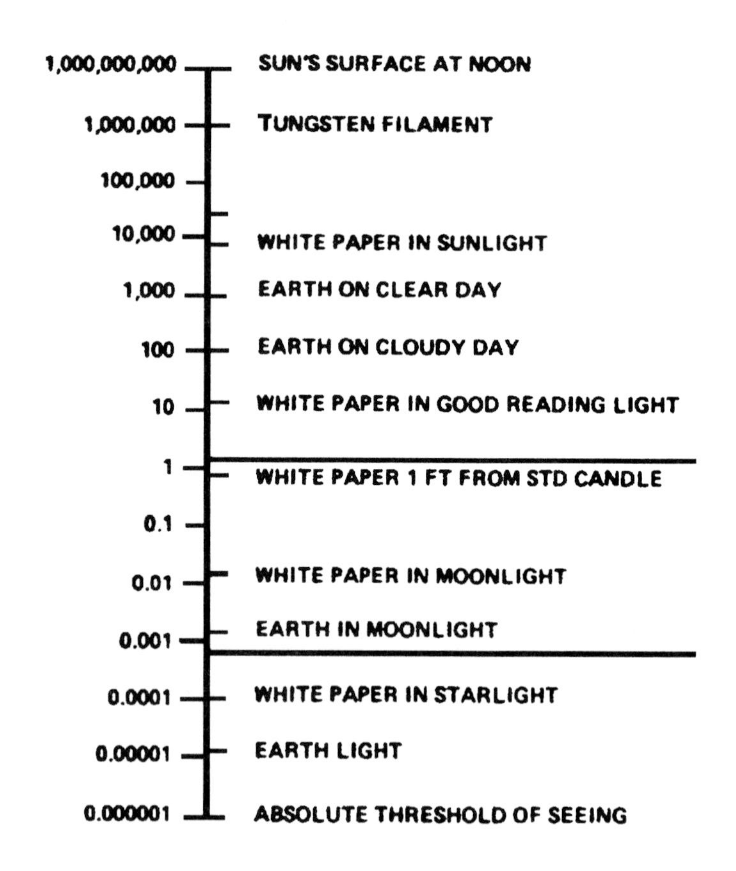

Table WRKSTN-D3: Recommended Maximum Luminance Ratios for Various Tasks[1]

Conditions	Luminance Ratios	
	Office	Industrial
Between tasks and adjacent darker surroundings	3 to 1	3 to 1
Between tasks and adjacent lighter surroundings	1 to 3	1 to 3
Between tasks and more remote darker surfaces	5 to 1	20 to 1
Between tasks and more remote lighter surfaces	1 to 5	1 to 20
Between luminaires (or windows, skylights) and surfaces adjacent to them	20 to 1	NC*
Anywhere within normal field of view	40 to 1	NC*

* NC means not controllable in practice.

The ratios of luminance between a task or display and its background (column 1) are given for offices (column 2) and production workplaces (column 3). These ratios are based on recommendations from the Illuminating Engineering Society (IES). No recommended ratios have been given for situations where it is not practical to try to control the brightness, as in contrasts between workplace and outdoor lighting. It is often difficult to keep the luminance ratios this low, especially in production workplaces where a multitude of tasks are done.

2 Workplace Tables

Table WRKSTN-D4: Recommended Illumination Levels (Footcandles) For a Variety of Different Tasks[1]

Task	Footcandles
Assembly	
Rough easy seeing	30
Rough difficult seeing	50
Medium	100
Fine	500
Extra fine	1000
Auditoriums	
Assembly only	15
Exhibitions	30
Social activities	5
Banks	
Lobby	
General	50
Writing areas	70
Teller's stations	150
Posting and keying	150
Conference rooms	
Critical seeing tasks	100
Conferring	30
Note-taking during projection	30
Hospitals	
Anesthetizing and preparation room	30
Lobby (or entrance foyer)	
During day	50
During night	20
Medical records room	100
Nurses station	
General—day	70
General—night	30
Patients' rooms	
General	20
Reading	30
Observation (by nurse)	2
Surgical suite	
Operating room, general	200
Operating table	2500
Inspection	
Ordinary	50
Difficult	100
Highly difficult	200
Very difficult	500
Most difficult	1000
Library	
Reading areas	
Reading printed material	30
Study and note taking	70
Card files	100
Circulation desks	70
Offices	
Reading handwriting in pencil, active filing, mail sorting	100
Reading handwriting in ink, intermittent filing	70
Reading high contrast or well printed materials	30
Conferring and interviewing	30
Corridors	20
Residences	
Specific visual tasks	
Dining	15
Grooming, shaving, make-up	50
Food preparation	50
Ironing	50
Sewing	
Dark fabrics	200
Medium fabrics	100
Light fabrics	50
General lighting	
Conversation, relaxation	10
Visual tasks	30

Table WRKSTN-D5: Recommended Illumination Levels for Interior Lighting[2]

Activity type	Illumination level (lx)
Rough orientation	75
Occasional rough visual tasks	150
Rough assembly	320
Rough toolmaking	550
Office work—simple	750
Bookkeeping—small character size	1,500
Difficult inspection	1,500
Technical drawing	2,200
Precise assembly work	5,000
Prolonged difficult visual task	7,500
Precise and delicate visual work	11,000
Very special visual tasks—extremely low contrast and small object size	15,000

Table WRKSTN-D6: General Illumination Levels and Types of Illumination for Different Task Conditions[3]

Task condition	Type of task or area	Illuminance level (Ft.-c)	Type of illumination
Small detail, low contrast, prolonged periods, high speed, extreme accuracy.	Sewing, inspecting dark materials, etc.	100	General plus supplementary, e.g., desk lamp.
Small detail, fair contrast, speed not essential.	Machining, detail drafting, watch repairing, inspecting medium materials, etc.	50–100	General plus supplementary.
Normal detail, prolonged periods.	Reading, parts assembly; general office and laboratory work	20–50	General, e.g., overhead ceiling fixture.
Normal detail, no prolonged periods.	Washrooms, power plants, waiting rooms, kitchens	10–20	General, e.g., random natural or artificial light
Good contrast, fairly large objects.	Recreational facilities..	5–10	General.
Large objects..........	Restaurants, stairways, bulk-supply warehouses.	2–5	General.

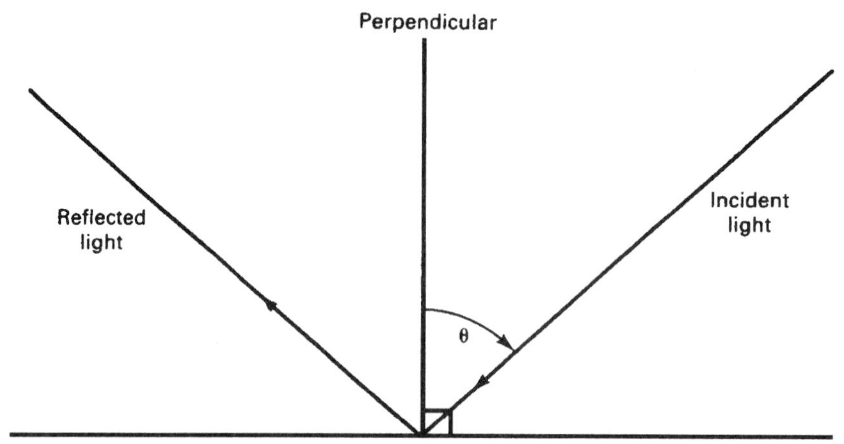

Figure WRKSTN-D1: Reflectance Relationships[2]

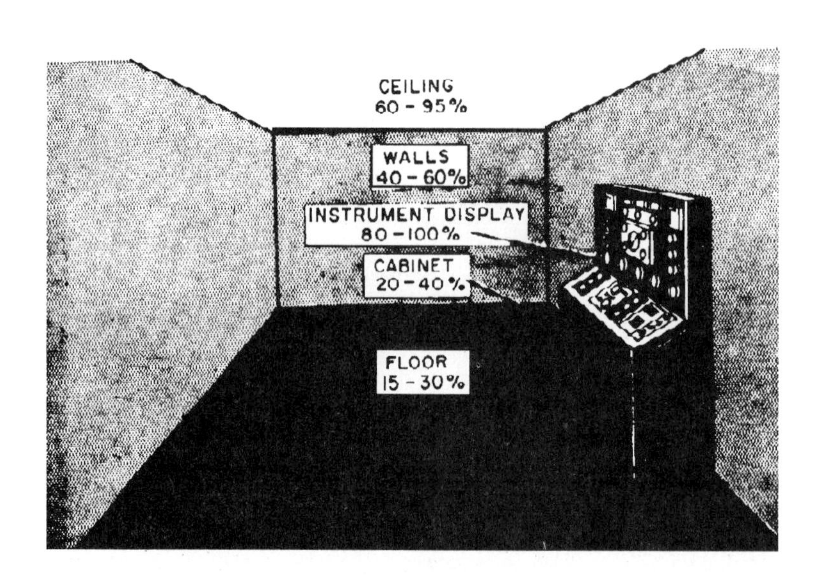

Figure WRKSTN-D2: General Recommendations for Workplace Reflectances[3]

Table WRKSTN-D7: Reflectance Factors for Surface Color[3]

Color	Reflectance	Color	Reflectance
White	85		
Light:		Dark:	
Cream	75	Gray	30
Gray	75	Red	13
Yellow	75	Brown	10
Buff	70	Blue	8
Green	65	Green	7
Blue	55		
Medium:		Wood Finish:	
Yellow	65	Maple	42
Buff	63	Satinwood	34
Gray	55	English Oak	17
Green	52	Walnut	16
Blue	35	Mahogany	12

Table WRKSTN-D8: Techniques for Controlling Glare[4]

To Control Direct Glare	To Control Indirect Glare (Veiling Reflections and Reflected Glare)
Position luminaires, the lighting units, as far from the operator's line of sight as is practical	Avoid placing luminaires in the indirect-glare offending zone (see Figure VC–2)
Use several low-intensity luminaires instead of one bright one	Use luminaires with diffusing or polarizing lenses
Use luminaires that produce a batwing light distribution*, and position workers so that the highest light level comes from the sides, not front and back	Use surfaces that diffuse light, such as flat paint, non-gloss paper, and textured finishes
Use luminaires with louvers or prismatic lenses	Change the orientation of a workplace, task, viewing angle, or viewing direction until maximum visibility is achieved
Use indirect lighting	
Use light shields, hoods, and visors at the workplace if other methods are impractical	

* The effectiveness of the batwing distribution varies with the orientation of the workplace and worker. It can also be used to control indirect glare, because maximum output is in the arc between approximately 35° to 45° angles.

Examples of ways to control direct glare (column 1) and indirect glare (column 2) at the workplace are given. These methods include design approaches that can be used when installing the lighting, as well as interventions that can be made after glare has been identified in a workplace.

Table WRKSTN-D9: Special Purpose Lighting for Inspection [4]

Column 1 describes fourteen improvement goals for inspection task performance. Aids that assist the inspector in detecting the defects are given in column 2; short explanations of how these aids work or descriptions of other actions that help the inspector are given in column 3. There is often more than one way to make a defect more visible; the nature of the material being inspected will help identify the most effective method. When more than one type of defect is being searched for, a combination of aids at the workplace may be appropriate.

Desired Improvement in Inspection Task	Special-Purpose Lighting or Other Aids	Techniques
1. Enhance surface scratches	Edge lighting, for a glass or plastic plate at least 1.5 mm, or 0.06 in., thick	Internal reflection of light in a transparent product; use a high-intensity fluorescent or tubular quartz lamp
	Spotlight	Assumes linear scratches of known direction; provide adjustability so that they can be aligned to one side of the scratch direction; use louvers to reduce glare for the inspector
	Dark-field illumination (e.g., microscopes)	Light is reflected off or projected through the product and focused to a point just beside the eye; scratches diffract light to one side
2. Enhance surface projections or indentations	Surface grazing or shadowing	Collimated light source with an oval beam
	Moiré patterns (to accentuate surface curvatures)	Project a bright collimated beam through parallel lines a short distance away from the surface; looking for interference patterns (Stengel, 1979); either a flat surface or a known contour is needed
	Spotlight	Adjust angle to optimize visualization of these defects

Table WRKSTN-D9: Special Purpose Lighting for Inspection[4] (Continued)

Desired Improvement in Inspection Task	Special-Purpose Lighting or Other Aids	Techniques
2. Enhance surface projections or indentations (cont.)	Polarized light	Reduces subsurface reflections when the transmission axis is parallel to the product surface
	Brightness patterns	Reflection of a high-contrast symmetrical image on the surface of a specular product; pattern detail should be adjusted to product size, with more detail for a smaller surface
3. Enhance internal stresses and strains	Cross-polarization	Place two sheets of linear polarizer at 90° to each other, one on each side of the transparent product to be inspected; detect changes in color or pattern with defects
4. Enhance thickness changes	Cross-polarization	Use in combination with dichroic materials
	Diffuse reflection	Reduce contrast of brightness patterns by reflecting a white diffuse surface on a flat specular product; produces an iridescent rainbow of colors that will be caused by defects in a thin transparent coating
	Moiré patterns	See item 2 in this table
5. Enhance non-specular defects in a specular surface, such as a mar on a product	Polarized light	A specular nonmetallic surface acts, under certain conditions, like a horizontal polarizer and reflects light; non-specular portions of the surface will depolarize it; project a horizontally polarized light at a 35° angle to the horizontal

Table WRKSTN-D9: Special Purpose Lighting for Inspection[4] (Continued)

Desired Improvement in Inspection Task	Special-Purpose Lighting or Other Aids	Techniques
5. Enhance non-specular defects in a specular surface, such as a mar on a product (cont.)	Convergent light	Project the light at a spherical mirror, reflect it off the product, and focus it at the eye; requires very rigid posture for inspectors, however; mirror should be larger than the area being inspected
6. Enhance opacity changes	Transillumination	For transparent products, such as bottles, adjust lights to give uniform lighting to the entire surface; use opalized glass as a diffuser over fluorescent tubes for sheet inspection; double transmission transillumination can also be used
7. Enhance color changes, as in color matching in the textile industry	Spectrum-balanced lights	Choose lighting type to match the spectrum of lighting conditions expected when the product is used; use 3000°K lights if the product is used indoors, 7000°K light if it is used outdoors
	Negative filters, as in inspecting layers of color film for defects	These filters transmit light mainly from the end of the spectrum opposite to that from which the product ordinarily transmits or reflects; this reversal makes the product surface appear dark except for blemishes of a different hue, which are then brighter and more apparent
8. Enhance unsteadiness, jitter	Parallel line patterns (Moiré)	Two sets of parallel lines, 3–5° offset; one set is mounted on the product and the other is stationary; this pattern can magnify the jitter 10–40 times

Table WRKSTN-D9: Special Purpose Lighting for Inspection[4] (Continued)

Desired Improvement in Inspection Task	Special-Purpose Lighting or Other Aids	Techniques
9. Enhance repetitive defects, as in rotating shafts or drums	Stroboscopic lighting	Adjust strobe frequency to the expected frequency of the defect
10. Enhance fluorescing defects	Black light	Use ultraviolet light to detect cutting oils and other impurities; may be used in clothing industry for pattern marking; fluorescing ink is invisible under white light, but very visible under black light
11. Enhance hairline breaks in castings	Coat with fluorescing oils	Use of ultraviolet light inspection will detect pools of oil in the cracks
12. Reduce surface glow under white light that hides defects; the surface appears to fluoresce	Complementary filter or light source, similar to a negative filter	Use a filter or light source with low transmission in wavelengths reflected by the object's surface, and high transmission in other parts of the spectrum, so as to create a gray appearance
13. Remove distracting reflections	Light shields	Place overhead or side shields on a workplace to eliminate reflections caused by room lighting
	Light traps	For VDUs mount a circular polarizer in front of the tube, set at a downward angle; the polarizer traps all incoming light from the room and allows only internally generated light back to the observer

Table WRKSTN-D9: Special Purpose Lighting for Inspection [4] (Continued)

Desired Improvement in Inspection Task	Special-Purpose Lighting or Other Aids	Techniques
13. Remove distracting reflections (cont.)	Reposition workplace	Rather than have operators face a wall, with ceiling lights behind them reflecting off the 45°–90° surfaces of their workpieces, have them sit with their backs to the wall so that workpieces reflect the low-luminance wall instead
14. Reduce blurring of fast-moving product, as in the printing industry	Synchronized moving images	Projected or reflected images on flat, otherwise formless webs can provide fixation points and reduction of streaming
	Stroboscopic lighting	Pulsed light above the fusion threshold, approximately 40+ Hz, will make a random spot type of defect appear as a string of pearls, even if the formless web itself is blurred (Taylor and Watson, 1972)
	Elongate the observation area	Rule of thumb: 0.3 m (1 ft) of observation area per 18.3 m/min (60 ft/min) of object speed at close inspection distances of 0.6–1.2 m (2–4 ft) allows proper fixation time, eye pursuit, and stopped images of the product (T. J. Murphy, 1981, Eastman Kodak Company)
	Group the product	For the same result, it is better to tighten the grouping and reduce the speed rather than to spread the product out and increase the speed

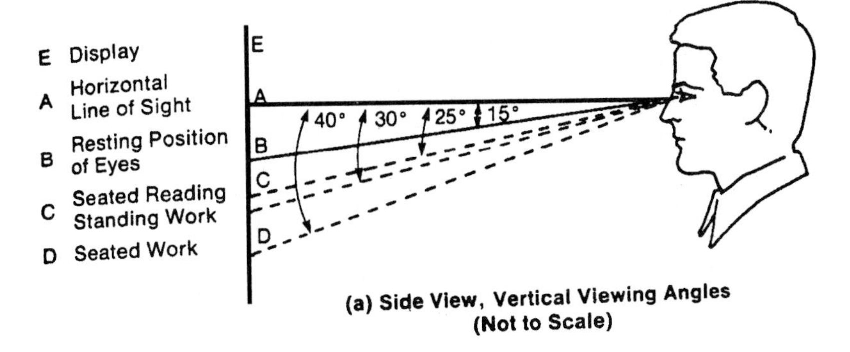

(a) Side View, Vertical Viewing Angles (Not to Scale)

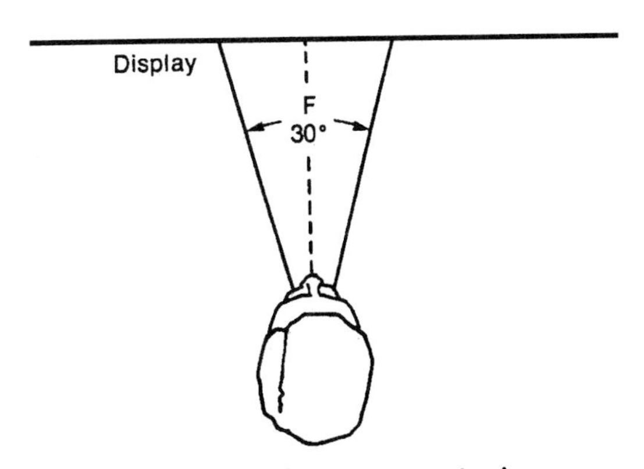

(b) Top View, Lateral Viewing Angles

Side view (part a) and top view (part b) of preferred viewing angles are shown. The vertical viewing angle is the angle between the operator's horizontal line of sight (A) and the actual line of sight (B, C, or D). The most comfortable angle is about 15° downward (B) and straight ahead. The preferred lateral viewing angles are within 15° of either side of the centerline (F). Observations of people at work have indicated that visual angles up to 40° downward are common for visual tasks (C, D). It is possible to keep most objects within these preferred viewing angles by moving the head and eyes.

Figure WRKSTN-D3: Preferred Viewing Angles[4]

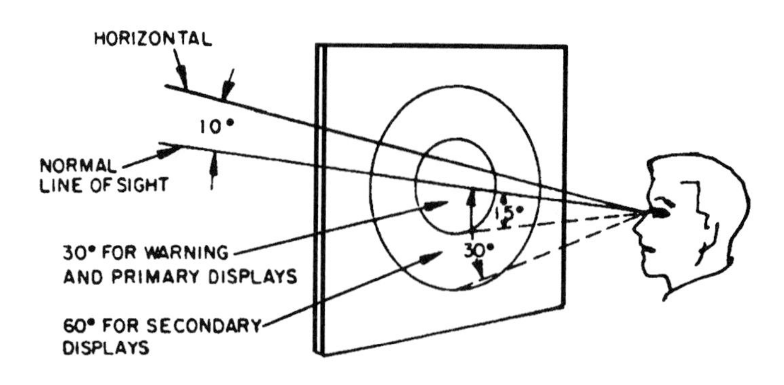

Figure WRKSTN-D4: Preferred Placement for Visual Displays[3]

REFERENCES

1. Bailey, R.W. 1989. *Human Performance Engineering. (2nd Ed)* Englewood Cliffs, NJ: Prentice-Hall. Reprinted with Permission.

2. Pulat, B.M. 1992. *Fundamentals of Industrial Ergonomics.* Englewood Cliffs, NJ: Prentice-Hall. Reprinted with Permission.

3. Van Cott, H.P., and Kinkade, R.G., 1972. *Human Engineering Guide to Equipment Design. (Revised)* Washington, D.C.: US Government Printing Office. Library of Congress Number: 72-600054.

4. Reprinted with permission from *Ergonomic Design for People at Work,* © Eastman Kodak Company, 1983, published by Van Nostrand Reinhold. Courtesy of Eastman Kodak Company.

CHAPTER 2: WORKPLACE TABLES

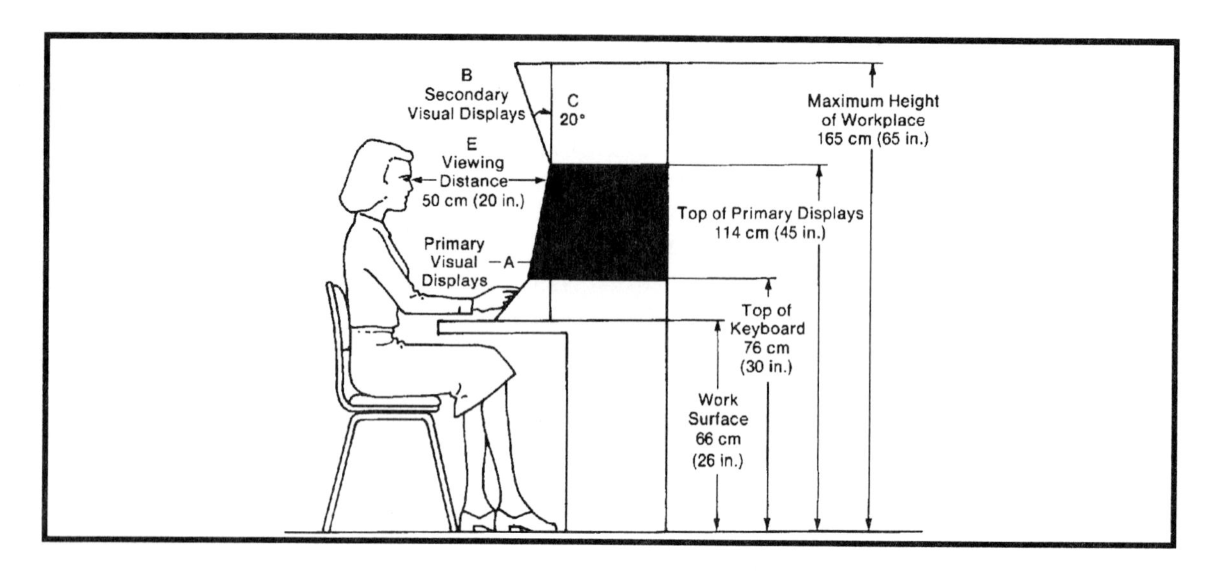

Section E: Workplace Environment

The work environment is as important to worker performance as the layout of the workstation or workplace. This section presents information concerning the working environment that need to be considered when designing jobs and workplaces.

WORKPLACE ENVIRONMENT

Table WRKSTN-E1: Impact of Work Environment on Health and Performance[1]

Environmental Factor	Physiological or Health Effects	Performance Effects
Electric Shock	Electrocution Electrical Burns Muscle Contraction	Distraction Interference with Manipulation Tasks
Noise	Hearing Loss Fatigue	Interference with Communication Interference with Signal Detection
Vibration	Muscle, Joint, Organ Pains Nausea Hand Circulatory Disturbances Sensory Loss in Fingers Reduced Gripping Strength	Interference with Manipulative Tasks Interference with Visual Tasks
Illumination	Headache Muscle Discomfort Fatigue Reduced Visual Acuity Eye Injury	Distraction Glare Interference Reduced Detection of Defects
Color	None Known	Influence on Appearance of Other Objects in the Area Influence on Mood
Heat/Humidity	Heat Illness: Circulatory Collapse Muscle Cramps Burns Discomfort	Distraction Interference with Manipulation Tasks (from sweaty hands, hot surface temperatures)
Cold	Hypothermia Frostbite Shivering Loss of Flexibility in the Fingers	Interference with Manipulation Tasks (through shivering, finger flexibility loss) Distraction

TEMPERATURE OF THE WORKPLACE

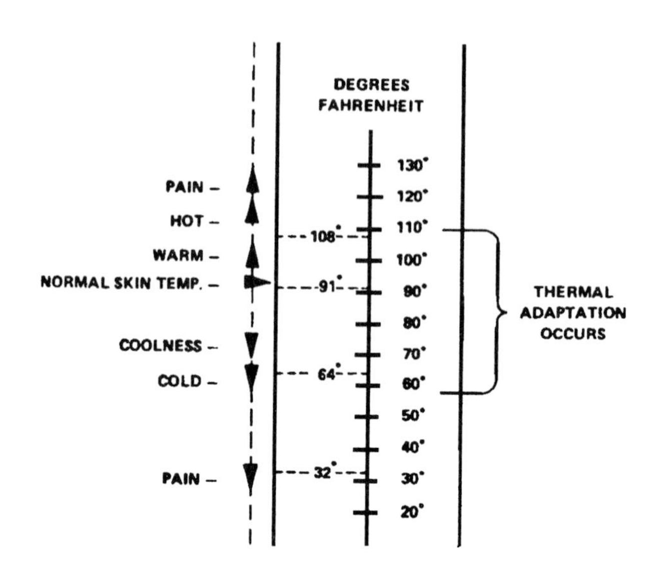

Cutaneous Senses

Figure WRKSTN-E1: Effects of Temperature on Skin[2]

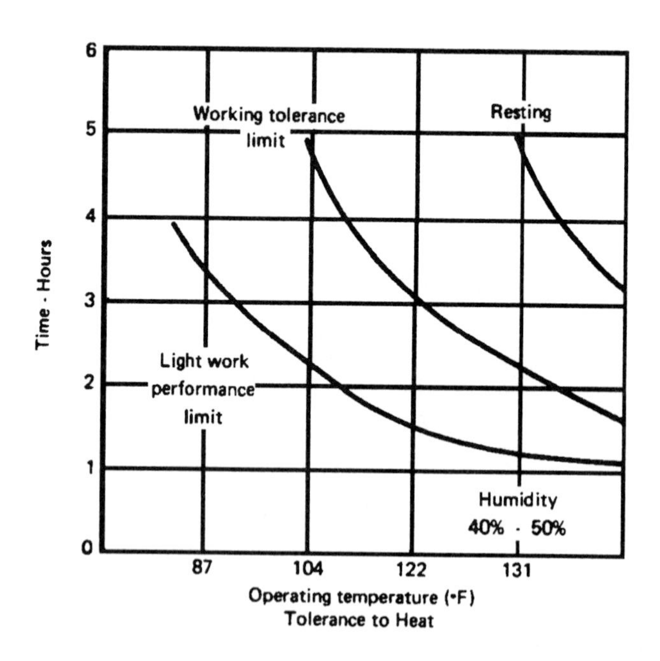

Figure WRKSTN-E2: Recommended Exposure Time to Heat as a Function of Activity and Temperature[2]

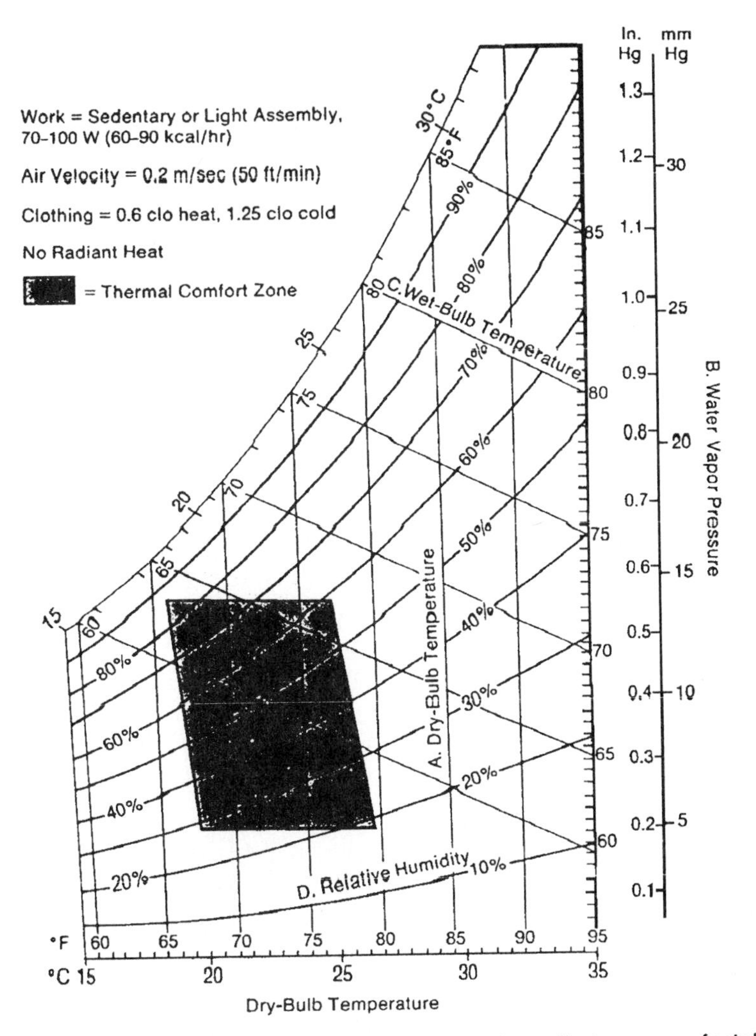

The dry bulb temperature and humidity combinations that are comfortable for most people doing sedentary or light work are shown as the shaded area on the psychometric chart. The dry bulb temperature range is from 19° to 26°C (66° to 79°F), and relative humidities (shown as parallel curves) range from 20 to 85 percent, with 35 to 65 percent being the most common values in the comfort zone. On this chart ambient dry bulb temperature (A) is plotted on the horizontal axis and indicated as parallel vertical lines; water vapor pressure (B) is on the vertical axis. Wet bulb temperatures (C) are shown as parallel lines with a negative slope; they intersect the dry bulb temperature lines and relative humidity curves (D) on the chart. In the definition of the thermal comfort zone, assumptions were made about the work load, air velocity, radiant heat, and clothing insulation levels. These assumptions are given in the top left corner of the chart.

Figure WRKSTN-E3: Thermal Comfort Zone[1]

Table WRKSTN-E2: Influences on Thermal Comfort Zone[1]

Factor	Level	Ambient Dry Bulb Temperature* Lower Limit °C	(°F)	Upper Limit °C	(°F)
Relative Humidity (%)	20	20	(68)	26	(79)
	50	19	(67)	25.5	(78)
	80	18.5	(66)	24	(76)
Air Velocity, m/sec (ft/min)	0.1 (20)	18	(65)	24	(76)
	0.25 (50)	19	(67)	25.5	(78)
	0.36 (70)	21	(70)	27	(80)
	0.51 (100)	22	(72)	28	(82)
	0.71 (140)	23	(74)	29	(84)
Work Load, 8-Hour Average, multiples of resting values	× 2	19	(67)	25.5	(78)
	× 3.5	17	(64)	23	(74)
	× 5	≈15.5	(≈60)†	20	(68)
Clothing Insulation (clo)	0.25	27	(80)	28	(83)
	1.25	19	(67)	22	(72)
	2.50	≈11	(≈52)	≈16	(≈62)
Radiant Heat, °C (°F), amount that globe temperature exceeds dry-bulb temperature	0	19	(67)	25.5	(78)
	1.1 (2)	17	(64)	24	(76)
	2.8 (5)	16	(62)	23	(74)
	5.6 (10)	13	(56)	20	(68)

* Unless otherwise noted, the following values have been used to calculate the thermal comfort zone limits: air velocity, 0.25 m/sec (50 ft/min); work load, sedentary, light assembly, up to two times resting metabolism; clothing insulation, 0.6 clo in heat, 1.25 clo in cold; no radiant heat load; humidity, 50 percent.

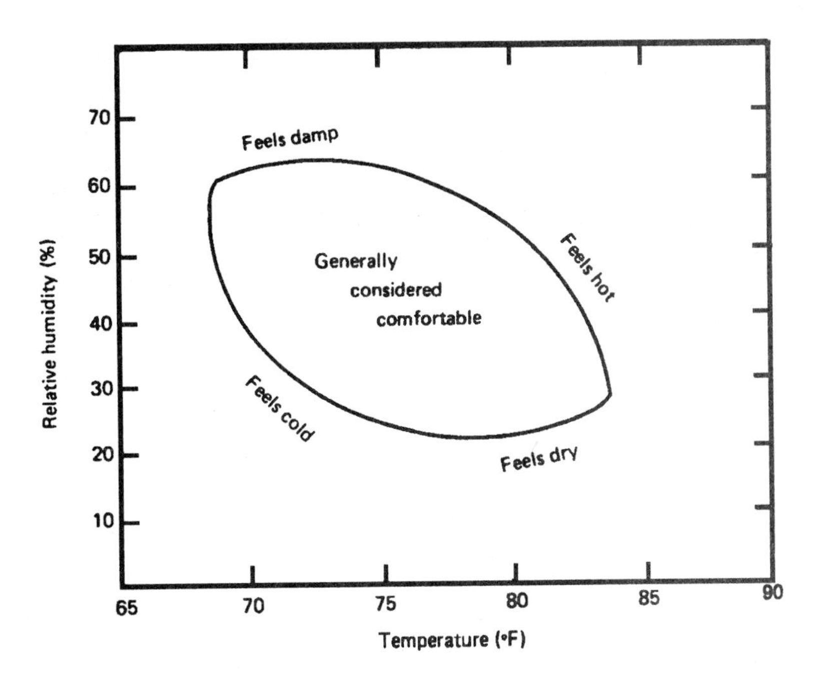

Figure WRKSTN-E4: Comfort Zone as a Function of Relative Humidity Versus Temperature[2]

Table WRKSTN-E3: Maximum Recommended Work Loads, Heat Discomfort Zone[1]

Maximum Recommended Work Load					
Ambient Temperature		Relative Humidity			
°C	°F	20%	40%	60%	80%
27	80	VH	VH	VH	H
32	90	VH	H	M	L
38	100	H	M	L	NR
43	110	M	L	NR	NR
49	120	L	NR	NR	NR

Note: Assumptions include 2-hour continuous exposure, 0.6 clo insulation, air velocity less than 0.5 m/sec (100 ft/min). Higher work loads may be sustained for shorter work periods. See Figure VD-3 for further information. Definitions of work load abbreviations: VH = very heavy, 350–420 W (300–360 kcal/hr); H = heavy, 280–350 W (240–300 kcal/hr); M = moderate, 140–280 W (120–240 kcal/hr); L = light, less than 140 W (120 kcal/hr). NR = not recommended for 2 hours of continuous exposure.

Table WRKSTN-E4: Maximum Temperatures for Short-Duration Exposures to High-Heat Environments (Up to 63°C or 136°F)[1]

Exposure Time (min)	Work Load*	Maximum Ambient Temperature, °C (°F)† Relative Humidity 20%	Relative Humidity 50%	Relative Humidity 80%
5	L	63 (146)	56 (133)	56 (133)
	M	59 (138)	48 (118)	46 (115)
	H	57 (135)	46 (115)	42 (108)
15	L	53 (128)	45 (113)	40 (104)
	M	52 (126)	43 (110)	38 (100)
	H	51 (124)	41 (106)	36 (97)
30	L	52 (126)	44 (112)	39 (102)
	M	47 (116)	38 (100)	34 (93)
	H	41 (106)	36 (97)	30 (86)
45	L	51 (124)	43 (110)	38 (100)
	M	41 (106)	36 (97)	31 (88)
	H	36 (97)	32 (90)	27 (81)

Note: For 5-min exposure times in high air velocities (2 m/sec, or 400 ft/min), the following maximum temperatures are recommended (L = light work load, M = moderate work load, H = heavy work load):

Workload	Relative Humidity, % 20%	50%	80%
L	56 (133)	50 (122)	48 (118)
M	54 (129)	49 (120)	44 (111)
H	52 (126)	48 (118)	42 (103)

* Work load abbreviations: L = light, up to 140 W (120 kcal/hr); M = moderate, >140 to 230 W (>120 to 240 kcal/hr); H = heavy, >230 to 350 W (>240 to 300 kcal/hr).

† These temperatures assume the following conditions: clothing insulation = 0.6 clo; air velocity = 0.1 m/sec (20 ft/min); radiant heat = 2°C (3.6°F), which is the difference between the globe and dry bulb temperature readings. Exposures are for unacclimatized workers.

The highest recommended dry bulb temperatures are shown for continuous exposures of 5 to 45 minutes (column 1) at 20, 50, and 80 percent relative humidity (columns 3 through 5). Light (L), moderate (M), and heavy (H) work loads are indicated in column 2 for each of the exposure time and relative humidity combinations. Maximum exposure temperatures decrease with increased exposure time, relative humidity, and work load. Air velocity, radiant heat, and clothing insulation assumptions are stated in a footnote. At temperatures above 50°C (122°F), especially in high humidities, high air velocity (up to 2 m/sec, or 400 ft/min) will increase body heat storage, so lower ambient temperatures would be required. At lower temperatures increased air velocity will lengthen exposure time. A summary of recommended maximum temperatures in high air flow for 5-minute exposures is given in the table note.

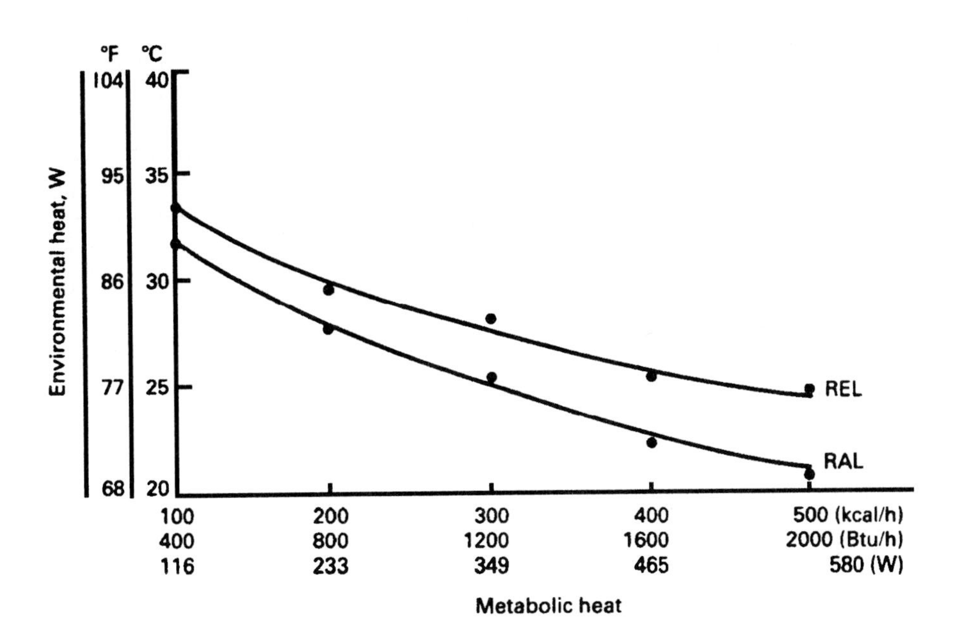

Figure WRKSTN-E5: Recommended Heat-Stress Exposure Limits (REL) and Alert Limits (RAL)[3]

Table WRKSTN-E5: Suggested Maximum Wet-Bulb Globe Temperatures for Various Working Conditions [°F(°C)][3]

Air velocity	Work intensity (kcal/h)		
	Light (≤ 200)	Moderate (201 < ≤300)	Heavy (> 300)
Low: < 1.5 m/s (< 300 ft/min)	86 (30)	82 (27.7)	79 (61.2)
High: ≥ 1.5 m/s (≥ 300 ft/min)	90 (32.2)	87 (30.5)	84 (66.2)

Table WRKSTN-E6: Insulation Value (clo) of Various Clothing Ensembles[1]

Clothing Ensemble	Icl (clo units)*
Sleeveless blouse, light cotton skirt, sandals	0.3
Shorts, open-neck shirt with short sleeves, light socks, sandals	0.3–0.4
Long lightweight trousers, open-neck shirt	
with short sleeves	0.5
with long sleeves	0.6
Cotton fatigues, lightweight underwear, cotton shirt and trousers, cushion-sole socks and boots	0.7
Typical light business suit; pant suit (with full jacket)	1.0
Typical light business suit and cotton coat (lab coat)	1.5
Heavy traditional European business suit, long cotton underwear, long-sleeved shirt, woolen socks, shoes; suit includes trousers, jacket, and vest	1.5–2.0

* Icl = Rcl/0.18, where Icl is the insulation value, in clo units, and Rcl is the total heat transfer resistance from skin to the outer surface of the clothed body, in degrees Celsius per kilocalorie per square meter of body surface area per hour. To express Icl in degrees Celsius per watt per square meter, change the constant from 0.18 to 0.16.

Column 2 shows the insulation values, or Icl in clo units (defined in the footnote), of several clothing ensembles that are described in column 1. Insulation increases with increased layers of clothes and with fabrics, like wool, that incorporate an air layer. Artificial fabrics often have higher insulation values but may not breathe. Their low moisture permeability can limit their usefulness because they reduce evaporative cooling in hot environments and trap the moisture near the skin in cold environments.

Table WRKSTN-E7: Contact With Hot Surfaces, Maximum Temperatures[1]

	Temperature Threshold	
Material	Pain Threshold, °C (°F)	First-Degree Burn Threshold, °C (°F)
Polystyrene GP	77 (171)	138 (281)
Wood (average)	76 (169)	135 (275)
ABS Resins	74 (166)	131 (268)
Phenolics (average)	60 (141)	99 (210)
Brick	59 (138)	95 (202)
Heat-resistant Glass	54 (129)	82 (180)
Water	53 (127)	80 (176)
Concrete	50 (122)	73 (164)
Steel	45 (113)	62 (143)
Aluminum	45 (112)	60 (141)

Note: Data are based on a 1-second contact by the finger.

The temperature of several materials (column 1) are shown at which pain (column 2) or a first-degree burn (column 3) will result from a 1-second finger contact. These values are maximum temperatures, so designs should not exceed them. Lower temperatures are desirable, since they accommodate longer contact times, contact with other skin such as the forearm, or the presence of breaks in the skin from cuts or abrasions.

ELECTRICITY

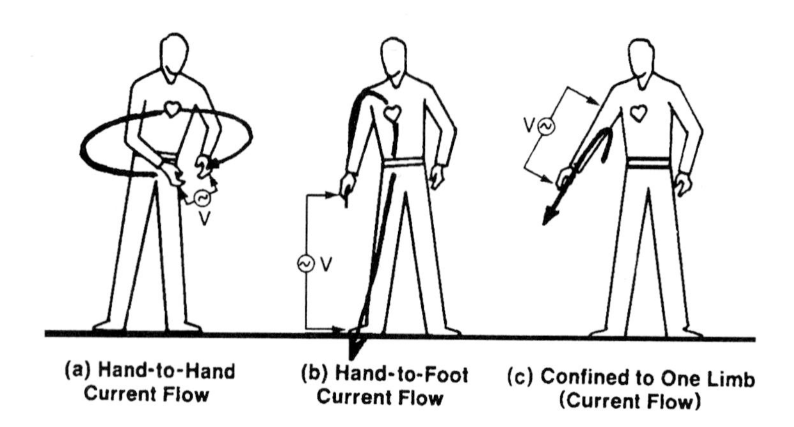

The pathways for current flow are shown for three types of contact with electricity: across the hands (part a), from hand to foot (part b), and within one limb (part c). The first two types of contact result in electric current passing through the body, making disorders of heart rhythms more possible. The voltage *(V)* applied is signified by a circle in which the symbol for current (~) is shown; the two arrows indicate the source and the sink (closing of the circuit) of the current flow.

Figure WRKSTN-E6: Electric Current Flow Through Body[1]

Table WRKSTN-E8: Skin Resistance[1]

Contact Condition	Measured Skin Resistance	
	Dry	Wet
Finger touch	40 kΩ–1 MΩ	4–15 kΩ
Hand holding wire	15–50 kΩ	3–6 kΩ
Finger-thumb grasp	10–30 kΩ	2–5 kΩ
Hand holding pliers	5–10 kΩ	1–3 kΩ
Palm touch	3–8 kΩ	1–2 kΩ
Hand around 4-cm ($1\frac{1}{2}$-in.) pipe	1–3 kΩ	0.5–1.5 kΩ
Two hands around 4-cm ($1\frac{1}{2}$-in.) pipe	0.5–1.5 kΩ	250–750 Ω
Hand immersed	—	200–500 Ω
Foot immersed	—	100–300 Ω
Human body, internal, excluding skin	—	200–1000Ω

The resistance in ohms (Ω), kilohms (kΩ), and megohms (MΩ) of skin when dry or wet, from sweating or contact with a liquid, is given in columns 2 and 3. Column 1 specifies the contact conditions. The lower the resistance, the more current will flow for a given voltage applied to the skin. The presence of moisture on the skin reduces resistance fivefold or more in most of these conditions.

Table WRKSTN-E9: Electical Shock Thresholds[1]

Effect	Threshold Current	Frequency	Examples of Situations Where Contact Could Occur
Perception			
warmth	0.2 mA/cm^2 dc	DC	Charging car battery
tingling or startle reaction	0.35–1.19 mA/cm^2 rms	60 Hz	Touching an improperly grounded appliance
Pain	<0.9 mA	DC	Contacting a charged surface with hand
	3–10 mA	60 Hz	Grasping an improperly grounded power tool
Muscle Contraction			
Let-go threshold (maximum current one can tolerate and still release an object by using the muscles being directly stimulated)	29 mA	DC	Shorting out a plating electrode with hand
	6 mA	60 Hz	Gripping a faulty tool equipped with a cheater plug
Respiratory paralysis	18–22 mA	60 Hz	Contacting a charge with a hand or foot and having it pass through the body to another limb

Table WRKSTN-E9: Electical Shock Thresholds[1] (Continued)

High voltage/high current	>5 A or 500 V	DC, AC, or impulse	Contacting a power transmission line or the high-voltage section of a television set
Cardiac Effects	54 J discharge	non-oscillating, $t < 1$ sec	Grabbing a high-voltage capacitor while standing in water
Ventricular fibrillation (uncoordinated heart activity); whole-body muscular contraction	54 J discharge	oscillatory, $t < 1$ sec	Discharging an electrical circuit across the body
	>67 mA	60 Hz	Using a badly grounded power tool in wet footing conditions
Cardiac Arrest	>5 A	60 Hz	Contacting a power transmission line
Convulsions, Head	100 mA	60 Hz	Contacting an electric power line

EXPOSURE TO NOISE

Table WRKSTN-E9: Intensity and Effects of Common Noises [2]

Common Sounds	Noise Level (dB)	Effect
Carrier deck jet operation Air raid siren	140	Painfully loud (blurring vision, nausea, dizziness)
Jet takeoff (200 feet) Thunderclap	130	Begin to "feel" the sound
Loud Disco Auto horn (3 feet)	120	Hearing becomes uncomfortable
Pile drivers	110	Cannot speak over the sound
Garbage truck	100	
Heavy truck (50 feet) City traffic	90	Very annoying
Alarm clock (2 feet) Hair dryer	80	Annoying
Noisy restaurant Freeway traffic Man's voice (3 feet)	70	Telephone use difficult
Air conditioning unit (20 feet)	60	Intrusive
Light auto traffic (100 feet)	50	Ouiet
Living room Bedroom Quiet office	40	
Library Soft whisper (15 feet)	30	Very quiet
Broadcasting studio	20	
	10	Just audible
	0	Hearing begins

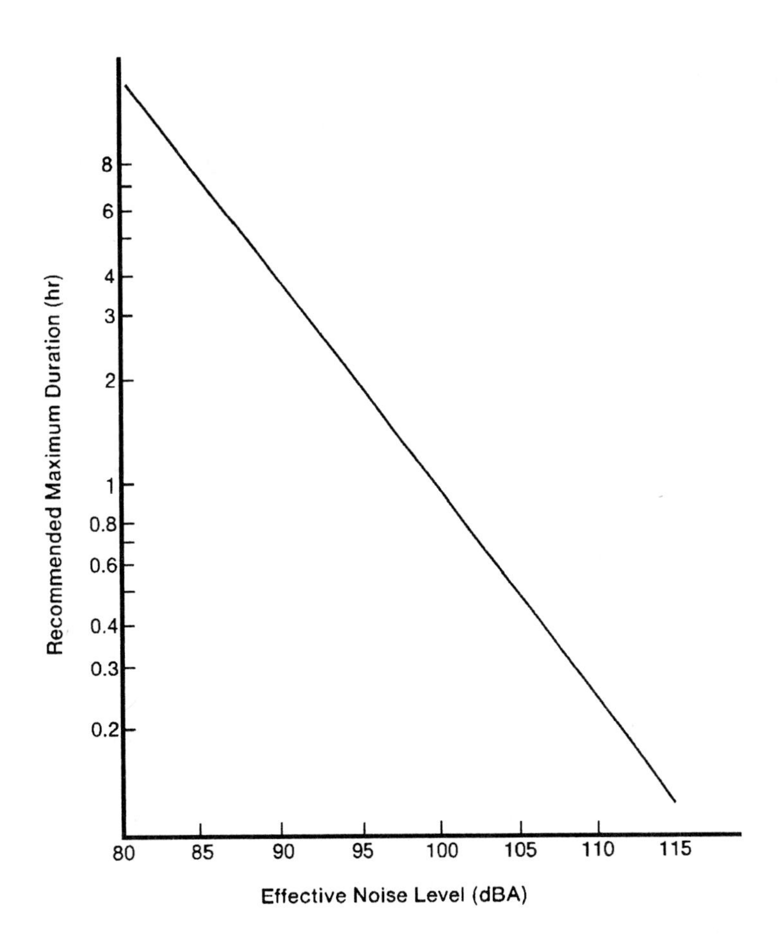

The recommended maximum duration of exposure (in hours, hr, on the vertical axis) to noise of different intensities (in decibels, dBA, on the horizontal axis) is given. The higher the noise level, the less time a person should be exposed to it in order to reduce the risk of hearing damage. Noise levels above 115 dBA should be avoided; levels below 80 dBA are not known to contribute to hearing loss over extended exposure times.

Figure WRKSTN-E7: Guidelines for Noise Exposure to Protect Hearing; Recommended Maximum Duration Versus Noise Level[1]

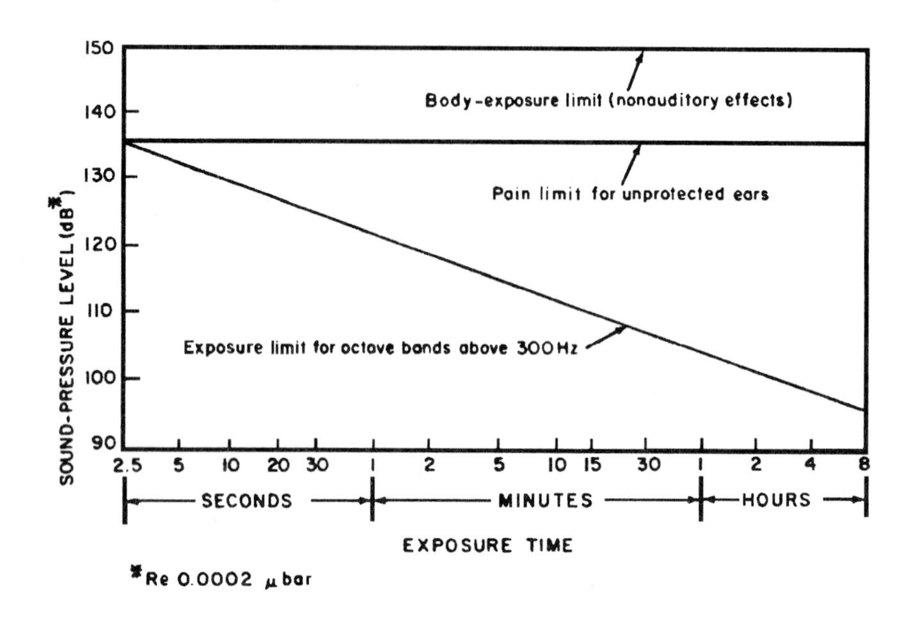

Figure WRKSTN-E8: Damage Risk Criteria for Exposure Times [4]

Pain limit for unprotected ears is shown at 135dB. When ear protectors are used, sound pressure level in sound field can exceed these criteria by amount of attenuation provided by the protectors. Body exposure limit at 150 dB is the point at which potentially dangerous non-auditory effects can occur.

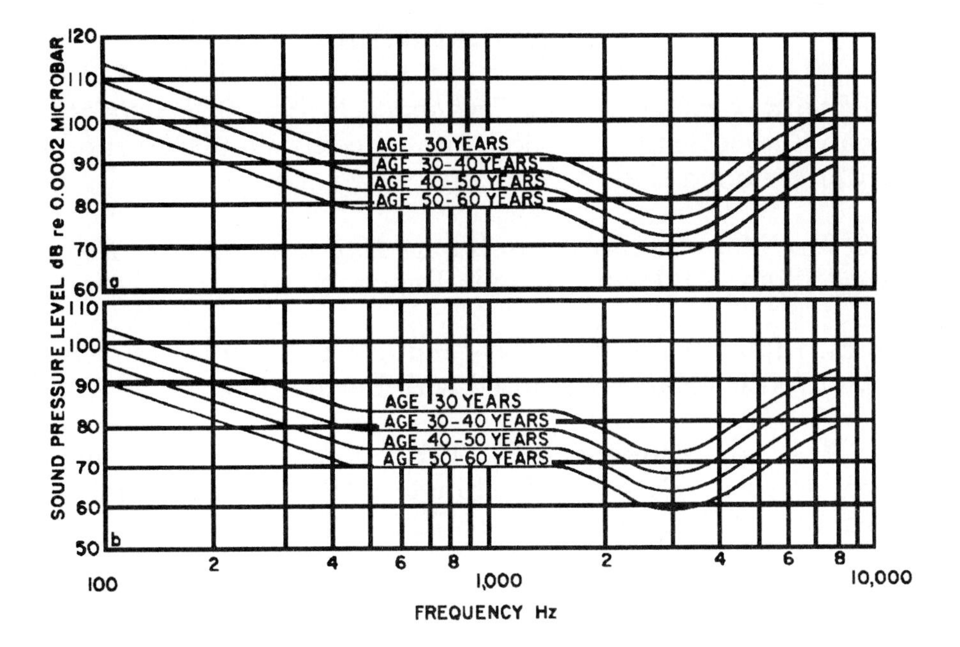

Figure WRKSTN-E9: Damage Risk Criteria By Age.[4]

Data shown for (a) Wide Band Noise Measured by Octave, 8 hr. Continuous Exposure; and (b) Pure Tones or Critical Bands of Noise.

REFERENCES

1. Reprinted with permission from *Ergonomic Design for People at Work*, © Eastman Kodak Company, 1983, published by Van Nostrand Reinhold. Courtesy of Eastman Kodak Company.

2. Bailey, R.W. 1989. *Human Performance Engineering. (2nd Ed)* Englewood Cliffs, NJ: Prentice-Hall. Reprinted with Permission.

3. Pulat, B.M. 1992. *Fundamentals of Industrial Ergonomics.* Englewood Cliffs, NJ: Prentice-Hall. Reprinted with Permission.

4. Van Cott, H.P., and Kinkade, R.G., 1972. *Human Engineering Guide to Equipment Design. (Revised)* Washington, DC.: US Government Printing Office. Library of Congress Number: 72-600054.

CHAPTER 3: CONTROL TABLES

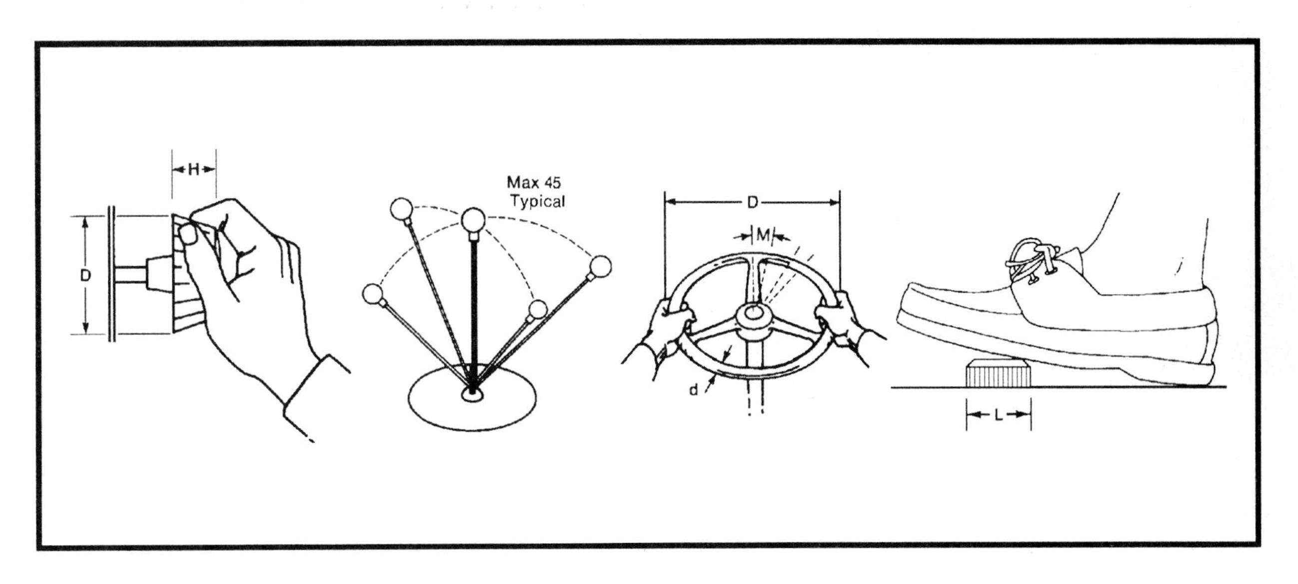

Section A: General Tables

This section presents generic information about the different varieties of displays. The information in this section is useful for the ergonomist or human factors engineer who must choose between a number of potential control systems.

GENERAL INFORMATION ABOUT CONTROLS

Table CNTRL-A1: General Rules for Designing, Selecting, and Arranging Controls[1]

1. Locate and orient controls so that their motion and location are compatible with the movement and location of associated display element or system response.
2. Functionally related controls are candidates for combining to reduce intermediate moves and panel space.
3. Control responses should be distributed to the limbs as equally as possible.
4. The hands should be assigned those controls that require precise setting.
5. Controls requiring large and continuous application of force should be assigned to the feet.
6. A positive indication of control activation should be provided.
7. The associated elements must be designed such that when a control is activated, the user will obtain feedback that a system response has been achieved.
8. Controls should be designed to withstand abuse. This is particularly important in emergency conditions.
9. Select multirotation controls when precise settings are required over a wide range of adjustments.
10. Select discrete adjustment controls when the system is to be adjusted for discrete positions or values over a continuum.
11. Select controls that can easily be identified based on touch or visual clues.
12. When precise adjustments over a continuum are required, continuous adjustment controls are the choice.
13. Control surfaces must prevent the activating hand from slipping accidentally.
14. Frequently used and critical controls must be located within easy reach of the user.
15. Control movements should conform to user expectations.
16. In designing and locating controls, consider the probability of accidental activation. Minimize this probability.
17. The least capable user must be able to apply the forces necessary to activate controls.

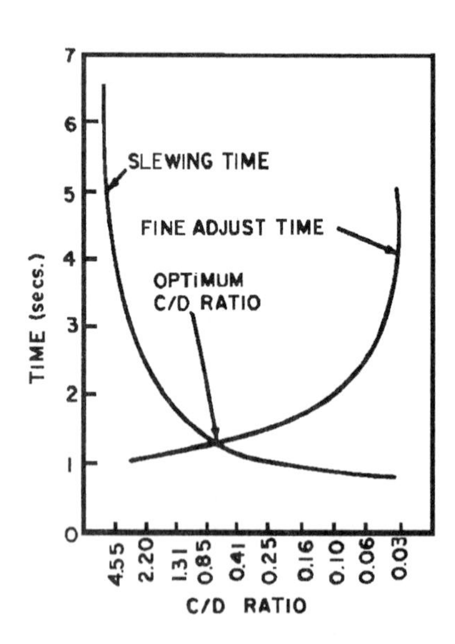

Figure CNTRL-A1: Control-Display Response Ratio[1]

Table CNTRL-A2: Mean Ratings on Task Dimensions of Control Selection[1]

	Control settings		Adjustment precision			Force required			Range of adjustment/discrete action			
	Disc	Cont.	Gross	Int.	Fine	Small	Med.	Large	$x = 2$	$10 \geqslant x > 2$	$24 \geqslant x > 10$	$x > 24$
Foot pushbutton	95.0	5.0	87.0	7.5	5.5	17.0	41.5	37.5	91.8	0.2	0.0	0.0
Foot pedal	28.7	71.3	51.1	35.5	13.4	14.7	48.7	36.6	62.8	35.0	2.2	0.0
Keyboard	98.5	1.5	34.2	9.9	59.9	90.5	6.5	3.0	64.3	7.4	9.6	18.7
Detent thumbwheel	90.0	10.0	21.7	46.3	32.0	73.0	24.5	2.5	17.2	62.3	16.0	4.5
Rotary selector switch	75.0	25.0	19.3	42.0	38.7	67.3	31.2	1.5	17.2	45.5	23.6	13.7
Hand pushbutton	98.0	2.0	78.0	10.5	11.5	67.0	27.3	5.7	90.3	7.3	2.1	0.3
Legend switch	99.5	0.5	76.0	7.5	16.5	72.8	23.2	4.0	88.7	10,3	1.0	0.0
Toggle switch	96.0	4.0	75.2	11.8	13.0	65.5	28.5	6.0	73.9	21.3	3.2	1.6
Ball controller	3.7	96.3	12.4	38.4	49.2	77.9	21.6	0.5	8.0	26.0	10.0	56.0
Joy stick	10.0	90.0	6.9	42.7	50.4	56.5	37.8	5.7	15.4	55.9	9.1	19.6
Round knob	16.4	83.6	11.3	32.4	56.3	74.2	24.2	1.6	25.0	39.0	11.5	24.5
Lever	37.7	62.3	37.6	44.2	18.2	21.9	53.4	24.7	45.6	42.4	6.1	5.9
Crank	2.8	97.2	47.5	36.7	15.8	21.2	43.3	35.5	14.3	20.0	15.7	50.0
Handwheel	2.9	97.1	31.6	49.7	18.7	18.2	47.6	34.2	15.0	35.0	17.5	32.5
Continuous adjustment thumbwheel	1.8	98.2	15.2	40.3	44.5	84.2	14.0	1.8	14.0	48.0	14.0	24.0

3
Control Tables

Table CNTRL-A3: Recommended Separations for Controls[2]

Control	Type of Use	Measurement of Separation	Recommended Separation Minimum mm	Minimum in.	Desirable mm	Desirable in.
Push Button	One Finger (Randomly)		12	$\frac{1}{2}$	51	2
	One Finger (Sequentially)		6	$\frac{1}{4}$	25	1
	Different Fingers (Randomly or Sequentially)		12	$\frac{1}{2}$	12	$\frac{1}{2}$
Toggle Switch	One Finger (Randomly)		20	$\frac{3}{4}$	51	2
	One Finger (Sequentially)		12	$\frac{1}{2}$	25	1
	Different Fingers (Randomly or Sequentially)		16	$\frac{5}{8}$	20	$\frac{3}{4}$
Crank and Lever	One Hand (Randomly)		51	2	100	4
	Two Hands (Simultaneously)		76	3	127	5
Knob	One Hand (Randomly)		25	1	51	2
	Two Hands (Simultaneously)		76	3	127	5
Pedal	One Foot (Randomly)	D, d	d = 100 D = 203	4 8	152 254	6 10
	One Foot (Sequentially)		d = 51 D = 152	2 6	100 203	4 8

The recommended distance between two similar controls on a panel or machine is determined by the type of control (Column 1) and the way it must be activated (Column 2). The minimum and optimum separations are given in Columns 4 and 5 for each type of control. Column 3 shows how the separation distances are defined, from the centers or the sides of the controls.

Table CNTRL-A4: Recommended Controls to Use Where Both Force and Range Settings are Important[3]

For *small* forces and—	Use —
Two discrete settings	Pushbutton or toggle switch.
Three discrete settings	Toggle switch or rotary selector switch.
Four to 24 discrete settings	Rotary selector switch.
Small range of continuous settings	Knob or lever.
Large range of continuous settings	Crank or multi-rotation knob.
For *large* forces and—	**Use—**
Two discrete settings	Detent lever, large hand pushbutton, or foot pushbutton.
Three to 24 discrete settings	Detent lever.
Small range of continuous settings	Handwheel, rotary pedal or lever.
Large range of continuous settings	Large crank.

Table CNTRL-A5: Recommended Control Movements[3]

Function	Control action
On	Up, right, forward, pull (switch knobs).
Off	Down, left, rearward, push (switch knobs)
Right	Clockwise, right.
Left	Counterclockwise, left.
Up	Up, rearward.
Down	Down, forward.
Retract	Rearward, pull, counterclockwise, up.
Extend	Forward, push, clockwise, down.
Increase	Right, up, forward.
Decrease	Left, down, rearward.

Table CNTRL-A6: A Comparison of Common Controls[1]

Control	Suitability Where Speed of Operation Is Required	Suitability Where Accuracy of Operation Is Required	Space Required to Mount Control	Ease of Operation in Array of Like Controls	Ease of Check Reading in Array of Like Controls
Toggle Switch (On-Off)	Good	Good	Small	Good	Good
Rocker Switch	Good	Good	Small	Good	Fair[1]
Push Button	Good	Unsuitable	Small	Good	Poor[1]
Legend Switch	Good	Good	Small	Good	Good
Rotary Selector Switch (discrete steps)	Good	Good	Medium	Poor	Good
Knob	Unsuitable	Fair	Small–Medium	Poor	Good
Crank	Fair	Poor	Medium–Large	Poor	Poor[2]
Handwheel	Poor	Good	Large	Poor	Poor
Lever	Good	Poor (H) Fair (V)	Medium–Large	Good	Good
Foot Pedal	Good	Poor	Large	Poor	Poor

[1] Except where control lights up for "on."
[2] Assumes control makes more than one revolution.
H = Horizontal
V = Vertical

A summary of the suitability of ten different controls (Column 1) for five different conditions or job requirements (across the top) is given. The ratings are based on typical examples of each control type, not on the extremes of performance in each range. The types of controls that should be considered when the design or work situation has certain requirements, such as speed, accuracy, ease of operation, ease of reading, or limited space on a control panel, can be determined from this summary.

Table CNTRL-A6: A Comparison of Common Controls (Continued)[1]

Characteristic	Type of Control: Discrete adjustment: Rotary selector switch	Thumb-wheel	Hand push-button	Foot push-button	Toggle switch	Continuous adjustment: Knob	Thumb-wheel	Hand-wheel	Crank	Pedal	Lever
Large forces can be developed	------	------	------	------	------	No	No	Yes	Yes	Yes	Yes
Time required to make control setting	Medium to quick.	------	Very quick.	Quick	Very quick.	------	------	------	------	------	
Recommended number of control positions (settings).	3 to 24	3 to 24	2	2	2 to 3	------	------	------	------	------	
Space requirements for location and operation of control.	Medium	Small	Small	Large	Small	Small to medium.	Small	Large	Medium to large.	Large	Medium to large.
Likelihood of accidental activation	Low	Low	Medium	High	Medium	Medium	High	High	Medium	Medium	High.
Desirable limits to control movement	270°	------	⅛" × 1¼".	½ × 4"	120°	Un-limited	180°	±60°	Un-limited	Small*	±45°.
Effectiveness of coding	Good	Poor	Fair to good.	Poor	Fair	Good	Poor	Fair	Fair	Poor	Good.
Effectiveness of visually identifying control position.	Fair to good.	Good	Poor†	Poor	Fair to good.	Fair‡ to good.	Poor	Poor to fair.	Poor§	Poor	Fair to good.
Effectiveness of non-visually identifying control position.	Fair to good.	Poor	Fair	Poor	Good	Poor to good.	Poor	Poor to fair.	Poor§	Poor to fair.	Poor to fair.
Effectiveness of check-reading to determine control position when part of a group of like controls.	Good	Good	Poor†	Poor	Good	Good‡	Poor	Poor	Poor§	Poor	Good.
Effectiveness of operating control simultaneously with like controls in an array.	Poor	Good	Good	Poor	Good	Poor	Good	Poor	Poor	Poor	Good.
Effectiveness as part of a combined control.	Fair	Fair	Good	Poor	Good	Good¶	Good	Good	Poor	Poor	Good.

*Except for rotary pedals which have unlimited range.
†Exception: when control is back-lighted and light comes on when control is activated.
‡Applicable only when control makes less than one rotation. Round knobs must also have a pointer attached.
§Assumes control makes more than one rotation.
¶Effective primarily when mounted concentrically on one axis with other knobs.

Table CNTRL-A7: A Comparison of Unconventional Controls[3]

Concept		Characteristic							
Controller	Type	Mechanization	Output	Location	Command capability	Provides hand freedom	Accessibility	Natural direction of operation	Accuracy
Hand	On chest or side	Electro-mechanical pencil stick.	Continuously variable or on-off.	On chest, stomach or hip.	Complete	One hand free	Good for both hands.	Yes	± 1° or less.
	In auxiliary glove.	Possibly push buttons.	Continuously variable or on-off.	Back of glove	Complete	One hand free	With one controller for each hand, accessibility good for either hand.	No	± 1° if continuously variable; on-off N/A.
Mouth	Voice	"Audrey," "Sceptron" "Shoe Box," etc.	On-off or incremental.	At throat	Complete	Both hands free	Good	No	N/A.
	Tone	Resonant transducers.	On-off or incremental.	At throat	Complete	Both hands free	Good, depending upon musical ability of operator.	No	N/A.
	Breath	Sensitive diaphragms in "mouth organ" configuration.	On-off or crenental.	In front of mouth.	Complete	Both hands free	Good	No	N/A.
	Tongue	Switches, ports, slide-wires, etc.	On-of or continuously variable.	On lips or in mouth.	Unknown	Both hands free	Good	Perhaps	N/A.
Eye	Reflected beam.	Light beam reflected from cornea.	Continuously variable.	At eye	Forward translation only.	One hand may be required for supplemental control and switching.	Good	Yes	±10 min at center of field, ±1° at edge of field.
	Corneal-retinal potentials.	Electrical potential across eye.	Continuously variable.	At eye	Forward translation only.	One hand may be required for supplemental control and switching.	Good	Yes	±1 – 2°.
	Muscle action potentials.	Electrical signals in eye muscles.	Continuously variable.	At eye	Forward translation only.	One hand may be required for supplemental control and switching.	Good	Yes	Unknown.

Table CNTRL-A7: A Comparison of Unconventional Controls (Continued)[3]

Concept		Characteristic								
Controller	Type	Cross coupling between rotational axes	Motion coupling	Acquisition actuation	Inadvertent actuation	Type of feedback	Response time (sec)	Reliability	Size*	Weight*
Body	Head	Electrical or mechanical pickoffs sensing position of head relative to body.	Continuously variable.	Sight at eye; pickoffs at neck or on head.	Forward translation only or rotation only.	One hand may be required for control or switching.	Good	Yes	With sight ± 3 mils, without sight ± 1°.	
	Limb motion	Force or displacement sensors attached to limb.	Continuously variable or on-off.	At controlling limb.	Can be complete.	One hand may be required for control or switching.	Good	Perhaps	± 1 – 2°.	
	Myoelectronics	Skin electrodes sensing muscle action potentials.	On-off or incremental.	At controlling limb.	Can be complete.	Both hands free.	Good	Perhaps	N/A.	
Hand	On chest or side	Not significant with small stick excursions.	Not significant at small accelerations.	Probable.	Not probable except in cramped quarters.	Visual, force	1.0 – 1.5 (0.5 – 1.0 reach plus 0.5 operate.)	Good. Can be simple, rugged.	1	1
	In auxiliary glove.	Not significant	Not significant	None if properly designed.	Not probable	Visual, possibly force.	1.5 – 2.5 (1.0 – 2.0 reach plus 0.5 operate).	Good. Simple, rugged, repeatable.	1	1
Mouth	Voice	Not possible	None	None	Not probable if unique sounds reserved for commands.	Visual	0.5	Good. Simple, repeatable.	5	4
	Tone	Not possible	None	None	Not probable	Visual	0.5	Good. Simple, repeatable.	3	2
	Breath	Not possible	None	None	Not probable if sensor thresholds are high enough.	Visual	0.5	Good. Simple, repeatable.	3	2
	Tongue	Not possible	None	Possible	Possible	Visual	0.5	Probably good	3	2

Table CNTRL-A7: A Comparison of Unconventional Controls (Continued)[3]

Concept		Characteristic								
Controller	Type	Cross coupling between rotational axes	Motion coupling	Acquisition actuation	Inadvertent actuation	Type of feedback	Response time (sec)	Reliability	Size*	Weight*
Eye	Reflectedbeam	Not possible	None	None	Not probable if lockout switch included.	Visual	0.1	Fair. Excessive equipment, tendency to misalignment.	5	3
	Corneal-retinal potentials.	Not possible	None	None	Not probable	Visual	0.1	Fair. Extraneous signals, spurious variations, frequent miscalculations.	5	3
	Muscle action potentials.	Not possible	None	None	Not probable	Visual	0.1	Fair. Variable contact resistance, excessive electronics.	5	3
Body	Head	Not possible	Not significant	None	Possible unless lockout is used.	Visual	0.5. Movement plus fixation time.	Probably good	1.5	1.5
	Limb motion	Possibly some	Possible; not probable at low thrust levels.	None	Possible	Visual	0.3 to 0.5	Probably good	1.5	1.5
	Myoelectronics	Not significant	Possible	None	Possible	Visual	0.2	Fair. Variable contact resistance, excessive electronics.	5	3

*Normalized with respect to hand controllers.

Table CNTRL-A8: United States Stereotypes for Up and Down Switches[2]

ON	OFF
START	STOP
HIGH	LOW
IN	OUT
FAST	SLOW
RAISE	LOWER
INCREASE	DECREASE
OPEN	CLOSE
ENGAGE	DISENGAGE
AUTOMATIC	MANUAL
FORWARD	REVERSE
ALTERNATING	DIRECT
POSITIVE	NEGATIVE

Some movement stereotypes for toggle switches used in equipment or production system control panels are given. The expected direction of activation of the switch (when mounted vertically) is given for 13 actions or conditions. These expectations are United States stereotypes and may vary in other countries.

REFERENCES

1. Pulat, B.M. 1992. *Fundamentals of Industrial Ergonomics.* Englewood Cliffs, NJ: Prentice-Hall. Reprinted with Permission.

2. Reprinted with permission from *Ergonomic Design for People at Work*, © Eastman Kodak Company, 1983, published by Van Nostrand Reinhold. Courtesy of Eastman Kodak Company.

3. Van Cott, H.P., and Kinkade, R.G., 1972. *Human Engineering Guide to Equipment Design. (Revised)* Washington, DC.: US Government Printing Office. Library of Congress Number: 72-600054.

4. Bailey, R.W. 1989. *Human Performance Engineering. (2nd Ed)* Englewood Cliffs, NJ: Prentice-Hall. Reprinted with Permission.

CHAPTER 3: CONTROL TABLES

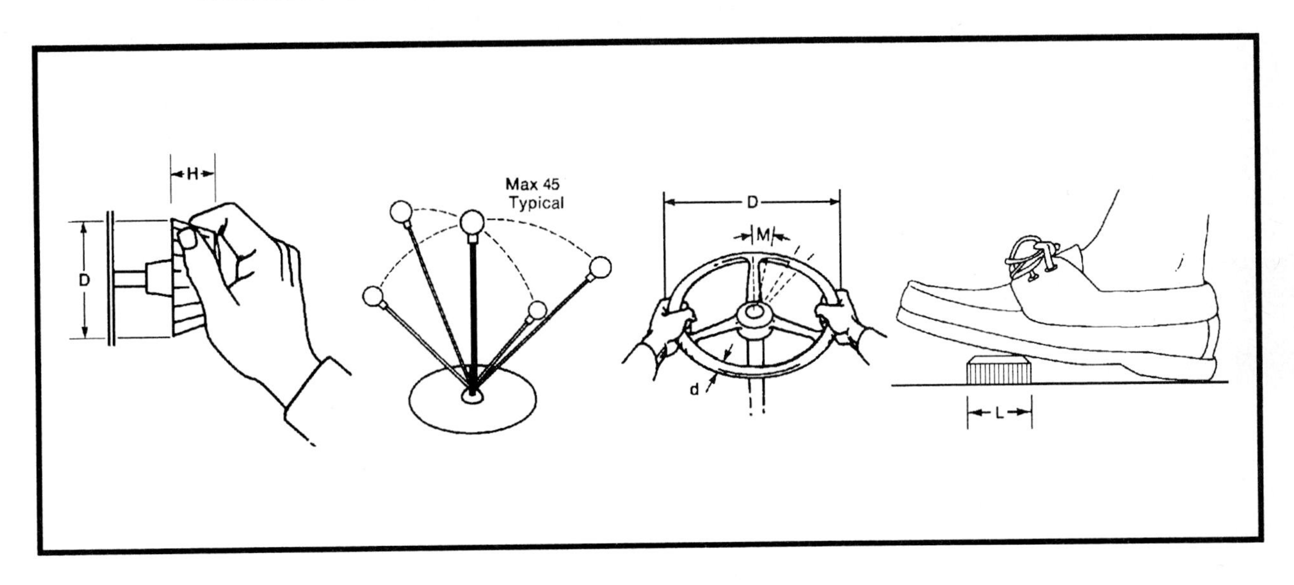

Section B: Knobs, Dials, and Switches

This section presents information to be used when designing systems that use knobs, dials, and switches as a means of control.

KNOB DESIGN

Table CNTRL-B1: Design Recommendations for Knobs[1]

Grasp	Fingertip	Palm of hand
Diameter:		
Minimum (in.)	0.375*	1.5
Maximum (in.)	4.0	3.0
Depth:		
Minimum (in.)	0.5	(†)
Maximum (in.)	1.0	
Displacement	Depends on C/D ratio.	
Resistance:		
Minimum	No limitation	
Maximum	4½–6 in.-oz.‡	
Control separation (in.):§		
One hand, randomly:		
Minimum	1.0	
Preferred	2.0	
Two hands, simultaneously:		
Minimum	3.0	
Preferred	5.0	

*0.25 when resistsnce is made very low.
†No limit set by operator performance.
‡First number is for 1 in. diameter; second number is for larger knobs.
§Edge to edge.

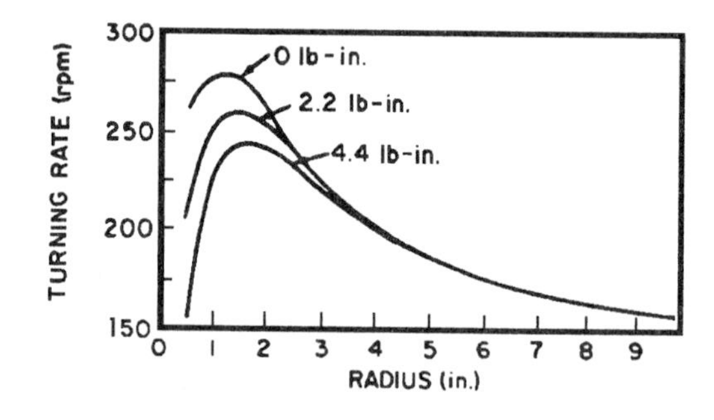

Figure CNTRL-B1: Turning Rate as a Function of Crank-Arm Radius and Load[1]

CLASS B

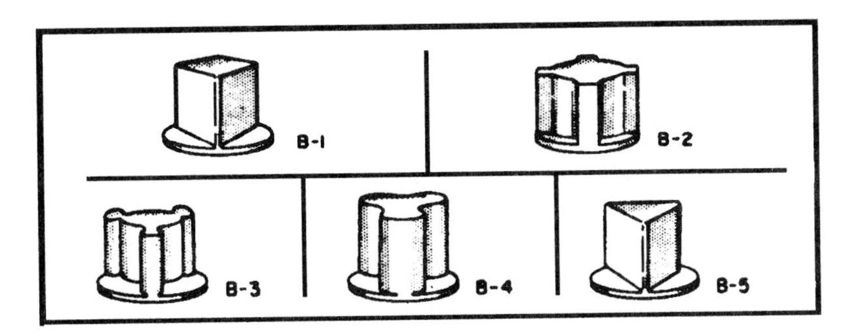

CLASS C

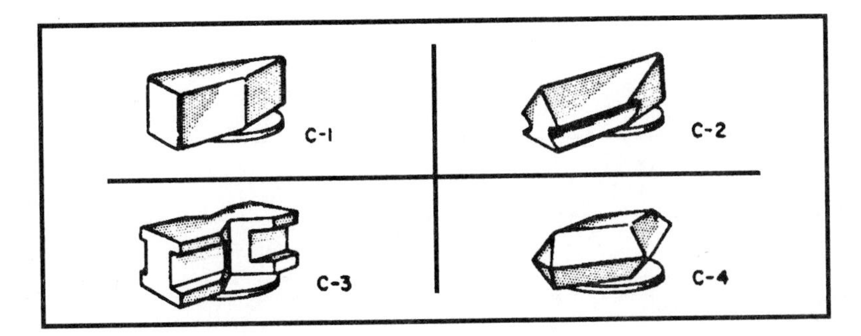

Figure CNTRL-B2: Examples of Three Classes of Knobs[1]

(A) Those for twirling or spinning; (B) those to be used where less than a full turn is required and position is not so important; and (C) those where less than a full turn is required and position is important.

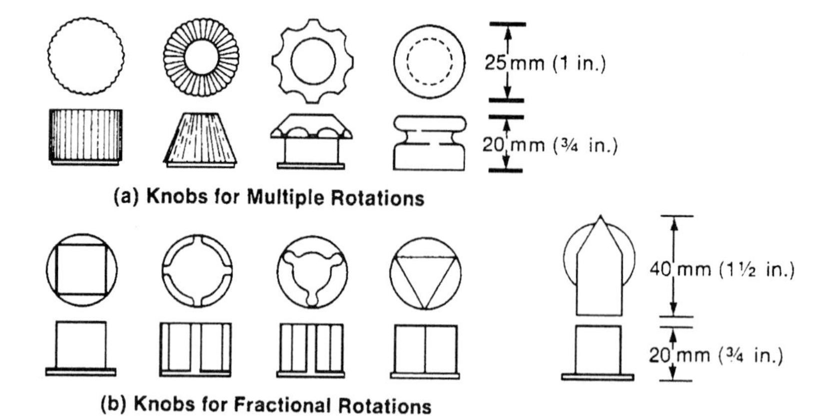

Figure CNTRL-B3: Shape Coding of Controls[2]

The nine knobs shown are shapes that will be distinguishable by touch alone even when an operator is wearing gloves. The knobs in part a are used when multiple rotations may be required. The first four knobs in part b are used in fractional rotation applications. The last knob in part b is a detent-positioning knob.

Table CNTRL-B2: Round Knob Characteristics[3]

	Fingertip grasp		Thumb and finger encircled		Palm Grasp	
	Height, h (mm)	Diameter, d (mm)	Height, h (mm)	Diameter, d (mm)	Height, h (mm)	Diameter, d (mm)
Minimum	12	10	12	25	15	35
Maximum	25	100	25	75	—	75

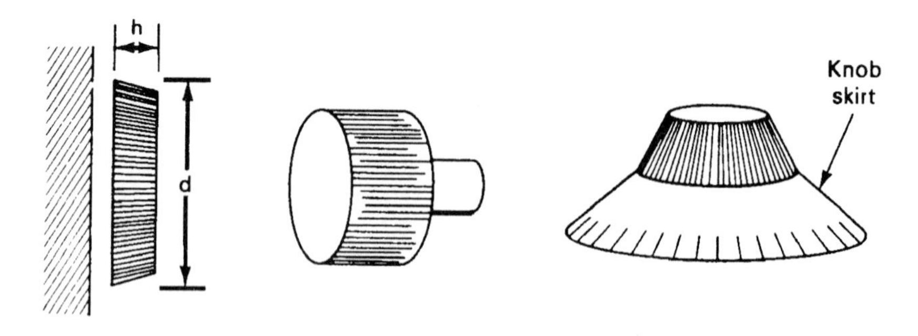

Table CNTRL-B3: Knob Design Recommendations[2]

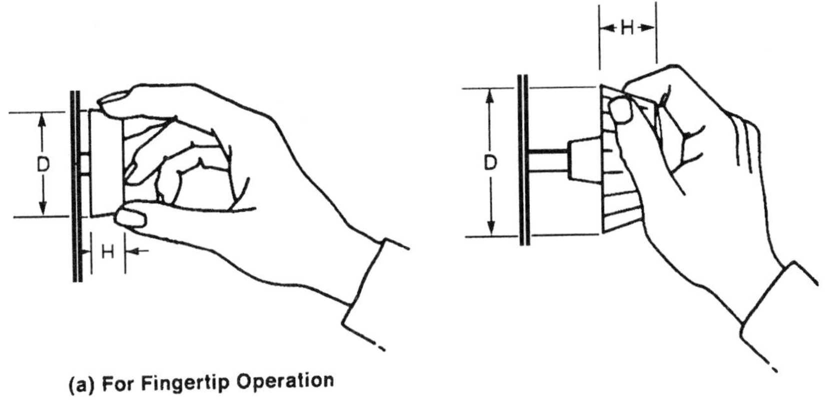

(a) For Fingertip Operation

(b) For Palm Grasp
(Star Pattern or Knurled Knob)

Fingertip Operation

Parameter	Recommended Design Values (Minimum–Maximum)	
Diameter (D)	10–100 mm	0.4–4.0 in.
Diameter minimum, for very low torque	6 mm	0.2 in.
Depth (H)	12–25 mm	0.5–1.0 in.

Palm Grasp

Parameter	Recommended Design Values (Minimum–Maximum)	
Diameter (D)	35–75 mm	1.5–3.0 in.
Depth (H), minimum	15 mm	0.6 in.

The recommended diameter (D) and depth (H) for knobs operated either by fingertips (part a) or with full palmar grasp (part b) are shown. The palmar grip design is appropriate for controls operating valves or other devices where fairly large forces (or torques) have to be developed.

SWITCHES

Table CNTRL-B4: Rotary Selector Switch Characteristics[3]

Bar Type					
	Length, *K* (mm)	Width, *W* (mm)	Depth, *D* (mm)	Displacement, α (visual positioning)	Resistance (kg)
Minimum	25	13	12	15°	0.25
Maximum	100	25	75	45°	1.4

Round Type				
	Diameter, *K* (mm)	Depth, *D* (mm)	Displacement, α (visual positioning)	Resistance (kg)
Minimum	25	12	15°	0.25
Maximum	100	75	45°	1.4

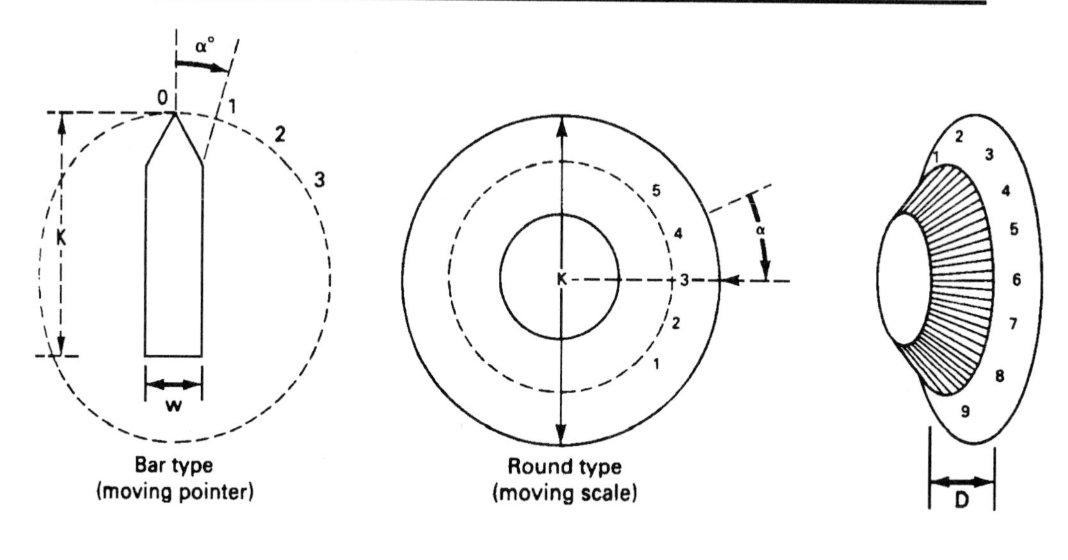

Bar type
(moving pointer)

Round type
(moving scale)

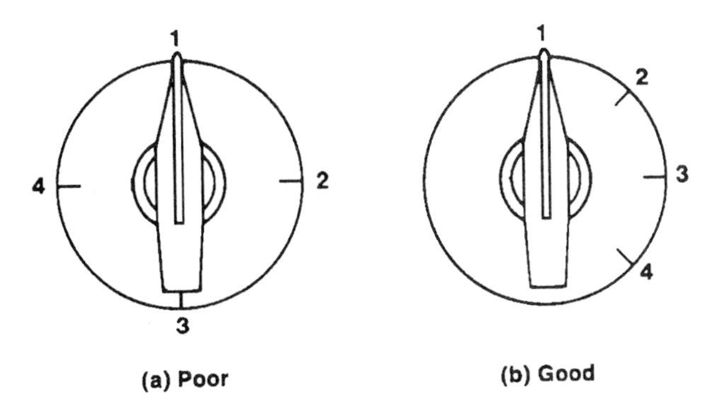

Figure CNTRL-B3: Examples of Good and Poor Rotary Selector Switch Movement[2]

The example of "poor" control movement on the left requires the operator to make a 270-degree rotation of the dial to cover three settings. The "good" control movement alternative requires only a 120-degree rotation to cover the same range. If the differences between settings are large and an error could be critical, however, the "poor" design may be appropriate.

Table CNTRL-B6: Recommended Design Criteria for Rotary Selector Switches[2]

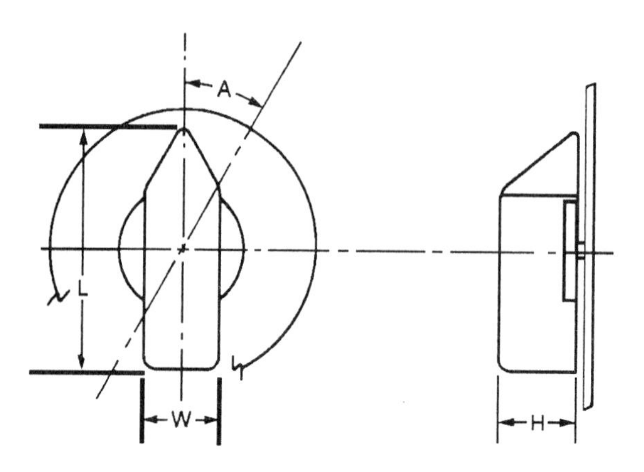

Parameter	Recommended Design Values (Minimum–Maximum)	
Dimensions		
Length (L)	25–100 mm	1.0–4.0 in.
Width (W)	NA–25 mm	NA–1.0 in.
Depth (H)	16–75 mm	0.6–3.0 in.
Displacement (A)		
Closely grouped controls	15°–40°	
Widely separated controls	30°–90°	
Resistance:	0.110–0.675 N · m	1–6 lbf · in.

Note: NA indicates data are not available.

Recommended dimensions (L, W, and H), displacement (A), and resistance ranges for bar-type rotary selector switches are given. Minimum widths (W) are not given since this value will vary with the characteristics of the material used to fabricate the switch. The marks at either end of the displacement path represent stops.

Table CNTRL-B6: Toggle Switch Characteristics[2]

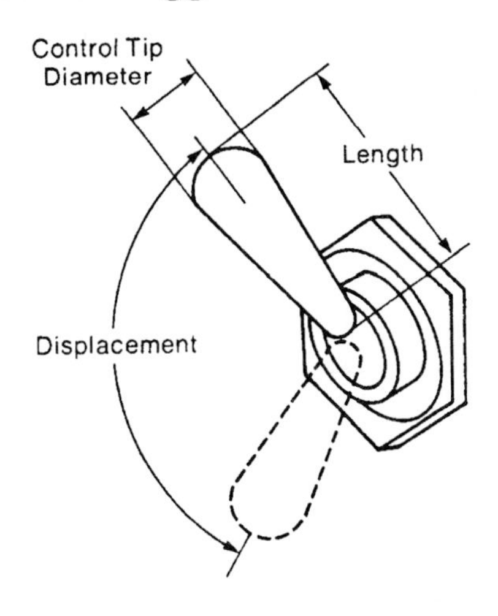

Parameter	Recommended Design Values (Minimum–Maximum)	
Control Tip Diameter	3–25 mm	0.12–1.00 in.
Length		
normal	12–50 mm	0.5–2.0 in.
if operator wears gloves	38–50 mm	1.5–2.0 in.
Displacement		
2-position switch	30°–120°	
3-position switch	18°–60°	
Resistance		
normal	3–11 N	10–40 oz
if control tip is small	3–5 N	10–16 oz

The recommended range of dimensions and resistances for two- and three-way toggle switches is given. The minimum resistance is specified to reduce the potential for accidental activation of the switch. The minimum and maximum displacements of the control are also specified.

3
Control Tables

Table CNTRL-B8: Rocker Switch Characteristics[3]

	Length, L (mm)	Width, W (mm)	Displacement, D
Minimum	13	5	30°
Maximum	—	—	30°

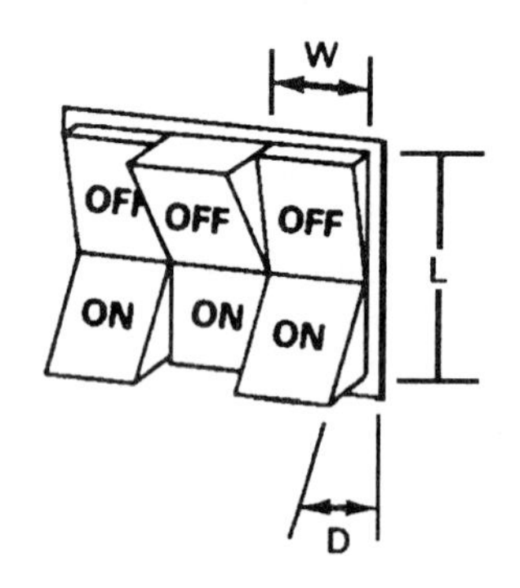

Table CNTRL-B9: United States Stereotypes for Up and Down Switch Settings[2]

UP	DOWN
ON	OFF
START	STOP
HIGH	LOW
IN	OUT
FAST	SLOW
RAISE	LOWER
INCREASE	DECREASE
OPEN	CLOSE
ENGAGE	DISENGAGE
AUTOMATIC	MANUAL
FORWARD	REVERSE
ALTERNATING	DIRECT
POSITIVE	NEGATIVE

Some movement stereotypes for toggle switches used in equipment or production system control panels are given. The expected direction of activation of the switch (when mounted vertically) is given for 13 actions or conditions. These expectations are United States stereotypes and may vary in other countries.

REFERENCES

1. Van Cott, H.P., and Kinkade, R.G., 1972. *Human Engineering Guide to Equipment Design. (Revised)* Washington, DC.: US Government Printing Office. Library of Congress Number: 72-600054.

2. Reprinted with permission from *Ergonomic Design for People at Work*, © Eastman Kodak Company, 1983, published by Van Nostrand Reinhold. Courtesy of Eastman Kodak Company.

3. Pulat, B.M. 1992. *Fundamentals of Industrial Ergonomics.* Englewood Cliffs, NJ: Prentice-Hall. Reprinted with Permission.

CHAPTER 3: CONTROL TABLES

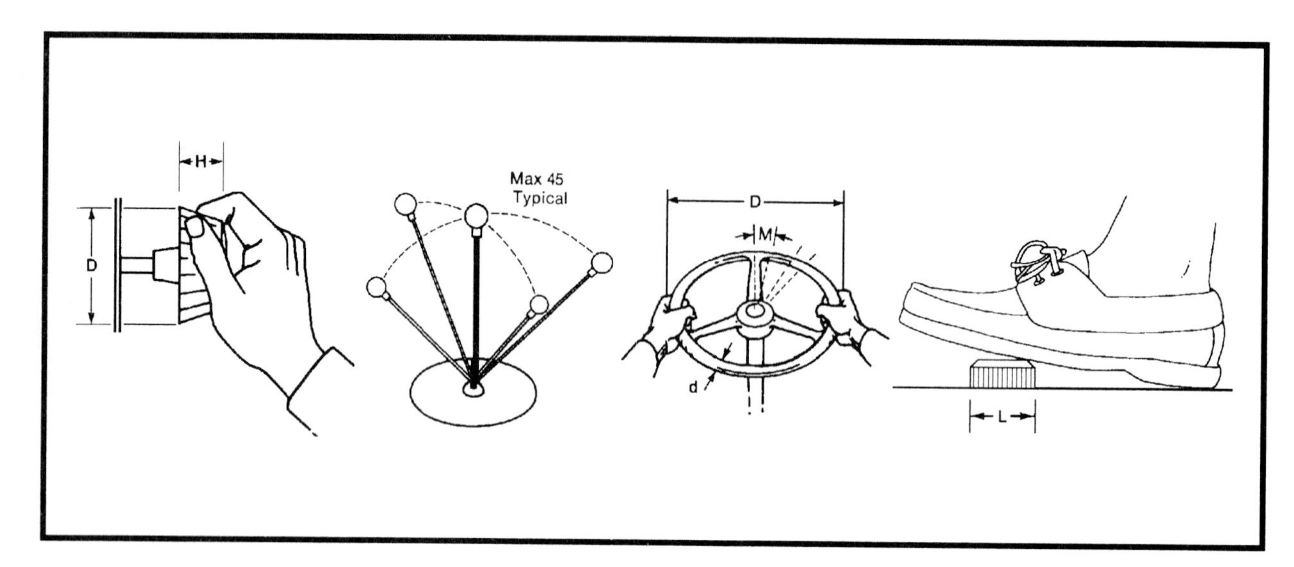

Section C: Levers, Joysticks, Wheels, and Cranks

This section presents information to be used when designing systems that use levers, joysticks, wheels, and cranks as a means of control.

LEVER AND JOYSTICK

Table CNTRL-C1: Criteria for Lever Design[1]

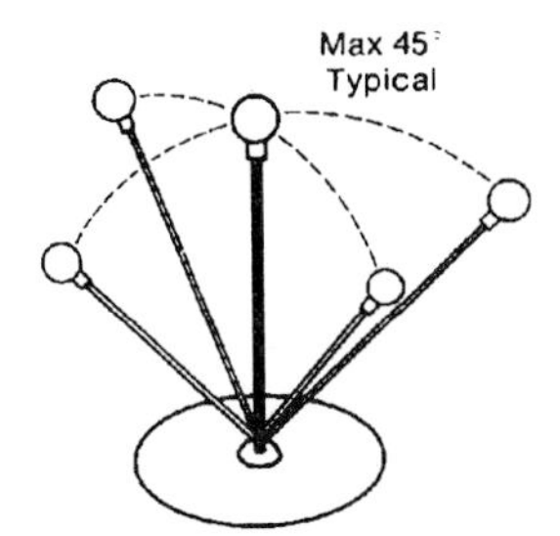

Maximum displacements and operating angles for a floor-mounted lever, such as a gearshift, are indicated, as well as the range of handle diameters for precision (finger) or power (palmar) grasp. The height of the lever handle above the floor (in neutral) is also given for seated and standing operations and the distance from the lever (in neutral) to the front of the body is specified. Maximum forces, in newtons and pounds-force, that can be developed by operators in one-handed operations from the seated position are shown for precision and power grasps.

3
Control Tables

Parameter	Recommended Design Value	
Operating angle, maximum, in each direction from neutral	45°	
Displacement, maximum		
From front to back	35 cm	14 in.
From side to side	95 mm	37 in.
Diameter of handle, minimum to maximum		
For finger grasp	12–75 mm	0.5–3.0 in.
For palm grasp	38–75 mm	1.5–3.0 in.
Height of lever handle above floor		
Seated operation	75 cm	30 in.
Standing operation	125 cm	49 in.
Distance range in front of body, lever in neutral (assumes optimal height for lever and that person can lean forward if necessary)	50–65 cm	20–26 in.
Recommended maximum forces (one-handed operation)		
Front to back, palm grasp	130 N	29 lbf
Front to back, finger grasp	9 N	2 lbf
Side to side, palm grasp	90 N	20 lbf
Side to side, finger grasp	3 N	0.8 lbf

Table CNTRL-C2: Lever Characteristics[2]

	Handle diameter, d (mm)	Grasp area height, h (mm)	Displacement, D (mm)		Resistance (kg)	
			Fore/aft	Right/left	Fore/aft	Right/left
Minimum	38	76	—	—	0.9	0.9
Maximum	70	—	350	950	13.5	9

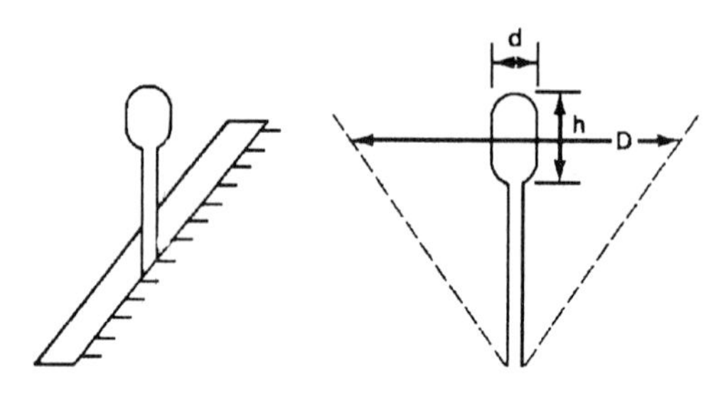

Table CNTRL-C3: Joystick Characteristics[2]

	Handle diameter, d (mm)	Displacement, D	Resistance (kg)
Minimum	5	—	0.34
Maximum	76	120°	0.9

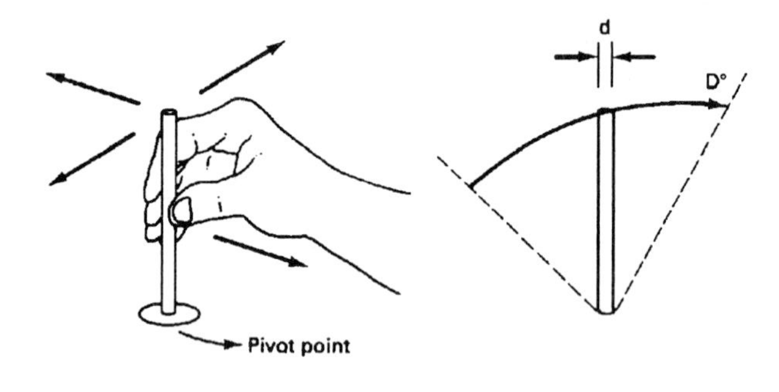

CRANK AND HANDWHEEL DESIGN

Table CNTRL-C4: Recommendation for Crank Design[1]

Torque Range		Minimum Crank Radius Orientation of Handles		
		Horizontal or Vertical on Side, 91 cm (36 in.) above floor	Vertical and Facing the Operator	
N · m	lbf · in.		91 cm (36 in.)	122–142 cm (48–56 in.)
0–2.3	0–20	3.8 cm (1.5 in.)	3.8 cm (1.5 in.)	6.4 cm (2.5 in.)
>2.3–4.5	>20–40	11.4 cm (4.5 in.)	6.4 cm (2.5 in.)	6.4 cm (2.5 in.)
>4.5–10.2	>40–90	19.0 cm (7.5 in.)	11.4 cm (4.5 in.)	11.4 cm (4.5 in.)
>10.2	>90	19.0 cm (7.5 in.)	11.4 cm (4.5 in.)	19.0 cm (7.5 in.)

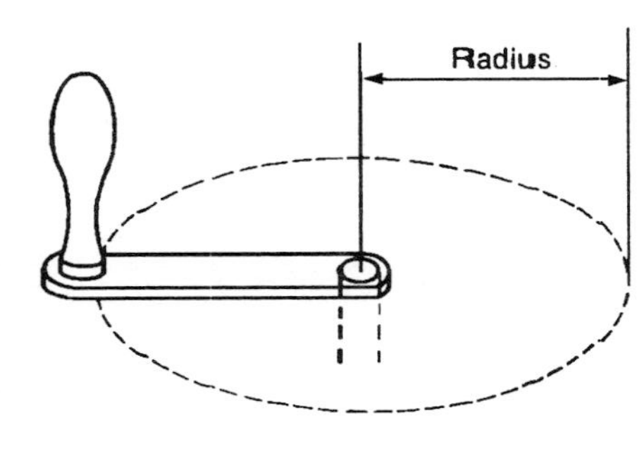

The minimum crank radius, in centimeters and inches, needed to exert force (torque) in four ranges (columns 1 and 2) and in different orientations (columns 3–5) is shown. Torque is expressed in newton meters and pound-force inches. Three different crank handle orientations or locations are shown. In column 3 the orientation is either horizontal or vertical on the side of the surface instead of in front of the operator, at 91 cm (36 in.) above the floor. In column 4 the crank is in a vertical position facing the operator, at 91 cm (36 in.) above the floor. In column 5 it is in the same position but 122–142 cm (48–56 in.) above the floor. Crank radii should not be less than the values given for each torque range, but larger cranks may be used. At radii greater than 25 mm (10 in.), vertical cranks mounted on the side of a piece of equipment may require excessive reaches for operation.

Table CNTRL-C5: Crank Characteristics[2]

	Handle			
	Diameter, d (mm)	Length, K (mm)	Radius, R (mm)	Resistance (kg)
Light loads $\leq$ 2.3 kg				
Minimum	9.5	25.4	12.7	0.9
Maximum	15.9	76.2	127	2.5
Heavy loads > 2.3 kg				
Minimum	25.4	76.2	127	1
Maximum	76.2	—	508	4

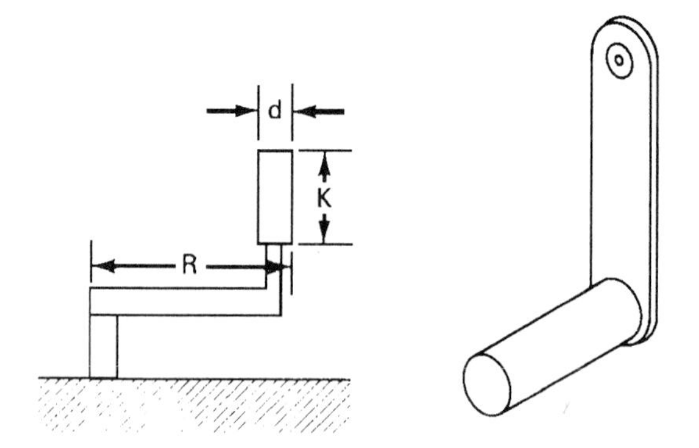

Table CNTRL-C6: Criteria for Handwheel Design[1]

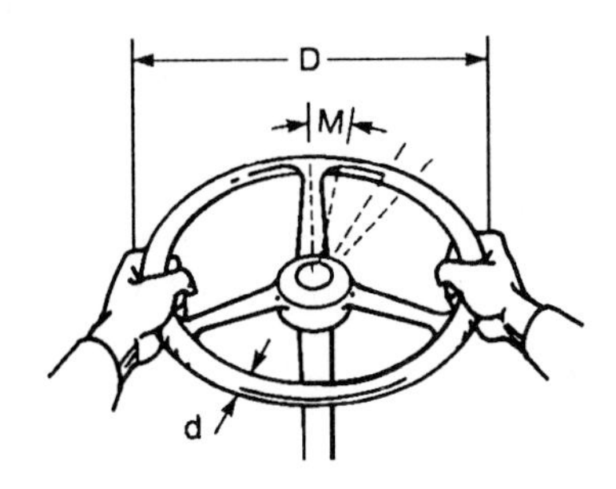

Parameter	Recommended Design Value (Minimum–Maximum)	
Handwheel diameter (D)	18–53 cm	7–21 in.
Rim diameter (d)	20–50 mm	0.8–2.0 in.
Displacement (M), from neutral	60°	
Resistance at rim (tangential force)		
One-hand operation	20–130 N	4–29 lbf
Two-hand operation	20–220 N	4–49 lbf

The recommended range of dimensions (handwheel diameter, D, and rim diameter, d) and displacement (M) for handwheels designed for operation by two hands are given. Minimum-to-maximum tangential forces needed to operate them, with one or both hands, are also shown.

REFERENCES

1. Reprinted with permission from *Ergonomic Design for People at Work*, © Eastman Kodak Company, 1983, published by Van Nostrand Reinhold. Courtesy of Eastman Kodak Company.

2. Pulat, B.M. 1992. *Fundamentals of Industrial Ergonomics.* Englewood Cliffs, NJ: Prentice-Hall. Reprinted with Permission.

3. Van Cott, H.P., and Kinkade, R.G., 1972. *Human Engineering Guide to Equipment Design. (Revised)* Washington, DC.: US Government Printing Office. Library of Congress Number: 72-600054.

CHAPTER 3: CONTROL TABLES

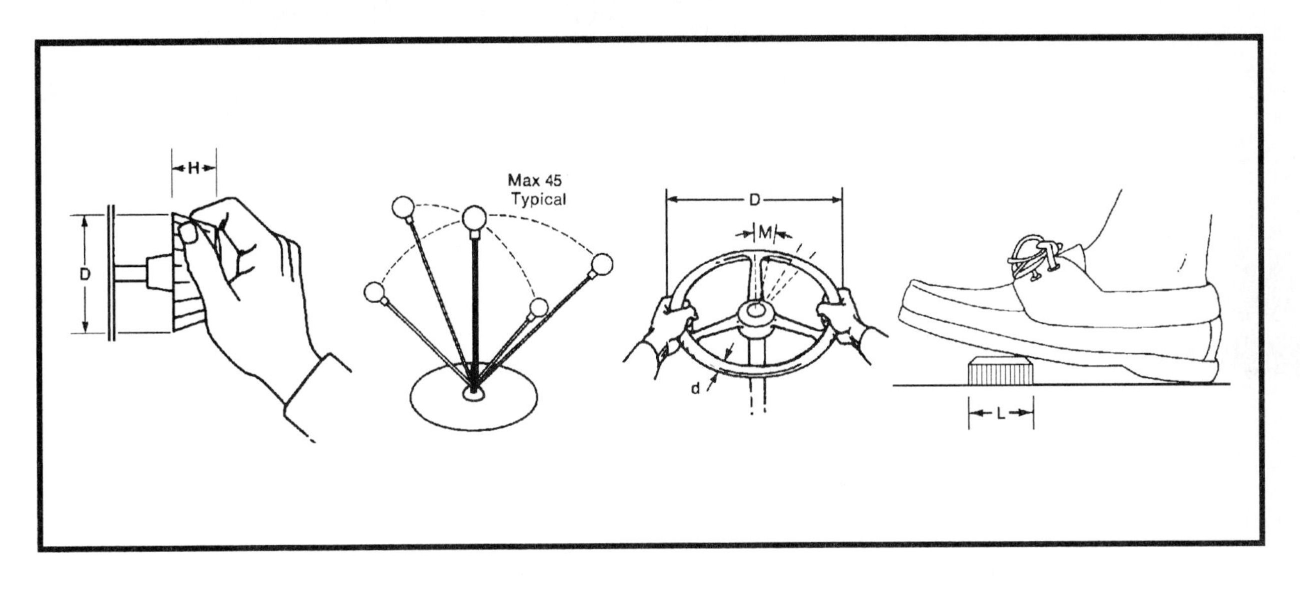

Section D: Thumbwheels, Buttons, Pedals, and Keys

This section presents information to be used when designing systems that use levers, Thumbwheels, buttons, pedals, and keys as a means of control.

THUMBWHEEL DESIGN

Table CNTRL-D1: Characteristics of Continuous Adjustment Thumbwheels[1]

	Diameter, D (mm)	Width, T (mm)	Protrusion, P (mm)	Resistance (torque) (cm–kg)
Minimum	38	6	3.2	1
Maximum	63	13	6.4	3

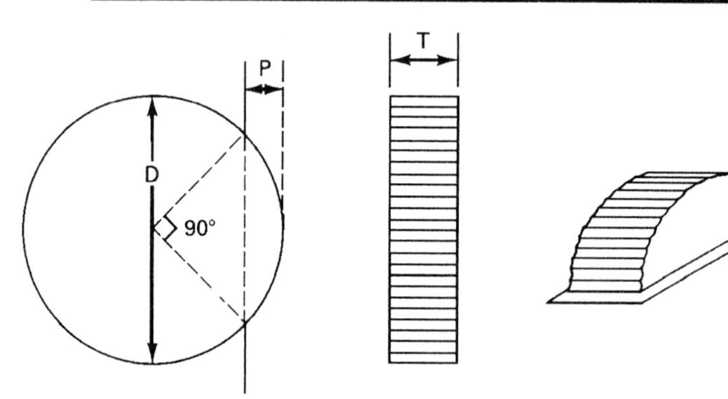

Table CNTRL-D2: Détente Thumbwheel Characteristics[1]

	Diameter, d (mm)	Width, w (mm)	Protrusion p (mm)	Separation, s (mm)	Resistance (kg)
Minimum	38	7	3	7	0.11
Maximum	63	14	6	20	0.34

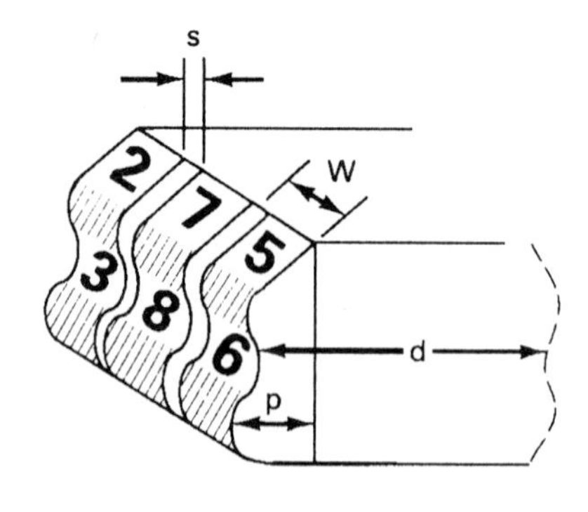

BUTTON AND PEDAL DESIGN

Table CNTRL-D3: Recommended Button and Pedal Design Characteristics[2]

Switching Pedal or Push Button (a) Parameter	Recommended Design Value	
Diameter (D)		
Minimum	12 mm	0.5 in.
Preferred	50–80 mm	2–3 in.
Displacement range (V)		
For ankle flexion	12–65 mm	0.5–2.5 in.
For whole leg movement	25–180 mm	1–7 in.
Maximum height of pedal above heel rest (H), lower leg vertical	8 cm	3 in.
Angle of ankle from neutral position, recommended minimum–maximum range, operator seated	20° up, 30° down	
Counterpressures, recommended minimum–maximum	15–75 N	3.3–16.5 lbf

Operating Pedal (b) Parameter	Recommended Design Value	
Minimum Length (L)		
Occasional use	8 cm	3 in.
Constant use	25 cm	10 in.
Minimum width (W)	9 cm	3.5 in.
Displacement Range (V)		
For ankle flexion	12–65 mm	0.5–2.5 in.
For whole leg movement	25–180 mm	1–7 in.
Angle of ankle from neutral position, recommended minimum–maximum range, operator seated	20° up, 30° down	
Counterpressures, recommended minimum–maximum	15–90 N	3.3–19.8 lbf

Pedals that are in constant use should be provided with an adjustable return spring to allow for differences in operator strength and variations in the nature of the work. Pedals that result in overstretching of the ankle joint (more than 25° around the resting position of the foot) are not recommended. The more frequently a foot pedal is operated, the nearer it should be to its minimum force limit.

If the operation of a foot pedal requires very high counterpressures, the pedal should be placed to allow the leg muscles, not just the ankle, to exert the force. Counterpressures greater than 400 N (90 lbf) should not be required on a frequent basis even when the leg is involved, as in operating a brake.

Table CNTRL-D4: Finger-Operated Pushbutton Characteristics[1]

	Diameter, *d* (mm)	Activation displacement, *v* (mm)	Resistance (kg)
Minimum	9	3	0.25
Maximum	19	38	1.1

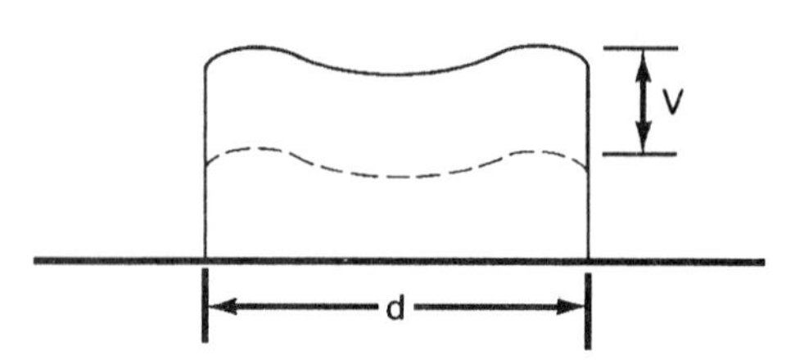

Table CNTRL-D5: Legend Button Characteristics [1]

	Size, *K* (mm)	Separators (mm)		Displacement activated/nonactivated (mm)	Resistance (kg)
		S_1	S_2		
Minimum	19	5	3	3	0.28
Maximum	38	7	7	7	1.27

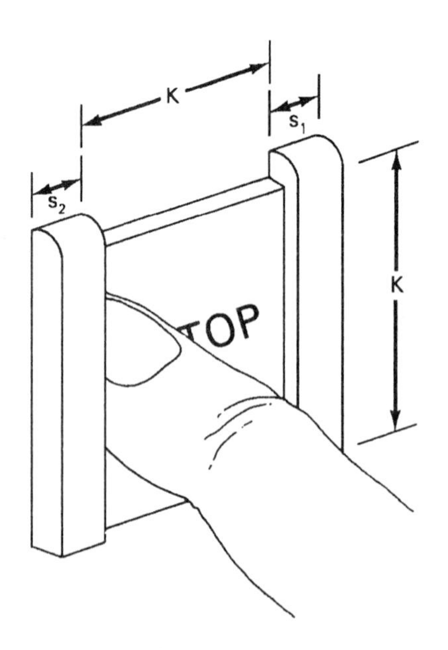

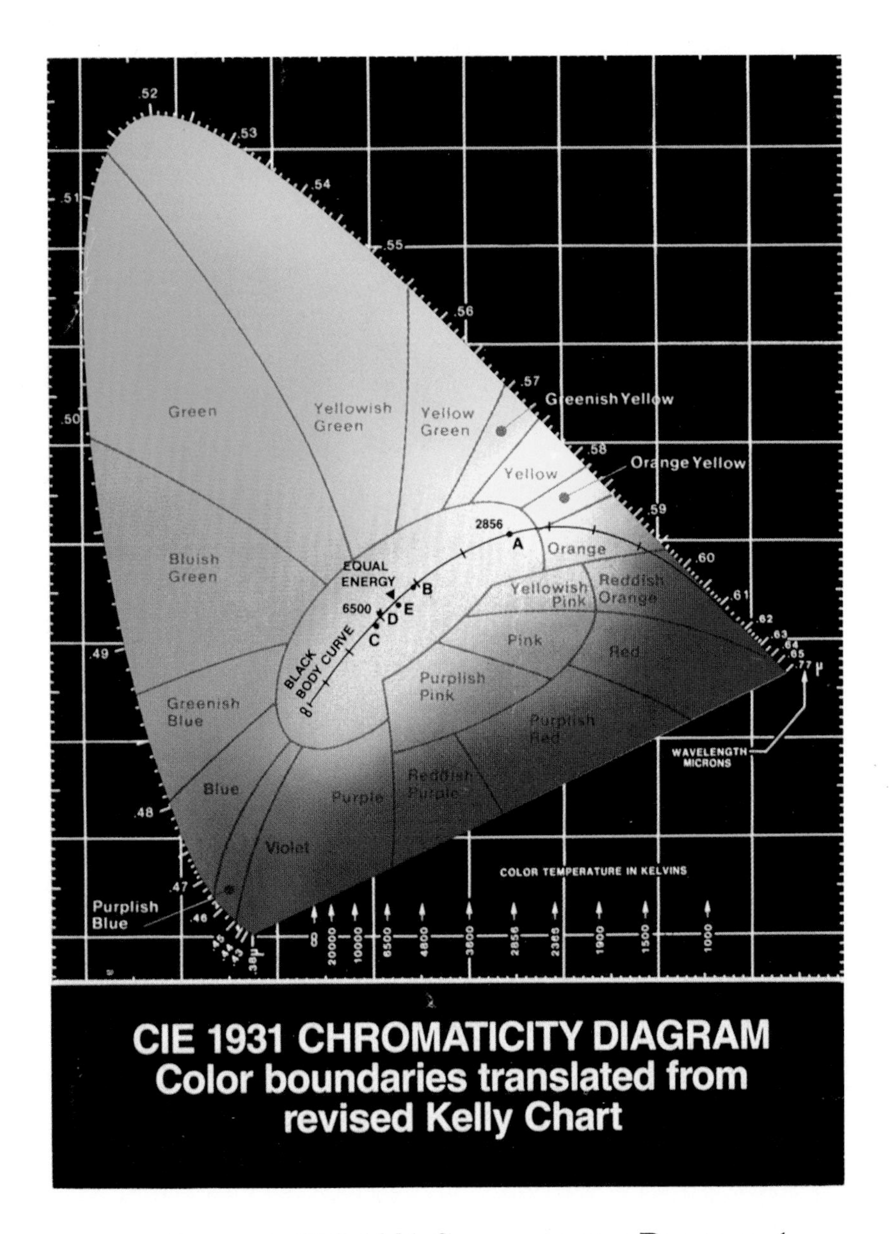

Color Plate 1: CIE 1931 Chromaticity Diagram [1]

The CIE 1931 chromaticity diagram relates the physical energy in light to its hue and saturation. Hue is arranged clockwise around the perimeter of the diagram, while saturation is shown from the center of the figure outward.

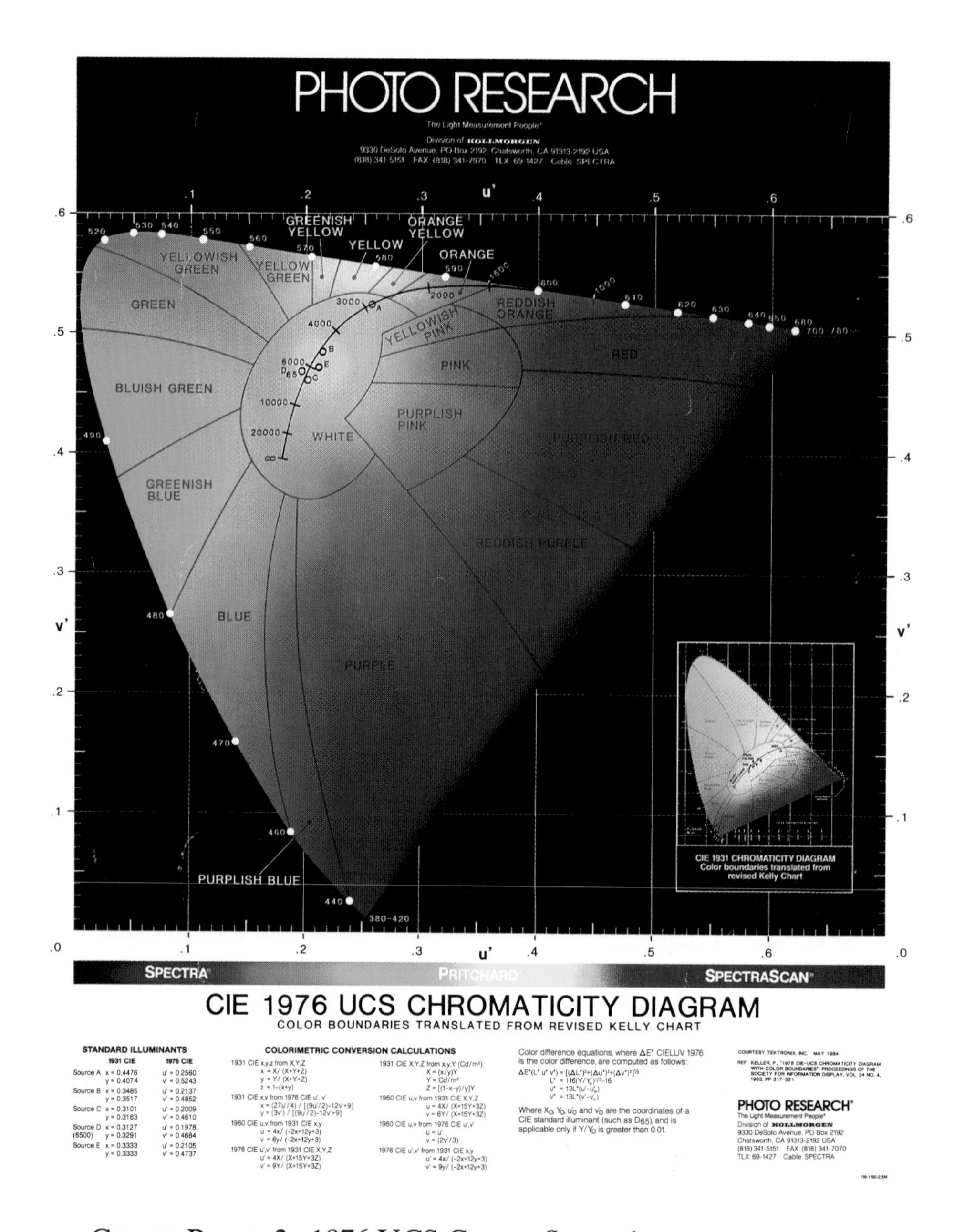

COLOR PLATE 2: 1976 UCS COLOR SPACE [1]

The 1976 UCS Color Space is a mathematical transformation of the CIE 1931 chromaticity diagram, linerarly representing the discrimination of hue and saturation. Equally spaced color differences are represented as equally spaced distances. This space can be used to predict perceptual relations among colors.

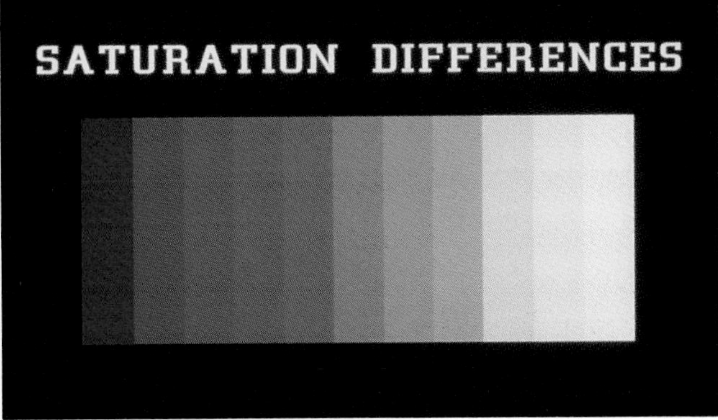

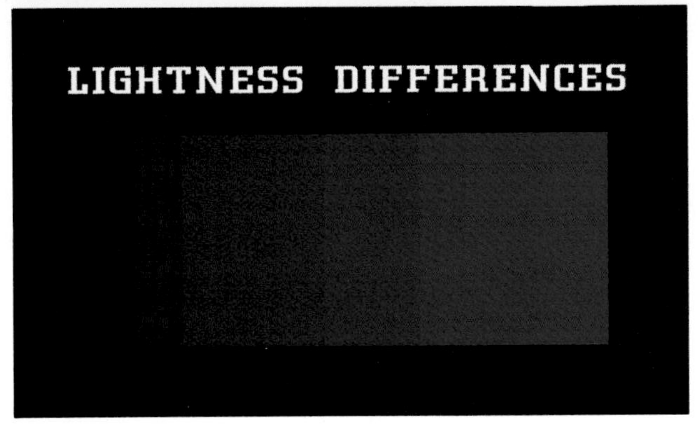

COLOR PLATE 3: PERCEPTUAL CORRELATES OF COLOR [2]

The differences between hue, lightness, and saturation are shown. Hue refers to the wavelength of light and is related to the perception of a given "color." Lightness is related to the perception of "brightness" of a color and refers to the amount of black or white in a given color. Saturation refers to the perception of the "richness" of the color and refers to the degree to which a hue differs from grey of the same lightness.

COLOR PLATE 4: IMAGE SIZE AND COLOR PERCEPTION [2]

As the squares decrease in size, it becomes increasingly more difficult for the observer to discriminate between the different colors. The blue squares are the most adversely affected.

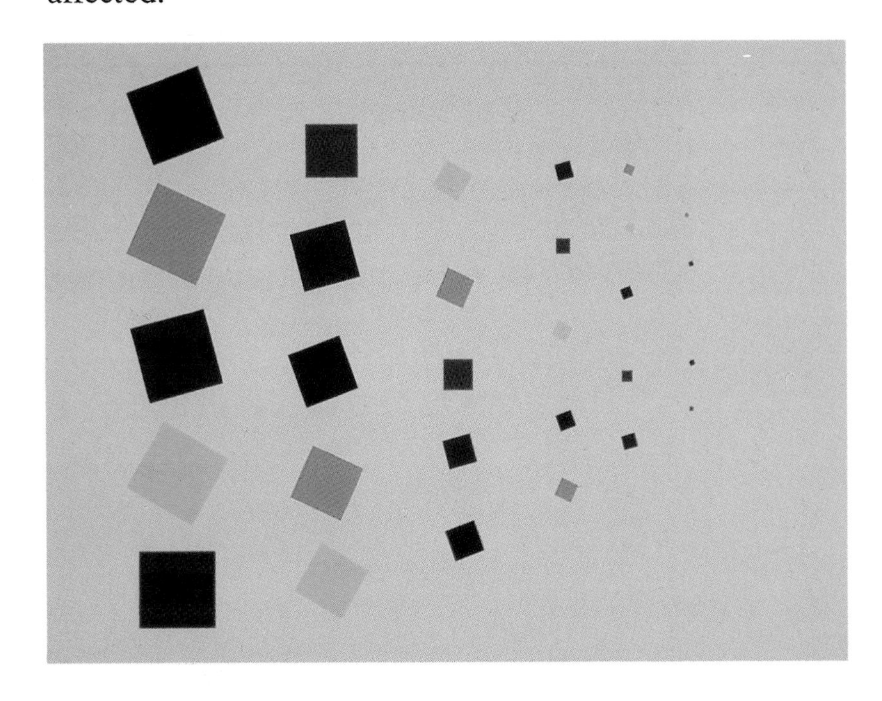

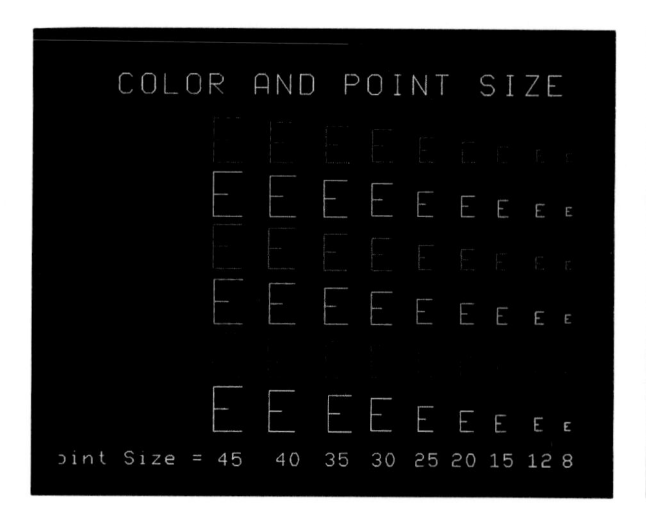

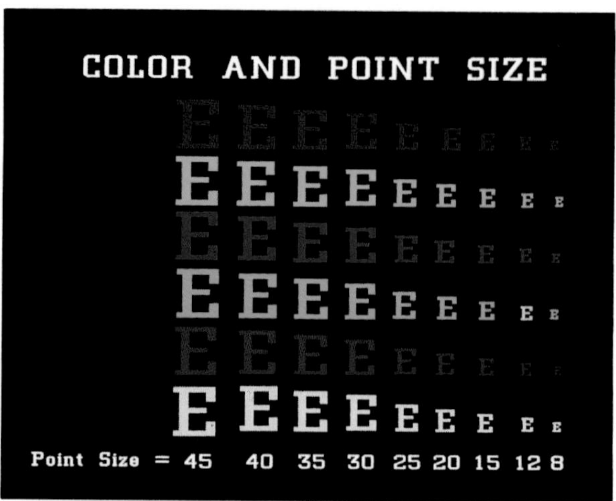

COLOR PLATE 5: TEST CHARTS FOR LEGIBILITY OF COLORED LETTERS [2]

The two charts can be used to determine the letters in the different fonts and sizes that can be seen correctly at different distances. (Note: one type point = .35mm = .0135 inches)

2: white yellow
3: white yellow cyan
4: white yellow cyan red
5: white yellow cyan red green
6: white yellow cyan red green magenta

2: green yellow
3: green yellow cyan
4: green yellow cyan magenta
5: green yellow cyan magenta red
6: green yellow cyan magenta red blue

COLOR PLATE 6: VISIBILITY OF COLORED FONTS [2]

The colors of thin fonts (top rows) are difficult to perceive on a white background. The colors of bold fonts (bottom rows) are easy to see on both white and black backgrounds.

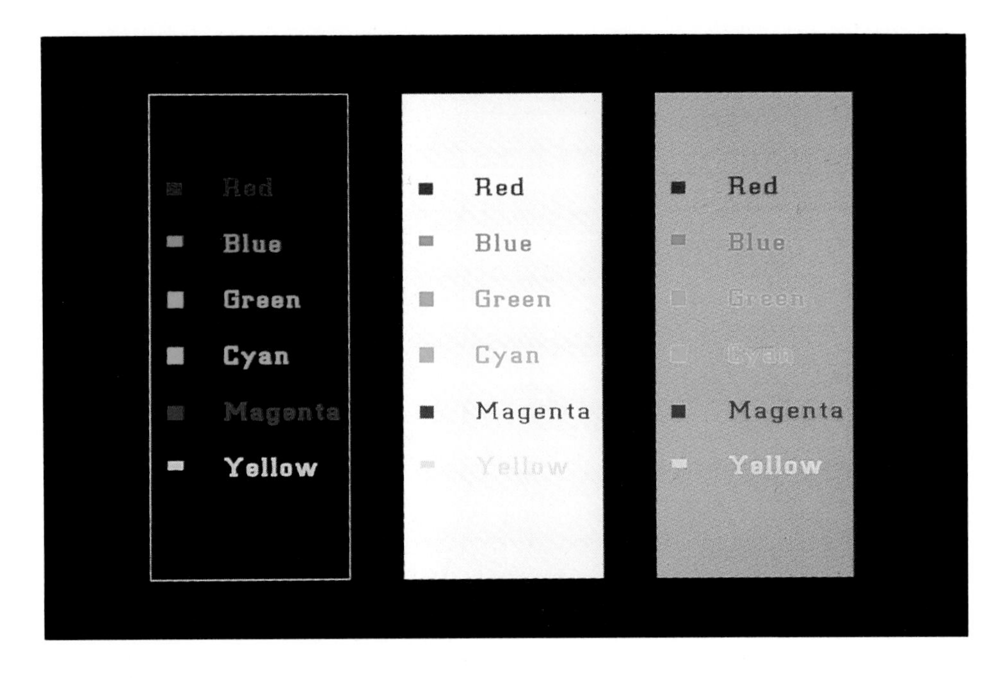

COLOR PLATE 7: VISIBILITY OF COLORS ON BLACK, WHITE, AND GREY [2]

The blocks show color combinations that have maximum visibility and discrimination. The words show the colors of text with high visibility, color discrimination, and legibility. The lighter colors have high visibility on a black background, while the darker colors have higher visibility on a white background.

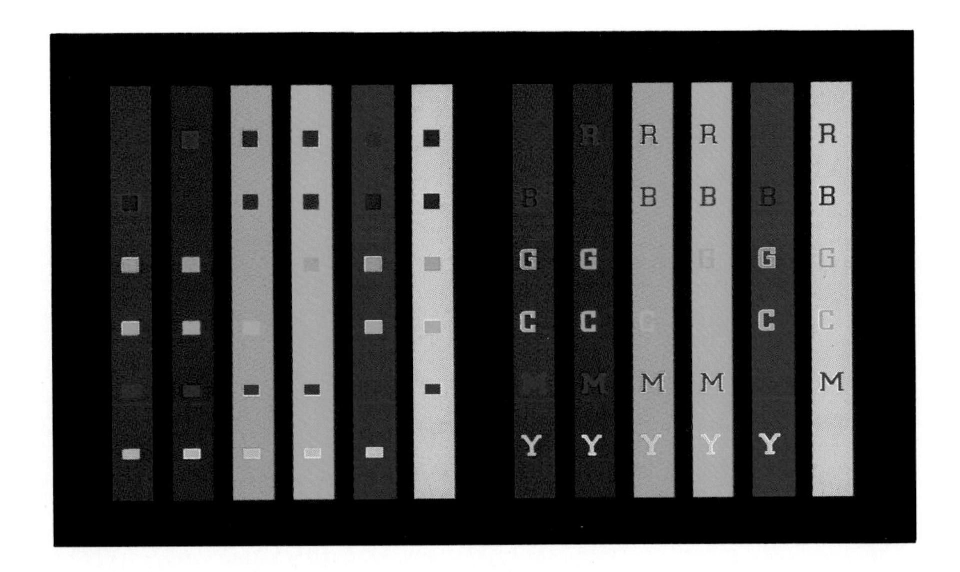

COLOR PLATE 8: VISIBILITY AND LEGIBILITY OF COLOR COMBINATIONS [2]

Each row shows two different figures (a square and a letter) in a constant color on different colored backgrounds. Visibility and legibility are best when hue and lightness contrast are maximized.

This text is the same color

COLOR PLATE 9: EFFECTS OF BACKGROUND ON NEUTRAL IMAGES [2]

All the color values of the text are identical. The characters look like lighter versions of their background colors. The effect depends upon the character brightness, ratio of the image size to the background, and area.

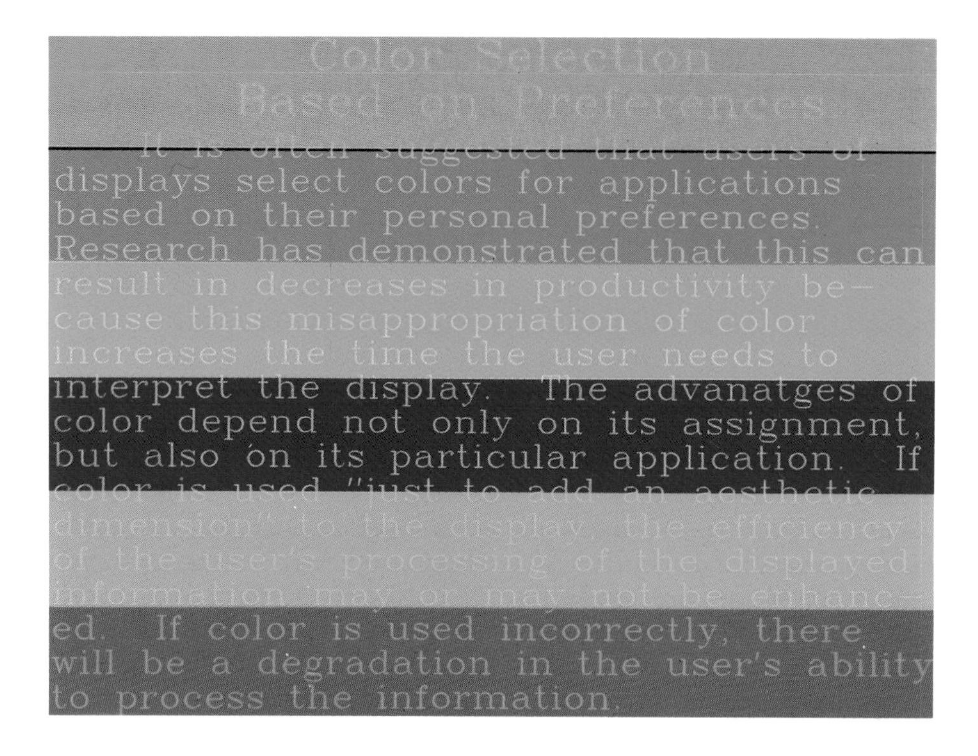

COLOR PLATE 10: SIMULTANEOUS CONTRAST [2]

The colors of the gray characters look like a desaturated value of the complement of their background color (e.g., grey text on blue appears yellowish, while grey on red appears greenish).

COLOR PLATE 11: VISIBILITY OF DIFFERENT COLOR CONTRASTS [2]

From top to bottom: The hue and lightness of the text color is constant, while that of the background becomes lighter, lowering the contrast and visibility of the text.

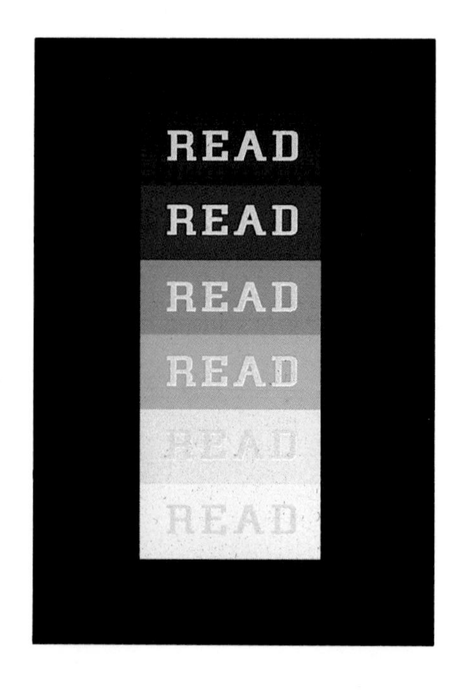

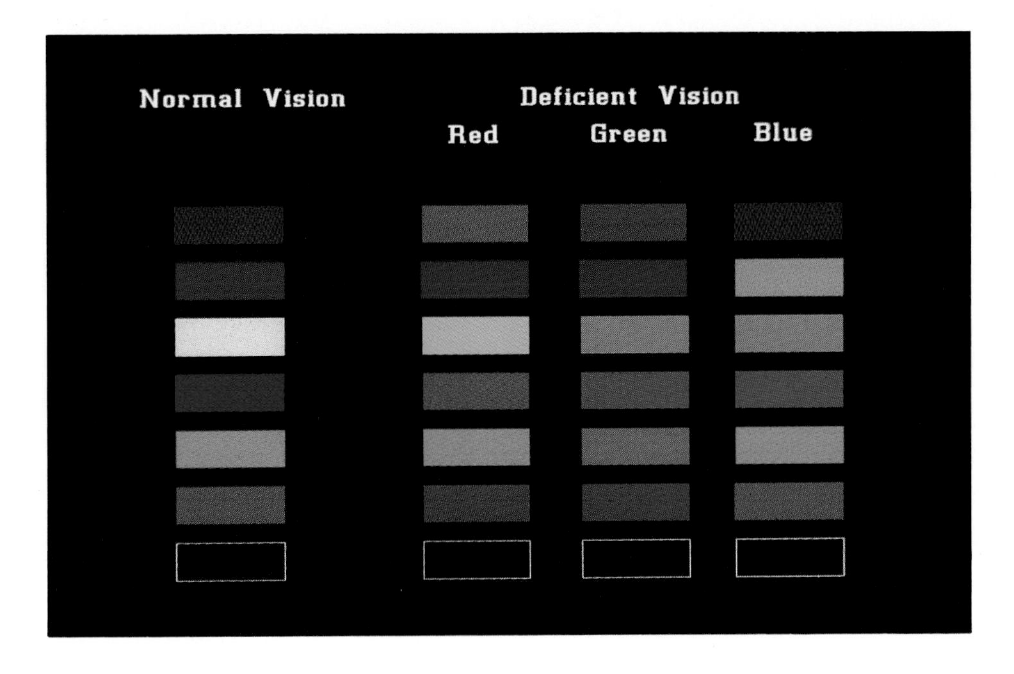

COLOR PLATE 12: COLOR DEFICIENCY TABLE [2]

The left column shows the appearance of colors to an individual with normal color vision. The three columns on the right show the appearances of colors to individuals with red, green, or blue color deficiencies.

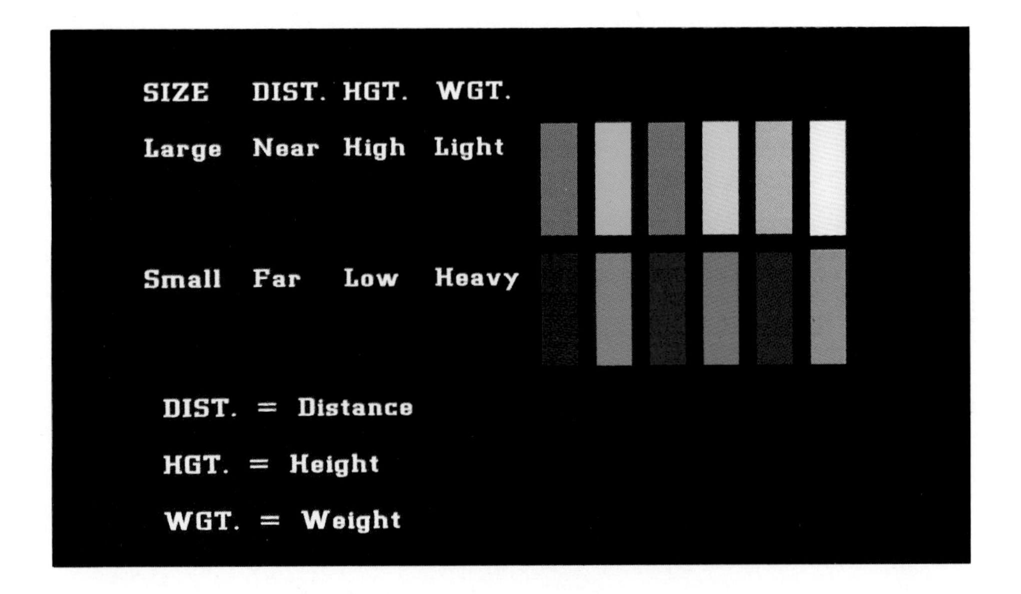

COLOR PLATE 13: COLORS AND PHYSICAL APPEARANCE [2]

The desaturated hues in the top row make images appear larger, nearer, higher, and lighter. The saturated hues in the bottom row make images appear smaller, further away, and heavier.

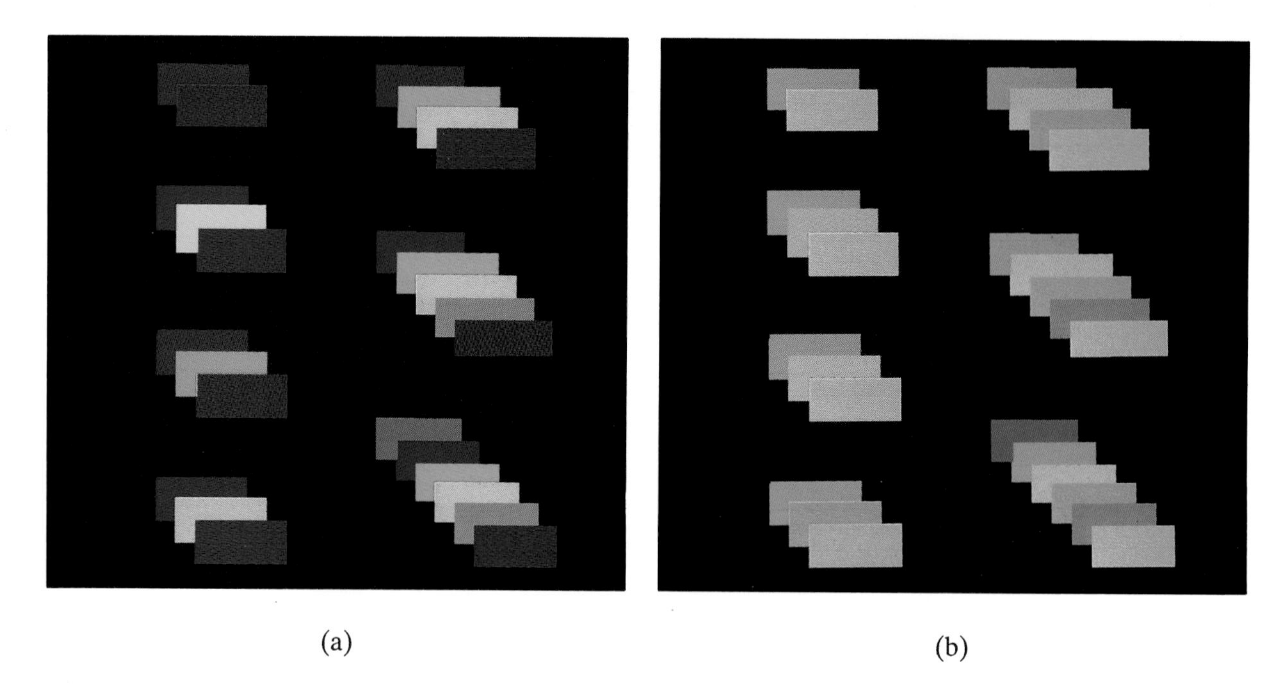

(a) (b)

COLOR PLATE 14: COLOR ORDERING CREATING 3D IMPRESSIONS [2]

Both of the pallettes are the same hue, but those shown in (a) are saturated values and (b) are unsaturated. The three dimensional effect is created by spectral ordering and balancing brightness (e.g., lowering the brightness of yellow to appear equal to the other colors).

COLOR PLATE 15: COLORS FOR SHOWING CHANGE [2]

Gradual saturation fo same hue or ordered spectrum shows continuous changes. Different hues represent discrete changes. Large changes in hues can represent the passing of critical points.

COLOR PALETTE: SHOWING CHANGES

CONTINUOUS

Same hues:
graduated saturations

Ordered Spectrum

DISCRETE

Different Hues

PASSING CRITICAL POINTS

Large Changes in hues

Table CNTRL-D6: Pushbutton Characteristics[2]

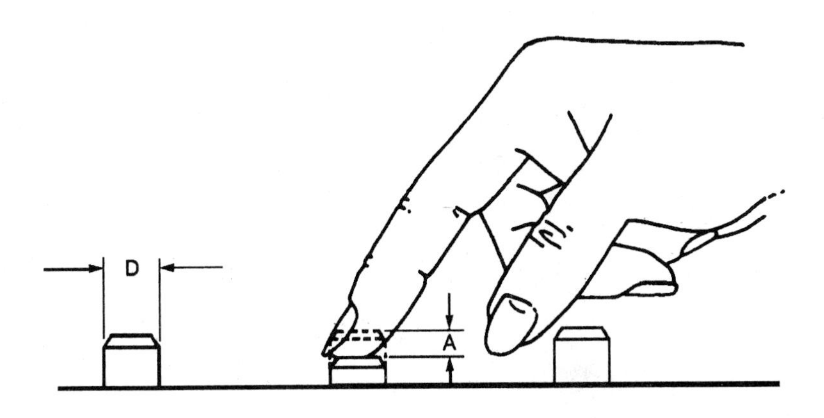

Parameter	Recommended Design Values (Minimum–Maximum)	
Diameter (D)		
Fingertip activation	10–19 mm	0.38–0.75 in.
Palm or thumb activation	19–NA mm	0.75–NA in.
Emergency push buttons	not less than 25 mm	not less than 1.0 in.
Displacement (A)		
Finger activation	3–6 mm	0.12–0.25 in.
Palm or thumb activation	3–38 mm	0.12–1.50 in.
Resistance		
Finger activation	2.8–11 N	10–40 oz
Thumb activation	2.8–22.7 N	10–80 oz

Note: NA indicates data are not available.

Recommended diameter (D), displacement (A), and resistance ranges for a standard push button are shown. Distinctions are drawn between push buttons operated with the index or middle finger and those activated by the thumb or palm. Maximum diameter is not indicated for the latter condition because it varies with the location of the push button in the workplace.

Table CNTRL-D7: Foot Pushbutton Characteristics[1]

	Diameter, L (mm)	Activation displacement, M (mm)	Resistance (foot resting) (newtons)
Minimum	12	12	44.5
Maximum	80	75	89

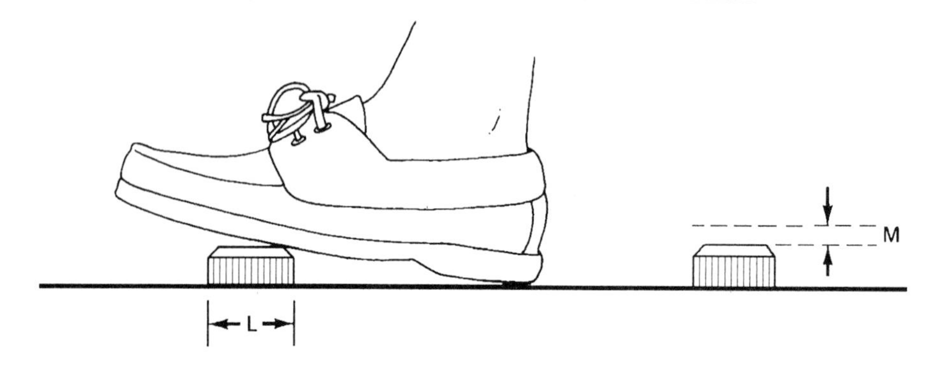

Table CNTRL-D8: Foot-Pedal Characteristics[1]

	Width (mm)		Length (mm)		Displacement range (mm)		Resistance (kg)	
	A	W	B	L	V	M	Ankle flexion	total leg movement
Minimum	75	108	25	230	25	12	3	4
Maximum	—	51	305	300	180	65	10	100

KEY DESIGN

Table CNTRL-D9: Recommended Key Design[2]

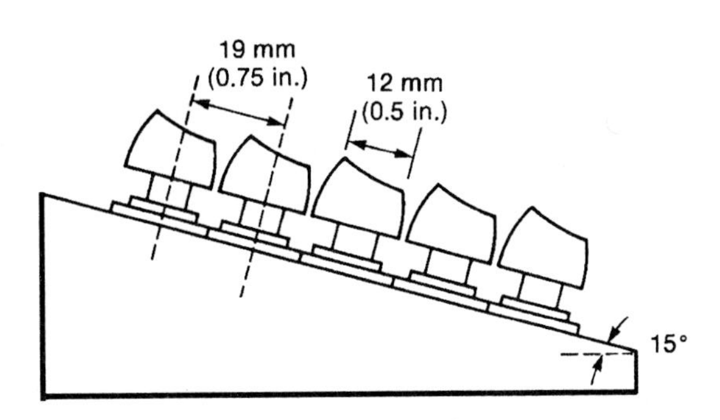

The minimum size of the square key surfaces and the recommended separation (center to center) of keys are indicated. The keyboard angle relative to the horizontal surface it sits on is also specified.

REFERENCES

1. Pulat, B.M. 1992. *Fundamentals of Industrial Ergonomics.* Englewood Cliffs, NJ: Prentice-Hall. Reprinted with Permission.

2. Reprinted with permission from *Ergonomic Design for People at Work,* © Eastman Kodak Company, 1983, published by Van Nostrand Reinhold. Courtesy of Eastman Kodak Company.

CHAPTER 4: SENSING TABLES

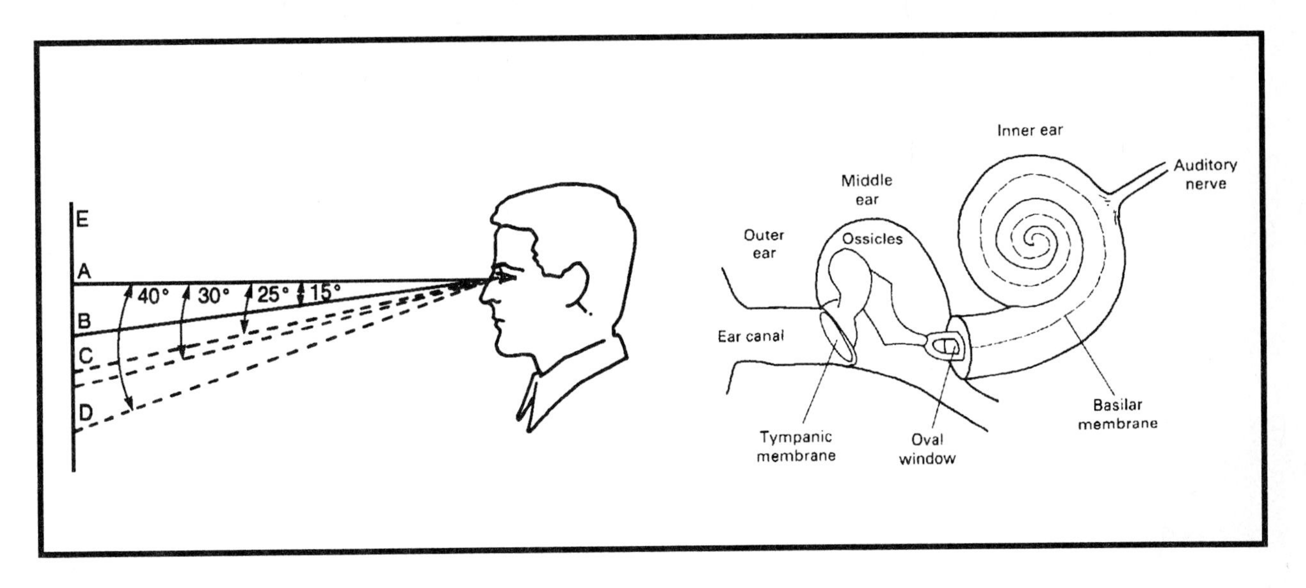

Section A: General Tables

This section presents generic information about the characteristics of the various senses. The information in this section is useful for the ergonomist or human factors engineer who must design display systems.

SENSING SUMMARY TABLES

Table SNSNG-A1: Some Approximate Sensory Thresholds[1]

Sense	*Detection Threshold*
Sight	Candle flame seen at 30 miles on a dark clear night
Hearing	Tick of a watch under quiet conditions at 20 feet
Taste	Teaspoon of sugar in 2 gallons of water
Smell	Drop of perfume diffused into the entire volume of a three-room apartment
Touch	Wing of a bee falling on your cheek from a distance of 1 centimeter

Table SNSNG-A2: Stimulation Intensity Ranges[2]

Sensation	Smallest detectable (threshold)	Largest tolerable or practical
Sight	10^{-6} mL.	10^{4} mL.
Hearing	2×10^{-4}dynes/cm^2.	$<10^{3}$ dynes/cm^2.
Mechanical vibration	25×10^{-5} mm average amplitude at the fingertip (Maximum sensitivity 200 Hz).	Varies with size and location of stimulator. Pain likely 40 dB above threshold.
Touch (pressure)	Fingertips, 0.04 to 1.1 erg (One erg approx. kinetic energy of 1 mg dropped 1 cm.) "Pressure," 3 gm/mm^2.	Unknown.
Smell	Very sensitive for some substances, e.g., 2×10^{-7} mg/m^3 of vanillin.	Unknown.
Taste	Very sensitive for some substances, e.g., 4×10^{-7} molar concentration of quinine sulfate.	Unknown.
Temperature	15×10^{-5} gm-cal/cm^2/sec. for 3 sec. exposure of 200 cm^2 skin.	22×10^{-2} gm-cal/cm^2/sec. for 3 sec. exposure of 200 cm^2 skin.
Position and movement	0.2–0.7 deg. at 10 deg./min. for joint movement.	Unknown.
Acceleration	0.02 g for linear acceleration 0.08 g for linear deceleration 0.12 deg./sec^2 rotational acceleration for oculogyral illusion (apparent motion or displacement of viewed object).	5 to 8 g positive; 3 to 4 g negative. Disorientation, confusion, vertigo, blackout, or redout.

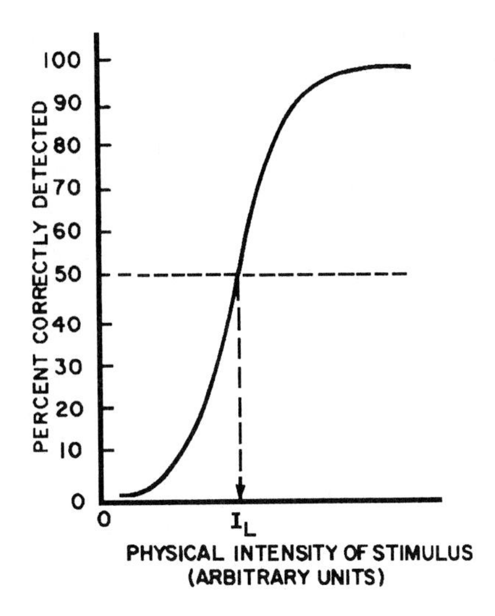

Figure SNSNG-A1: Determining the Lower Sensory Threshold (I_L)[2]

The lower threshold for a sensory modality is usually defined as the value that is detected 50% of the time under ideal conditions. This figure shows how the lower threshold for a given modality can be determined using a cumulative distribution of percentages of detection. For practical uses, lower thresholds should be set at a detection rate of 75% or better.

Table SNSNG-A3: Frequency Sensitivity Ranges of Senses[2]

Stimulus	Lower Limit	Upper Limit
Color (hue)	300 nm (300 × 10^{-9} m.)	800 nm.
Interrupted white light	Unlimited	50 interruptions/sec. at moderate intensities and duty cycle of 0.5.
Pure tones	20 Hz	20,000 Hz.
Mechanical vibration	Unlimited	10,000 Hz at high intensities.

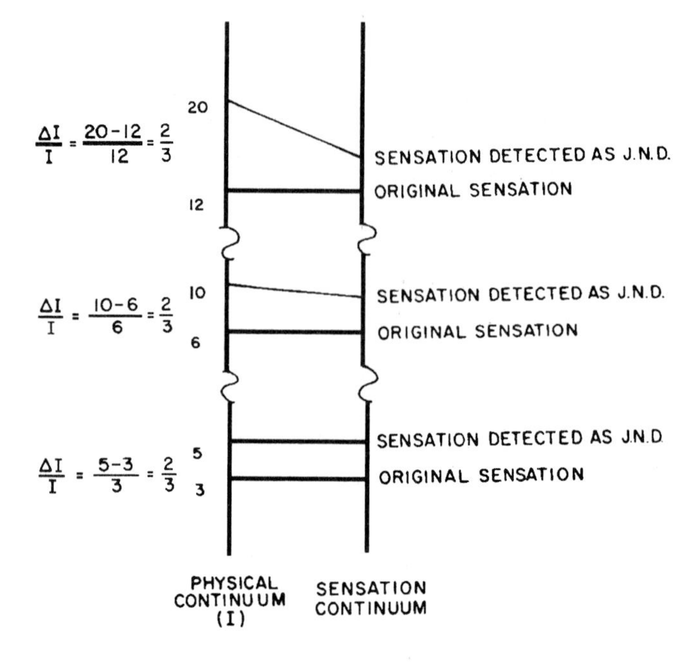

Figure SNSNG-A2: Relationship Between I, ΔI, and JND[2]

The Just Noticeable Difference (JND) between two stimuli, within the range of useful stimuli, the change in intensity (ΔI) corresponding to a JND bears a constant relationship to the reference magnitude (I). That is, as the reference stimuli increases in magnitude, increasing increments in stimulus intensity are required to be detected.

Table SNSNG-A4: Relative Discrimination of Physical Intensities[2]

Sensation	Number discriminable
Brightness	570 discriminable intensities, white light.
Loudness	325 discriminable intensities, 2,000 Hz.
Vibration	15 discriminable amplitudes in chest region using broad contact vibrator with 0.05-0.5 mm amplitude limits.

Table SNSNG-A5: Relative Discrimination of Frequency[2]

Sensation	Number discriminable
Hues	128 discriminable hues at medium intensities.
White light	375 discriminable interruption rates between 1–45 interruptions/sec. at moderate intensities and duty cycle of 0.5.
Pure tones	1,800 discriminable tones between 20 Hz and 20,000 Hz at 60-dB loudness.
Interrupted white noise	460 discriminable interruption rates between 1–45 interruptions/sec. at moderate intensities and duty cycle of 0.5.
Mechanical vibration	180 discriminable frequencies between 1 and 320 Hz.

Table SNSNG-A6: Absolute Identification of Intensity[2]

Sensation	Number identifiable
Brightness	3 to 5 discriminable intensities with white light of 0.1–50 millilamberts.
Loudness	3 to 5 discriminable intensities with pure tones.
Vibration	3 to 5 discriminable amplitudes.

Table SNSNG-A7: Absolute Identification of Frequency[2]

Sensation	Number identifiable
Hues	12 or 13 discriminable hues.
Interrupted white light	5 or 6 discriminable interruption rates.
Pure tones	4 or 5 discriminable tones.

REFERENCES

1. Bailey, R.W. 1989. *Human Performance Engineering. (2nd Ed)* Englewood Cliffs, NJ: Prentice-Hall. Reprinted with Permission.

2. Van Cott, H.P., and Kinkade, R.G., 1972. *Human Engineering Guide to Equipment Design. (Revised)* Washington, DC.: US Government Printing Office. Library of Congress Number: 72-600054.

CHAPTER 4: SENSING TABLES

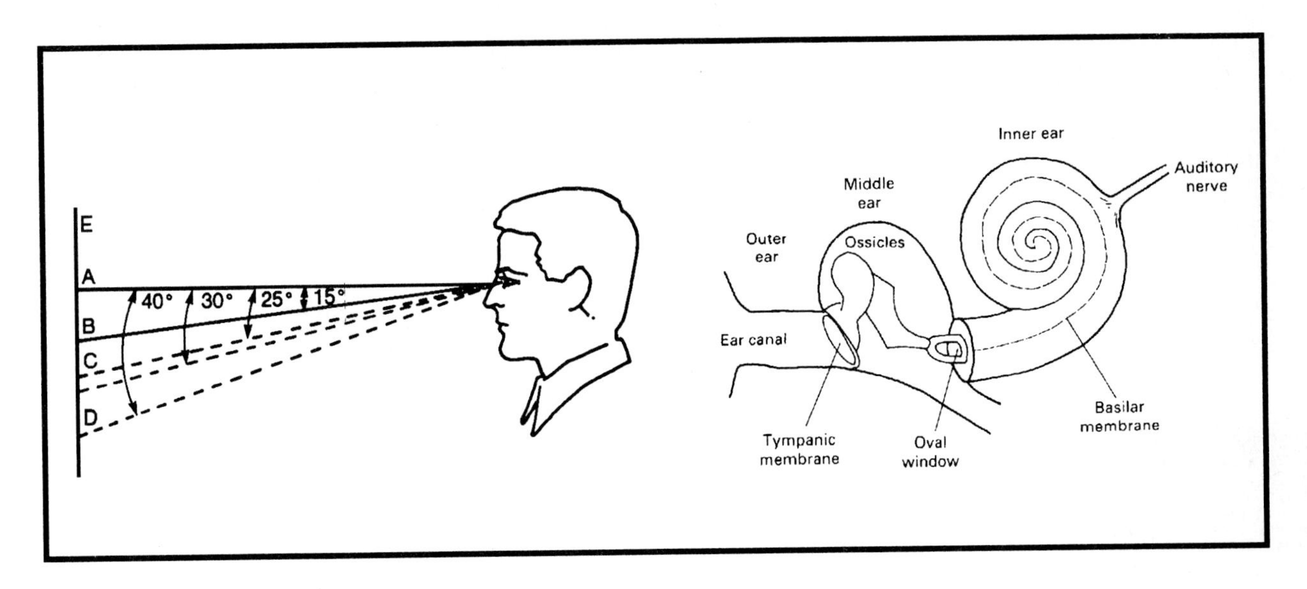

Section B: Visual System

This section presents information about the human visual systems, and should prove invaluable when designing visual display systems.

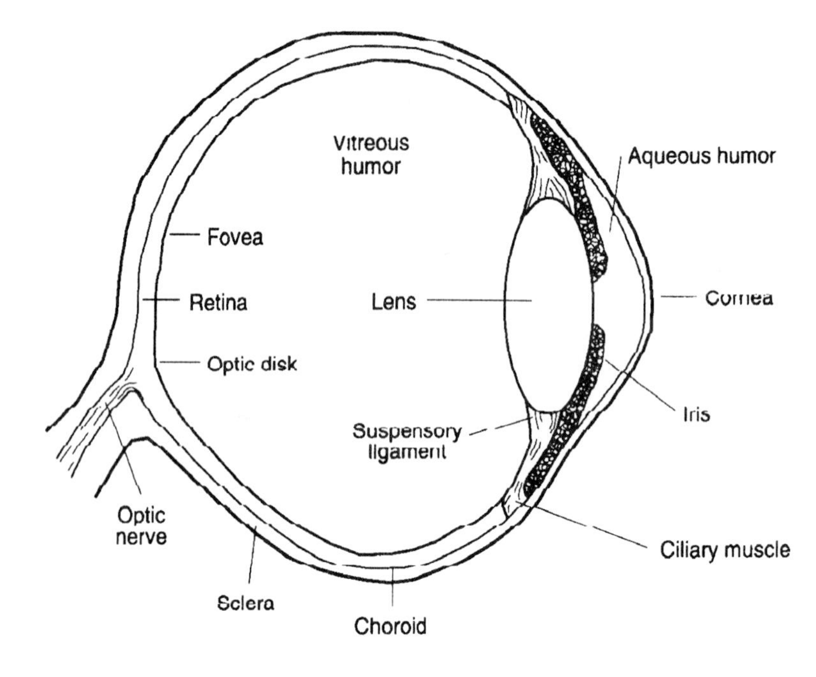

Figure SNSNG-B1: Anatomy of the Human Eye[1]

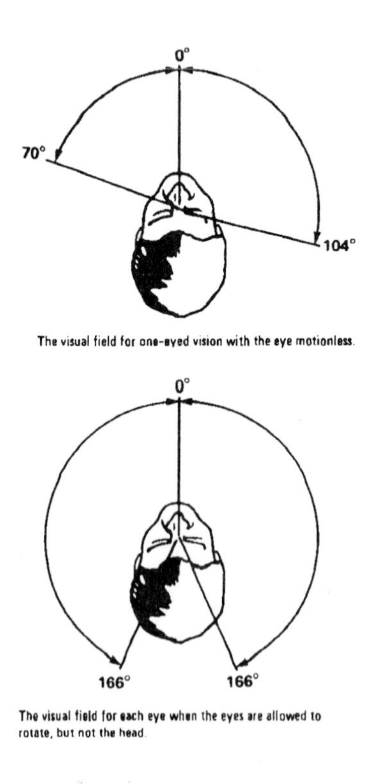

Figure SNSNG-B2: Visual Field[2]

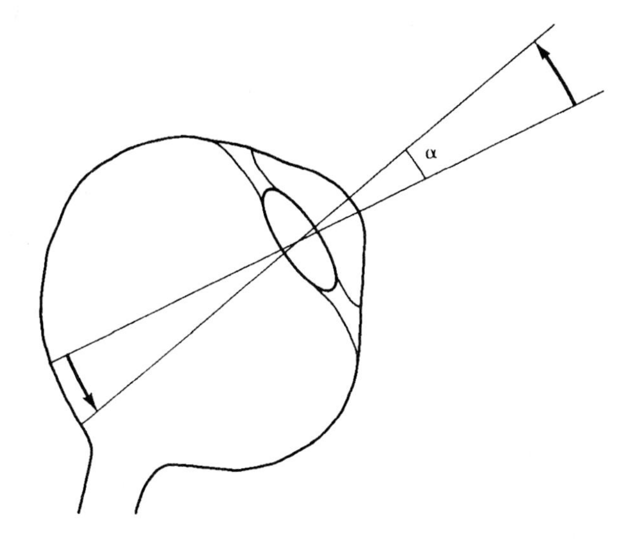

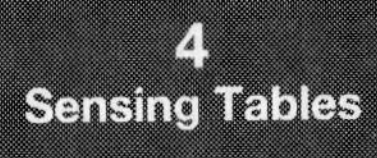

Figure SNSNG-B3: The Visual Angle α [1]

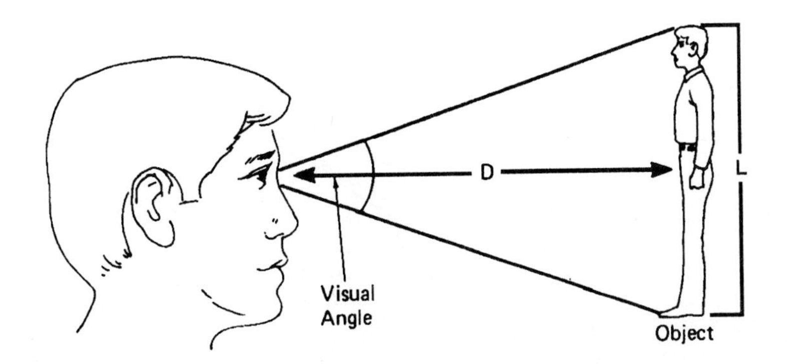

Figure SNSNG-B4: The Meaning of Visual Angle[2]

Table SNSNG-B1: The Cumulative Probability of Target Detection as a Function of Visual Angle [3]

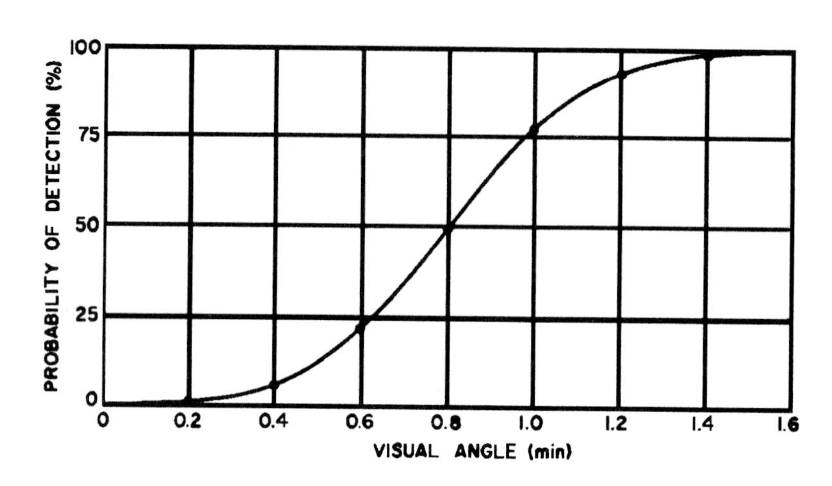

Table SNSNG-B2: Change of Required Target Size With Distance From Fovea for Targets of Varying Luminance[3]

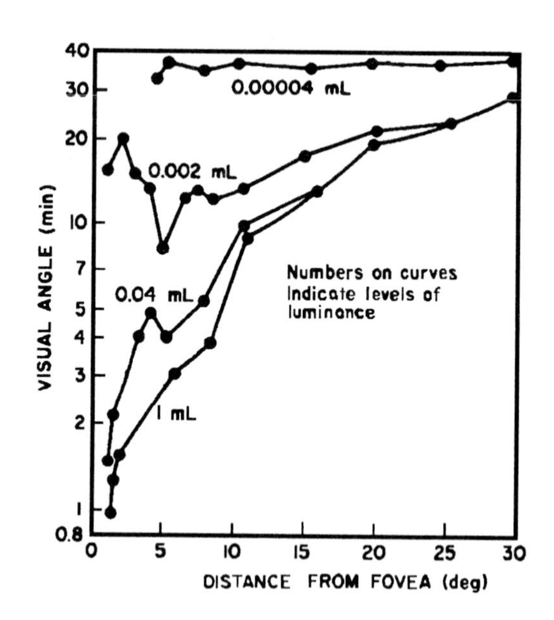

Table SNSNG-B3: Effect of Target Size on Search and Identification Tasks[3]

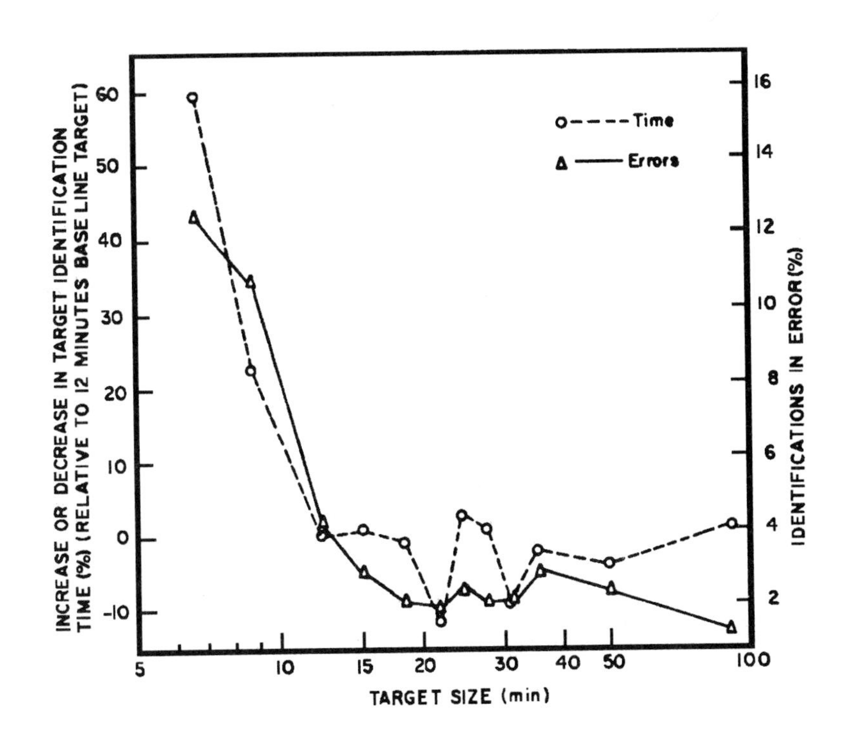

Table SNSNG-B4: Relative Loss of Sensitivity to Visual Stimuli, Immediately After Exposure to Higher Luminances[3]

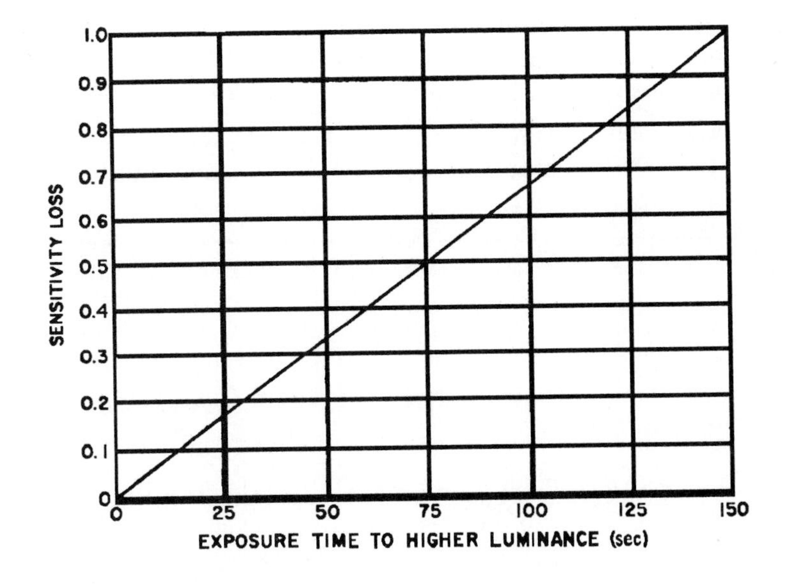

Table SNSNG-B5: Examples of Various Levels of Luminance[3]

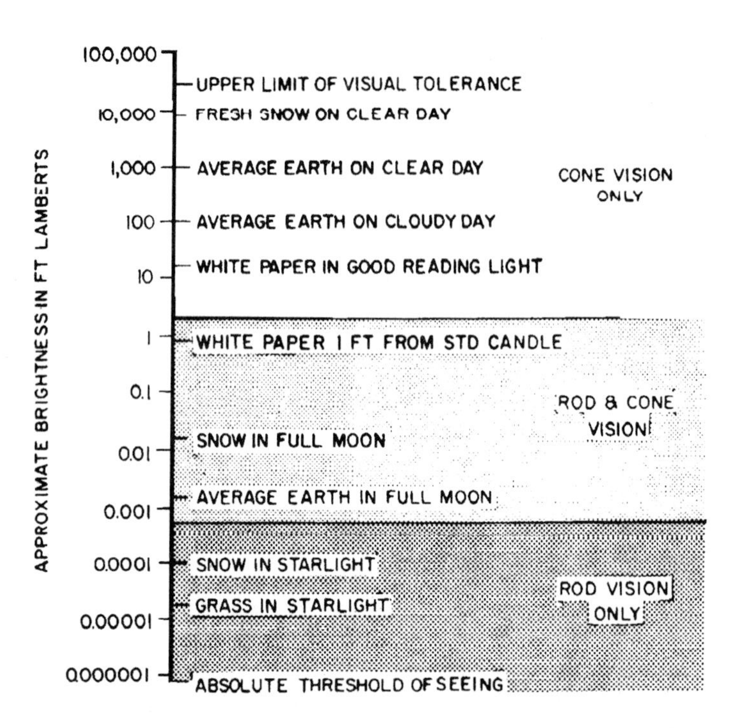

Table SNSNG-B6: Conversion Factors for Luminance Units[3]

Units	Foot-lamberts	Lamberts	Milli-lamberts	Candles per square inch	Candles per square foot	Candles per square centimeter
ft.-L.	-------------	1.076×10^{-3}	1.076	2.21×10^{-3}	3.18×10^{-1}	3.43×10^{-4}
L.	9.29×10^{2}	-------------	1.0×10^{3}	2.054	2.96×10^{2}	3.18×10^{-1}
mL.	9.29×10^{-1}	1.0×10^{-3}	-------------	2.054×10^{-3}	2.957×10^{-1}	3.183×10^{4}
c/in².	4.52×10^{2}	4.87×10^{-1}	4.87×10^{2}	-------------	1.44×10^{2}	1.55×10^{-1}
c/ft².	3.14	3.38×10^{-3}	3.38	6.94×10^{-3}	-------------	1.076×10^{-3}
c/cm².	2.92×10^{3}	3.14	3.14×10^{3}	6.45	9.29×10^{2}	-------------

Note: Value in units in left-hand column times conversion factor equals value in units shown at top of column.

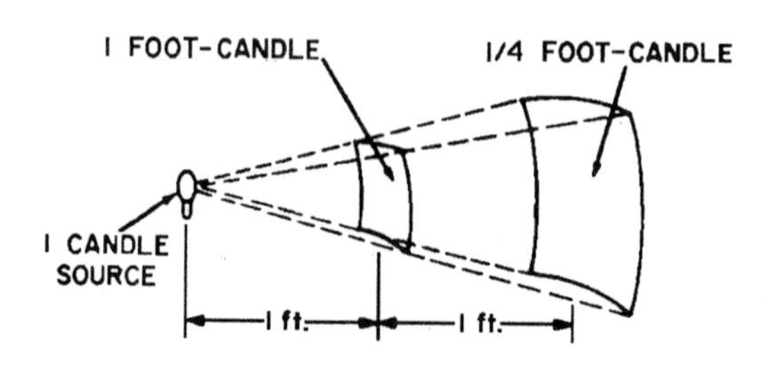

Figure SNSNG-B5: Inverse Square Law of Illumination[3]

VISUAL ACUITY

Figure SNSNG-B6: Common Types of Test Targets for Minimum Separable Acuity[3]

Table SNSNG-B7: Minimal Separable Acuity as a Function of Background Luminance[3]

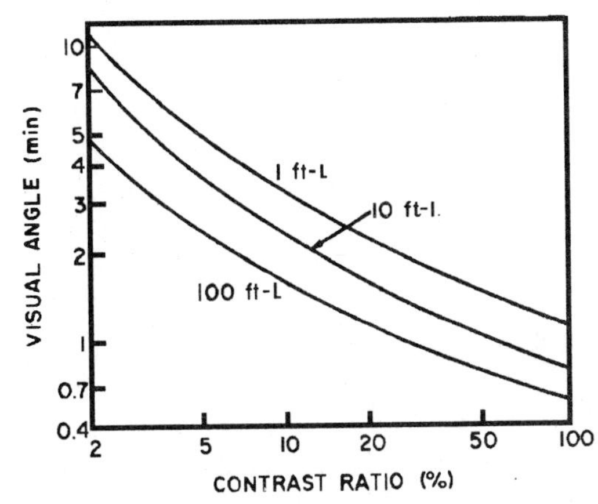

These measurements where made using a Landholt ring.

Table SNSNG-B8: Minimal Perceptible Detection for Circular Targets as a Function of Contrast, and Background Luminance[3]

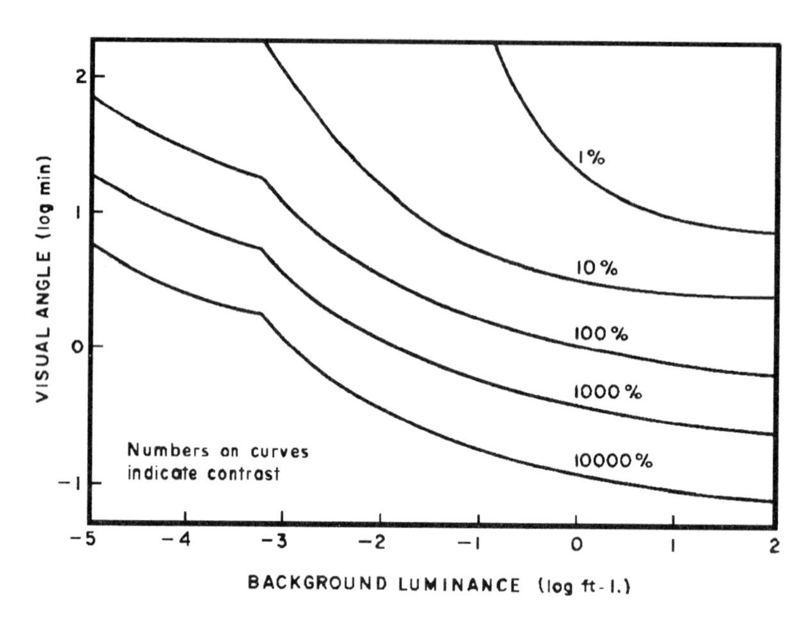

Minimal perceptible acuity is the measurement of the eye's ability to detect the smallest possible target that is lighter or darker than the background. Curves for contrast above 100% are for targets brighter than their backgrounds. The threshold here is 99%.

Table SNSNG-B9: Vernier Acuity as a Function of Background Luminance[3]

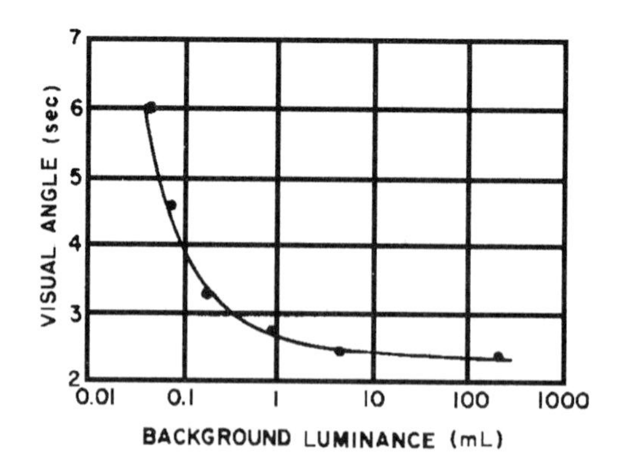

Vernier acuity is measured by having subjects view lines that are separated by progressively smaller spaces. The threshold is measured by the smallest lateral displacement that can be detected can be detected by the observer. This type of acuity is used for reading instruments and optical range finders. Vernier acuity can be increased by increasing the available light.

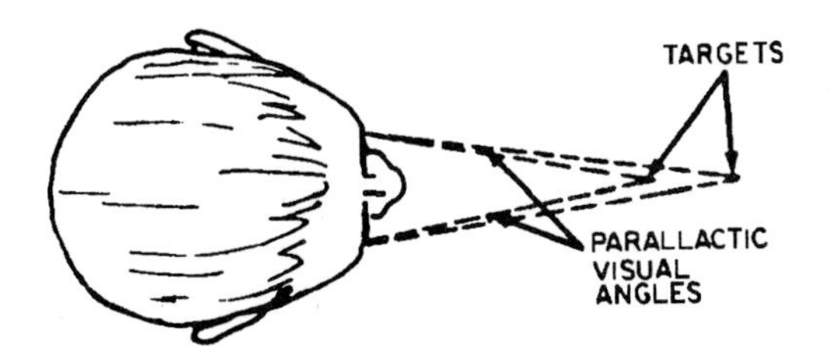

Figure SNSNG-B6: Meaning of Stereoscopic Acuity[3]

To determine the threshold for stereoscopic acuity, take the difference between the parallactic visual angles of two targets located at the JND distance from each other. As with other types of acuity, stereoscopic acuity increases with increasing light.

Table SNSNG-B10: Stereoscopic Acuity as a Function of Background Luminance[3]

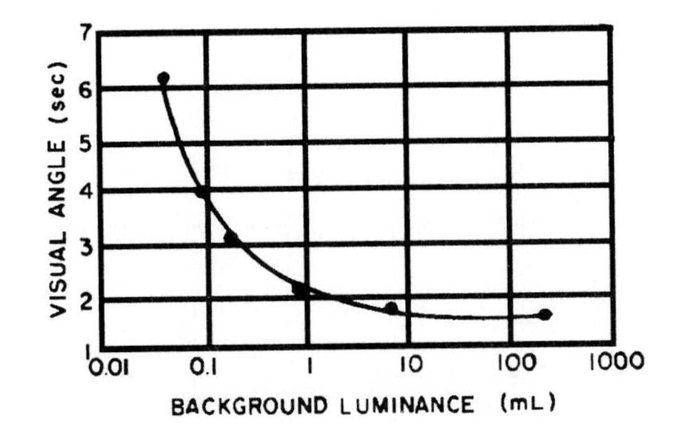

Table SNSNG-B11: Effect on Visual Acuity of Surround Luminance[3]

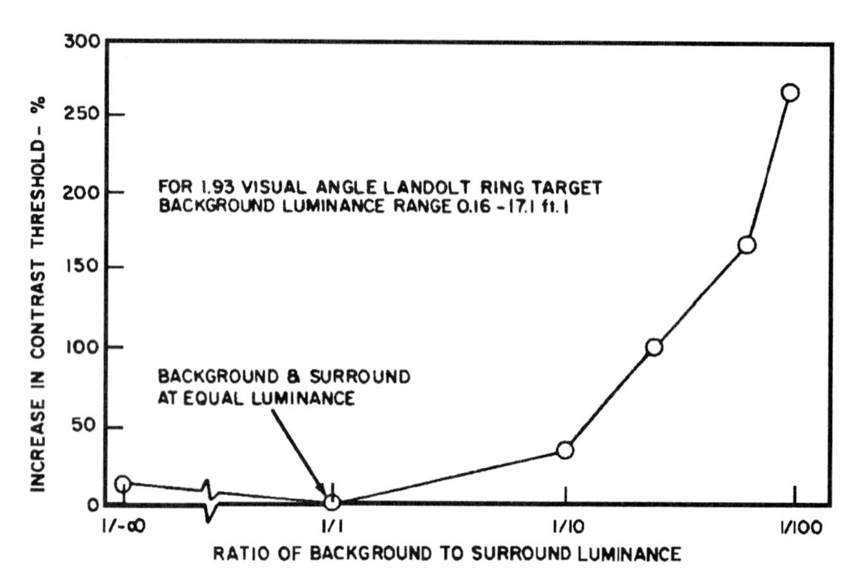

The data in this table are for a dark Landholt ring acuity target on a lighter background with a gap of 1.93 min. visual angle. As background luminance increases over surround, threshold increases. Thus, if you are trying to maintain maximum acuity, your targets should not be in shadows or near large areas of greater brightness.

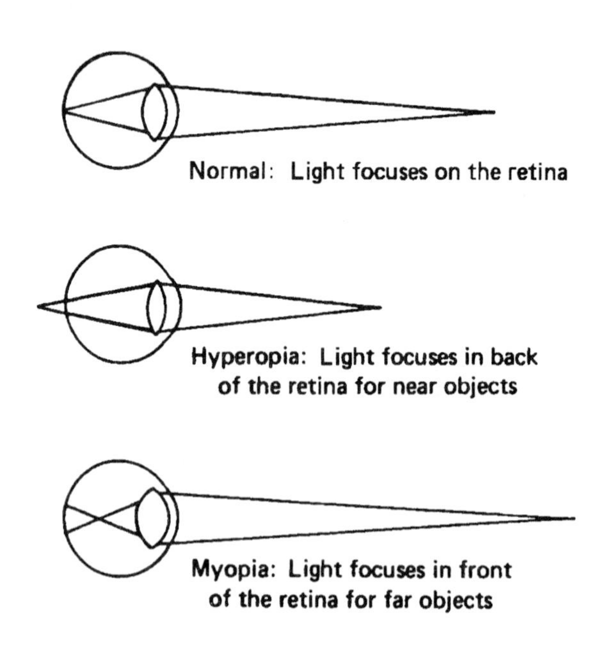

Figure SNSNG-B7: Problems of Visual Acuity[1]

Table SNSNG-B12: Change of Visual Acuity With Distance From the Fovea at High Luminance, 76 ft.-Lamberts.[3]

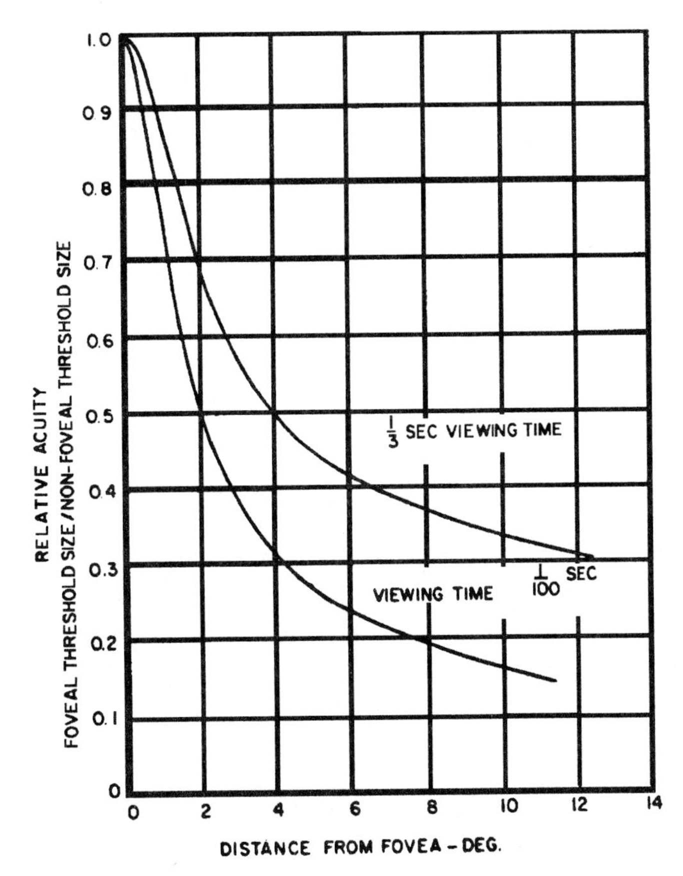

Data are for circular targets of varying size. From the table, you can see that visual acuity diminishes rapidly with increasing distance from the fovea. Thus, if an observer is to see a near-target, their eyes must be fixated within an angle as small as 1°; this also means that if critical details are to be observed at greater peripheral angles then the size of the target must be increased.

Table SNSNG-B13: The Variation of Static and Dynamic Visual Acuity With Age[3]

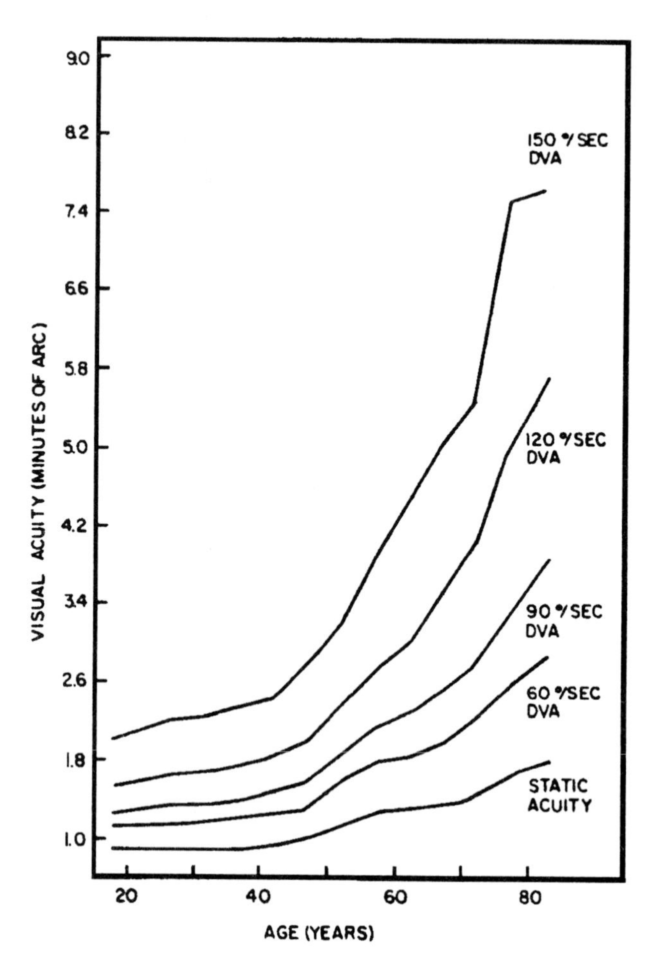

Dynamic visual acuity refers to situations in which either the target or the observer are moving, and is usually defined as the smallest detail that can be detected when either the target or the observer is moving. This table shows the threshold for minimal separable acuity for targets moving through 180°. Acuity diminishes rapidly as age rate of motion exceeds 60°. Acuity also diminishes as the observer exceeds age 50.

Table SNSNG-B14: Rod-Cone Distribution on the Retina[1]

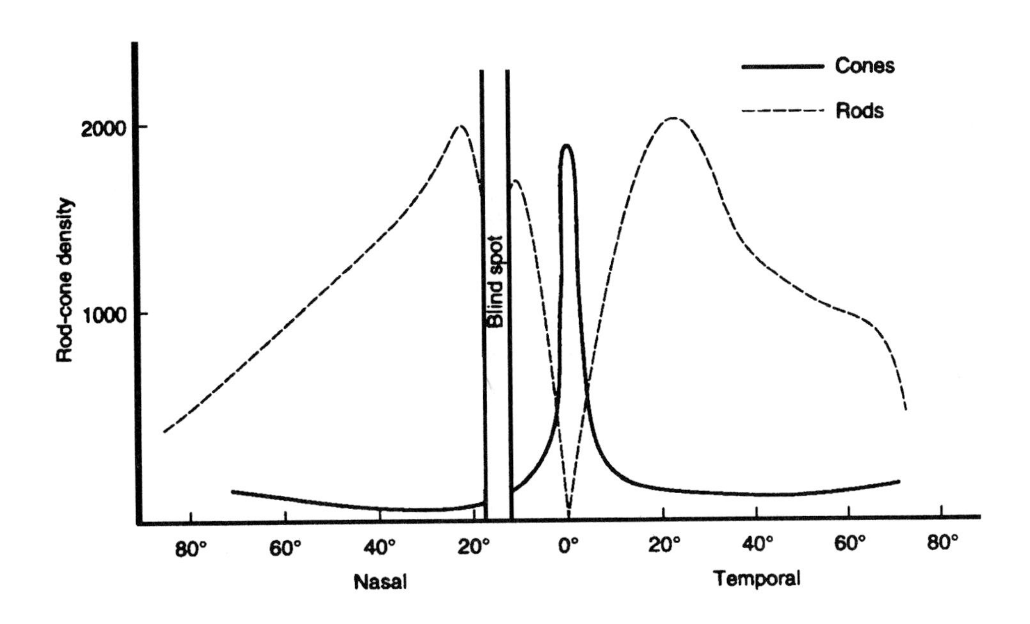

Table SNSNG-B15: Dark Adaptation Curve.[1]

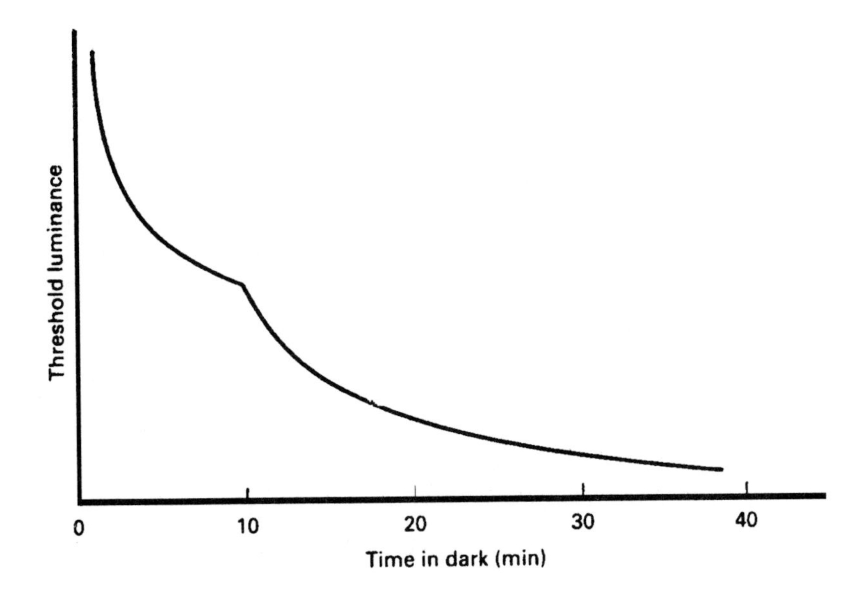

The breakpoint denotes the shift between the cones and the rods (i.e., the change from photopic to scotopic vision).

Table SNSNG-B16: Data on the Temporal Summation of the Eye[3]

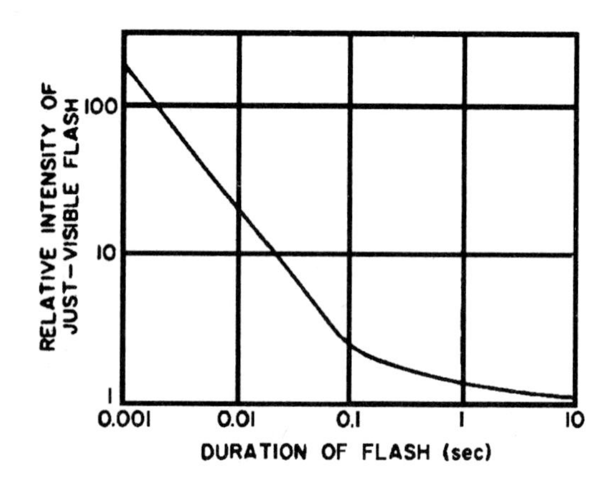

Intensity x times below 0.1 second equals a constant value for threshold visibility. Thus, for extremely short flashes of light, high intensities are required for visibility.

REFERENCES

1. Pulat, B.M. 1992. *Fundamentals of Industrial Ergonomics.* Englewood Cliffs, NJ: Prentice-Hall. Reprinted with Permission.

2. Bailey, R.W. 1989. *Human Performance Engineering. (2nd Ed)* Englewood Cliffs, NJ: Prentice-Hall. Reprinted with Permission.

3. Van Cott, H.P., and Kinkade, R.G., 1972. *Human Engineering Guide to Equipment Design. (Revised)* Washington, DC.: US Government Printing Office. Library of Congress Number: 72-600054.

CHAPTER 4: SENSING TABLES

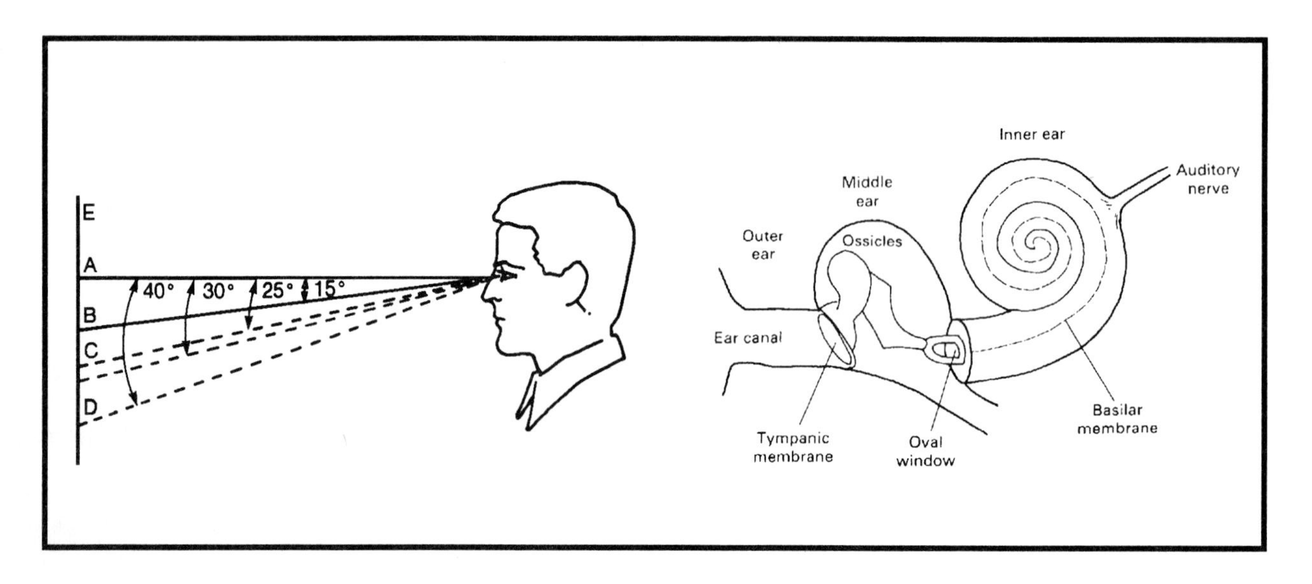

Section C: Color Vision

This section presents information regarding the chromatic (photopic) portion of human vision to be used when designing systems with chromatic displays.

COLOR VISION

Table SNSNG-C1: The Electromagnetic Spectrum Showing the Visible Wavelengths[1]

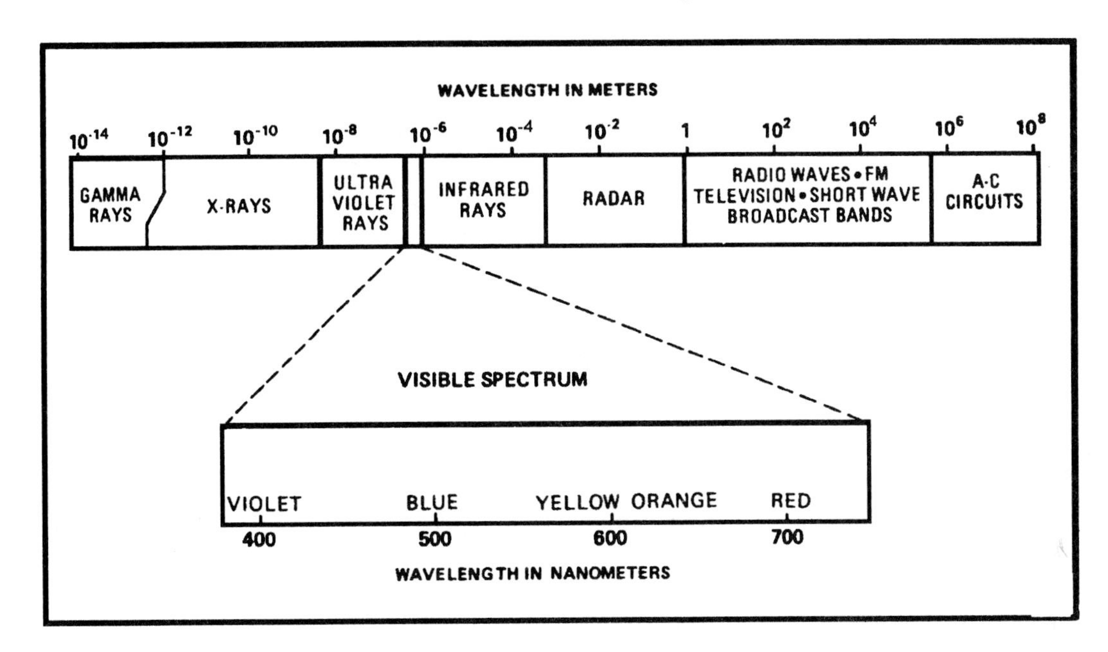

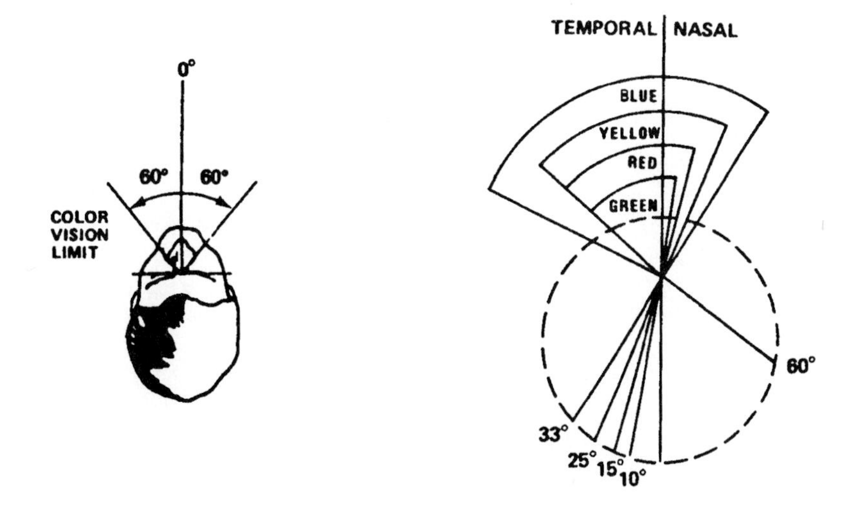

Figure SNSNG-C1: The Visual Field for Color With Eye Motionless[1]

Table SNSNG-C2: Intensity Thresholds of Point-Source Signal Lights as a Function of the Background Luminance for Signal-Light Color[2]

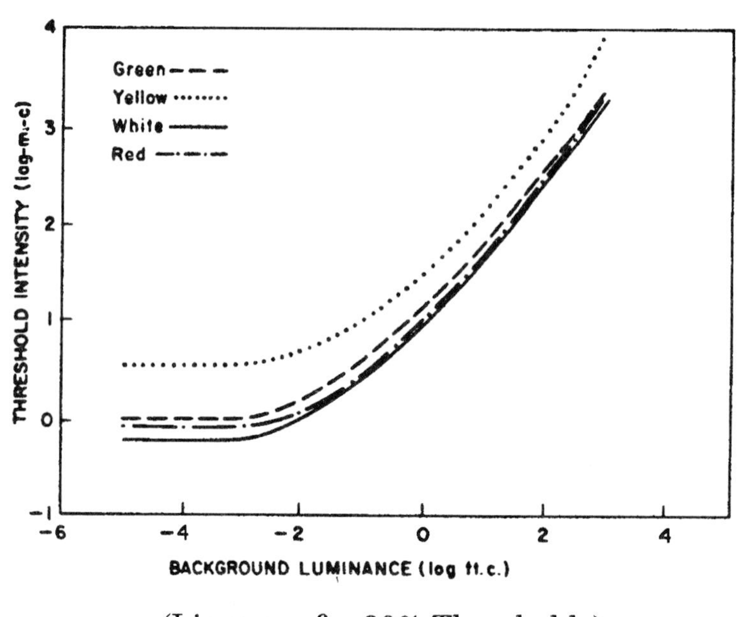

(Lines are for 90% Thresholds)

Table SNSNG-C3: Ten Spectral Colors that Can be Easily Identified With Little Training[2]

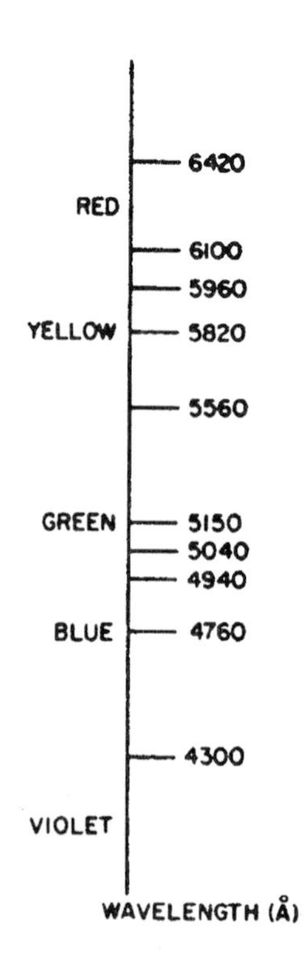

4 Sensing Tables

Table SNSNG-C4: Visible Colors for Different Viewing Angles[2]

Peripheral Viewing	*Red*	*Blue*	*Green*	*Cyan*	*Magenta*	*Yellow*
0–40 degrees	X	X	X			X
40–50 degrees		X				X
50–60 degrees		X				
>60 degrees						

Table SNSNG-C5: Easily Perceivable and Discriminable Colors for Thick Lines and Area-Fill [3]

	Background	
Number of Colors	*White*	*Black*
1	Red Green Blue Purple Magenta	Yellow Cyan Green Magenta
2	Green and magenta Red and blue Red and green	Magenta and cyan Magenta and green
3	Red, green, and blue Magenta, green, and blue	Magenta, cyan, and yellow Magenta, green, and yellow

Table SNSNG-C6: Easily Perceivable and Discriminable Colors for Thin Lines[3]

	Background	
Number of Colors	*White*	*Black*
1	Red Green	Yellow Cyan Green
2	Red and green Magenta and cyan Red and blue	Green and magenta Yellow and magenta Cyan and magenta
3	Red, blue, and green	Cyan, magenta, and yellow

Table SNSNG-C7: Ratings of Edge Sharpness of Color Images on Color Backgrounds[3]

	Background Color							
Image Color	*Red*	*Blue*	*Green*	*Cyan*	*Magenta*	*Yellow*	*Black*	*White*
Red	O	L	L	H	L	H	L	H
Blue	L	O	M	H	M	H	L	H
Green	M	M	O	L	M	L	M	L
Cyan	M	M	L	O	M	L	M	L
Magenta	L	M	M	M	O	H	M	M
Yellow	M	M	L	L	M	O	M	L
Black	M	L	H	H	M	H	O	H
White	M	H	L	L	M	L	H	O

H = high edge sharpness
M = medium edge sharpness
L = low edge sharpness (blurry)
O = no edge visibility

Table SNSNG-C8: Discriminable and Legible Color Combinations on Achromatic Backgrounds[3]

	Image Color					
	Red	*Blue*	*Green*	*Cyan*	*Magenta*	*Yellow*
On Black/Dark Gray Backgrounds						
Color Discrimination Combinations					X	X
				X	X	
			X		X	
				X		X
Legibility Combinations			X	X		X
On White Background						
Color Discrimination Combinations		X			X	
	X	X				
	X		X			
	X			X		
Legibility Combinations	X	X				
		X			X	
On Medium Gray						
Color Discrimination Combinations	X					X
		X			X	
Legibility Combinations	X					X

Table SNSNG-C9: Relations Between Physical, CIE, and Perceptual Descriptions of Color[3]

Properties of Color		
Physical	*← CIE Link →*	*Perceptual*
Wavelength	Dominant wavelength	Hue
Purity	Excitation purity	Saturation
Intensity	Luminance	Brightness and Lightness

Table SNSNG-C10: Comfort Ratings for Color CRT Text and Background Combinations[3]

	Background Color					
Text Color	*Red*	*Blue*	*Green*	*Cyan*	*Magenta*	*Yellow*
Red	—	P	P	F	P	F
Blue	P	—	F	F	P	F
Green	P	F	—	F	F	P
Cyan	P	F	P	—	F	P
Magenta	P	P	F	F	—	F
Yellow	F	G	P	P	F	—

G = Good
F = Fair
P = Poor

Table SNSNG-C11: Ratings of Display Characteristics that Can Affect Color Perceptions[3]

	CRT	*LCD*	*ELD*	*PPD*	*LED*
All Colors	good	good	fair	fair	good
Color Quality	good	poor	fair	poor	fair
Purity	fair	poor	good	good	good
Brightness	good	fair*	good	fair	good
Gray Scale Levels	good	fair	poor	fair	good
Power Consumption /Energy Emissions	fair	good	fair	poor	good
Resolution	good	fair	good	good	good
Contrast	good	poor	good	good	fair
Flicker	poor	good	fair	good	good
Glare/Reflections**	poor	poor	poor	poor	good
Viewing Range	good	poor	good	fair	good
Size	poor	good	good	fair	good
Weight	poor	good	good	fair	good

* Depends on ambient illumination
** Untreated screen

Table SNSNG-C12: Variables Contributing to Computer Display Color Quality[3]

Visual	*Hardware*	*Software*
Color	Chromaticity of display primaries	Color space algorithm
Hue	Wavelength	
Saturation	Purity	
Lightness or brightness	Light intensity	Gamma correction
Perceptual difference between colors	DAC levels	Frame buffer and color map
Contrast sensitivity	Spatial and luminance contrast	Virtual color controls
Legibility	Resolution	Anti-aliasing
Flicker	Refresh rate Phosphor decay Brightness	

Table SNSNG-C13: Comfort Ratings for Color CRT Text and Background Combinations[3]

	Background Color					
Text Color	*Red*	*Blue*	*Green*	*Cyan*	*Magenta*	*Yellow*
Red	—	P	P	F	P	F
Blue	P	—	F	F	P	F
Green	P	F	—	F	F	P
Cyan	P	F	P	—	F	P
Magenta	P	P	F	F	—	F
Yellow	F	G	P	P	F	—

G = Good
F = Fair
P = Poor

Table SNSNG-C14: Input Device Color Usability Comparison[3]

Feature	*Mouse*	*Tablet*	*Thumbwheel*	*Keyboard*	*Trackball*
Discrete color selection	X	X		X	
Continuous color selection	X	X	X	X	X
Drawing ability	X	X			
Tracing ability		X			
Good for simple graphics (such as pie charts)	X	X	X	X	X
Good for complicated graphics (such as integrated circuit designs)	X	X			
Requires extra work surface	X	X			X
Accuracy	X	X	X	X	X
Right or left hand use	X	X			X
Fast pointing speed	X	X			X

Table SNSNG-C15: Recommended Colors for Signal Lights[2]

For maximum distance		For intermediate and near distance
Two colors required	Three colors required	Select number required, beginning at top
Red Green	Red Green White or Yellow	Red Green Yellow Blue White Purple

REFERENCES

1. Bailey, R.W. 1989. *Human Performance Engineering. (2nd Ed)* Englewood Cliffs, NJ: Prentice-Hall. Reprinted with Permission.

2. Van Cott, H.P., and Kinkade, R.G., 1972. *Human Engineering Guide to Equipment Design. (Revised)* Washington, DC.: US Government Printing Office. Library of Congress Number: 72-600054.

3. Thorell, L.G., and Smith, W.J. 1990. *Using Computer Color Effectively: An Illustrated Reference.* Englewood Cliffs, NJ: Prentice-Hall. Page 25. Reprinted by permission of Prentice-Hall.

CHAPTER 4: SENSING TABLES

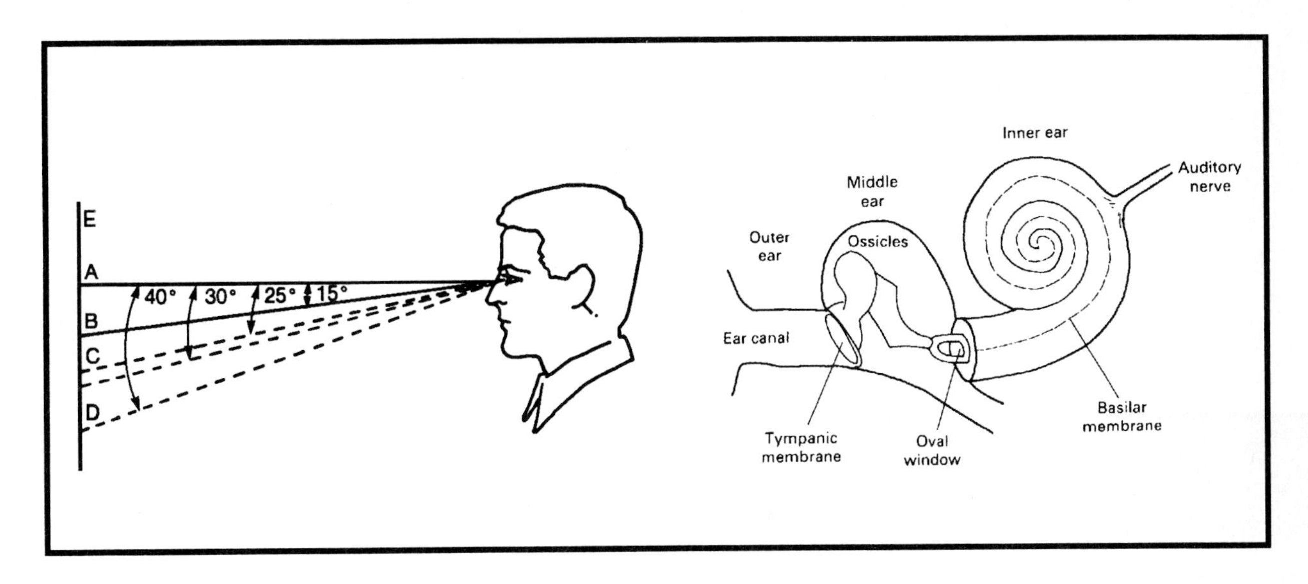

Section D: Audition

This section presents information about the human auditory and vestibular systems to be used when designing systems with auditory displays or that might disrupt the operator's sense of balance or proprioception.

AUDITION

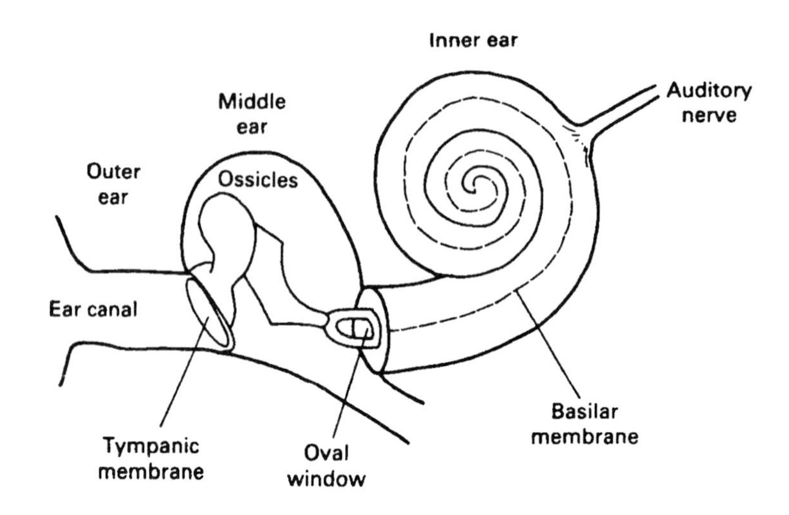

Figure SNSNG-D1: Schematic Diagram of the Ear[1]

Table SNSNG-D1: Sound Pressure Level[1]

Source	Intensity [dB(A)]
Quiet residence	42
Dictation	67
Conference speaking	70
Quiet factory	76
Loud shouting	82
18-in. automatic lathe	87
Wire drawer	89
Sawmill	90
Chain saw	105
Pneumatic bore hammer	120
Rifle shot	130

Table SNSNG-D2: Typical Sound Pressure Levels (SPL's) for Some Common Sounds[2]

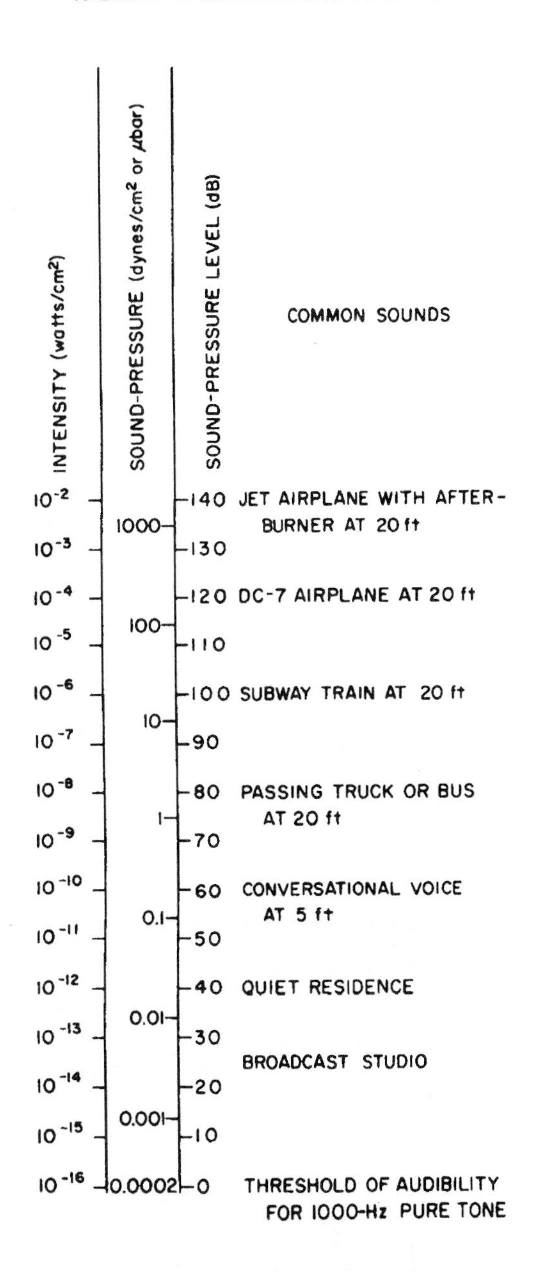

Table SNSNG-D3: Sound Pressure Level Attenuation as a Function of the Distance of Transmission in Calm Air[2]

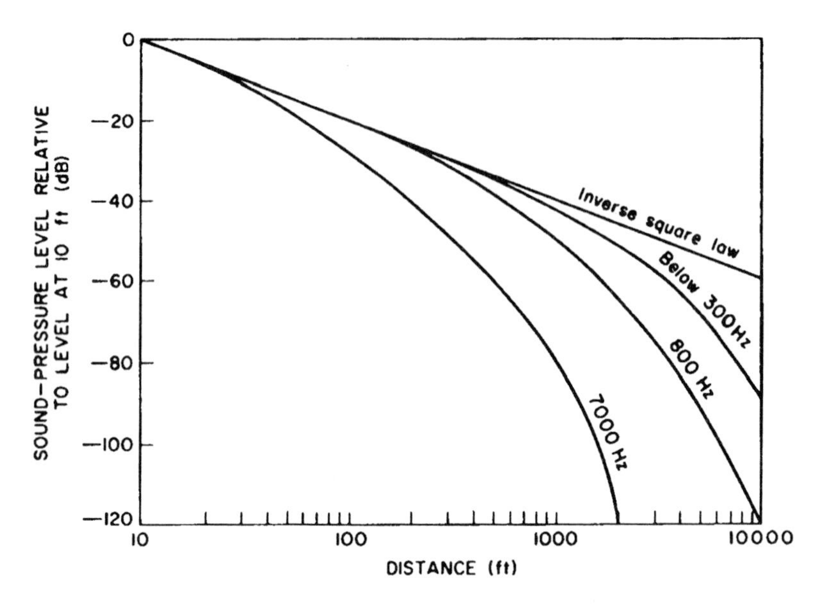

When designing alarms and warnings, sounds with frequencies below 1000 Hz should be used when the sounds must travel long distances, because of the absorption of high frequencies over long distances.

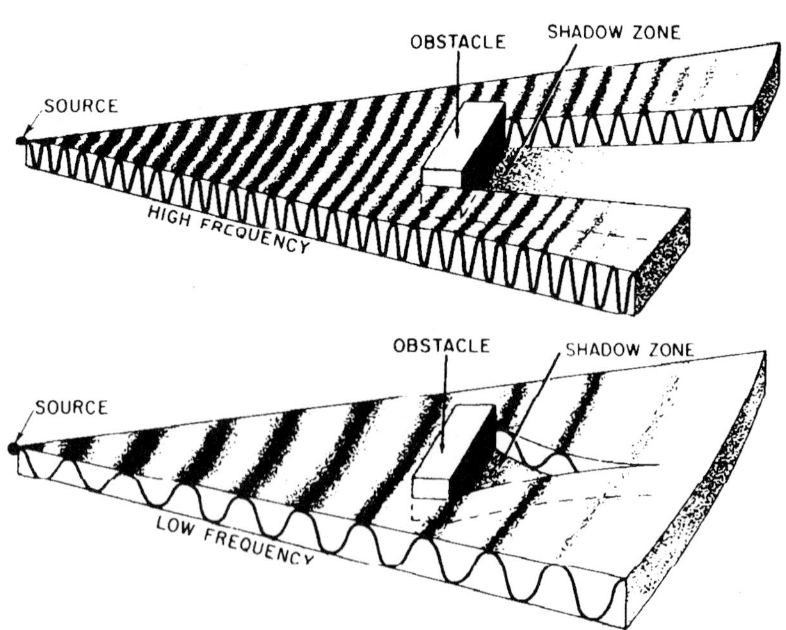

Figure SNSNG-E2: Attenuation of High-Frequency Sounds (Wavelength Shorter than Width of Obstacle) in the Shadow Zone[2]

Barriers based on this principle are sometimes used for control of high frequency-noise.

Table SNSNG-D4: Dependency of Pitch on Frequency[2]

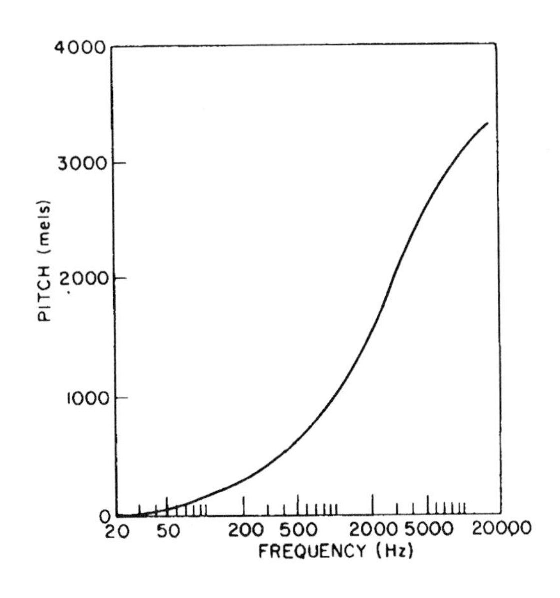

Use this table when picking a number of pure tones so that the intervals between given tones are perceived to be equal in pitch.

Table SNSNG-D5: Three Absolute Threshold Curves for Pure Tones[2]

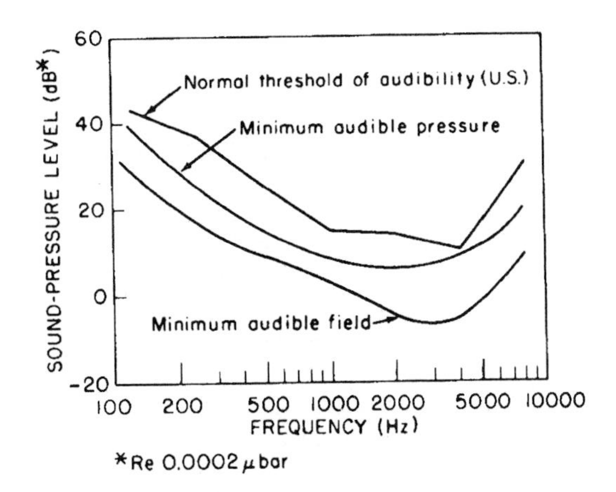

Minimal Audible Field (MAF), Minimal Audible Pressure (MAP), and Normal Threshold of Audibility (NTA) are the three accepted absolute thresholds for pure tones. Minimal Audible Field is the sound pressure level (SPL) at the absolute threshold of a young, trained listener sitting in an anechoic room. MAP is the sound pressure level the sound pressure level at the absolute threshold of a young, trained listener, presented through headphones. NTA is measured at the ear canal's orifice and is the modal value of the minimum sound pressure level that can be heard by young, untrained listeners.

Table SNSNG-D6: Equal Loudness Contours for Pure Tones[2]

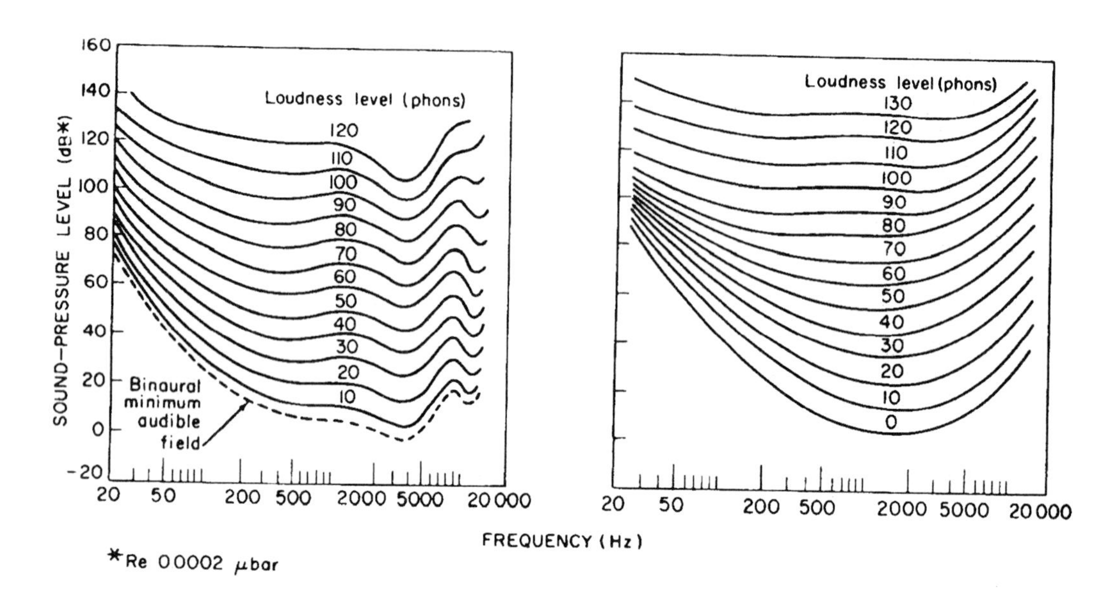

"For the left position data, the listener is in a free field facing source; sound pressure is measured at the listener's head center. For the right position, the listener is wearing headphones; sound pressure is measured under the headphones" (Van Cott and Kinkade, 1972).

Table SNSNG-D7: The Relationship Between Loudness (Sones) and Loudness Level (Phons)[2]

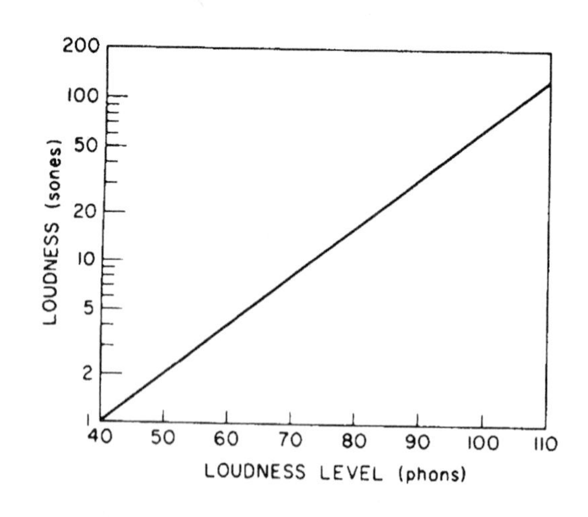

Table SNSNG-D8: Hearing Loss For Females and Males as a Function of Aging [2]

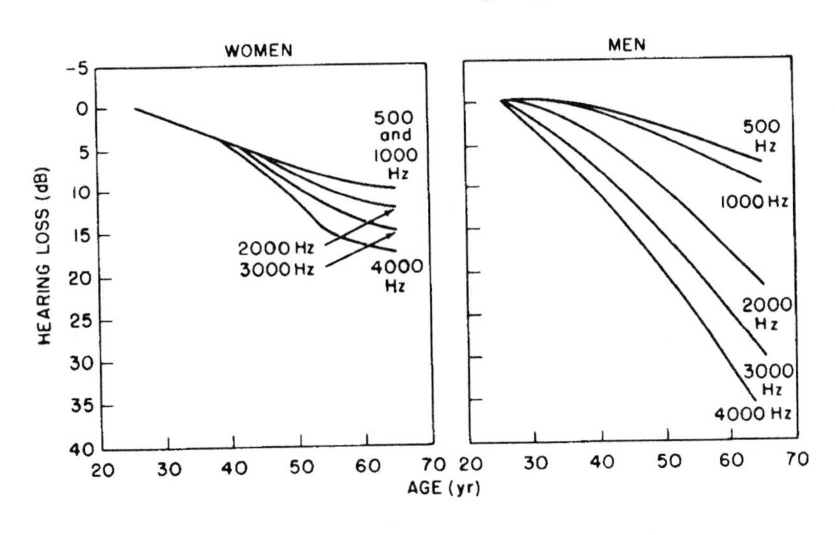

Table SNSNG-D9: JND in Frequency for Pure Tones at Various Levels Above Threshold[2]

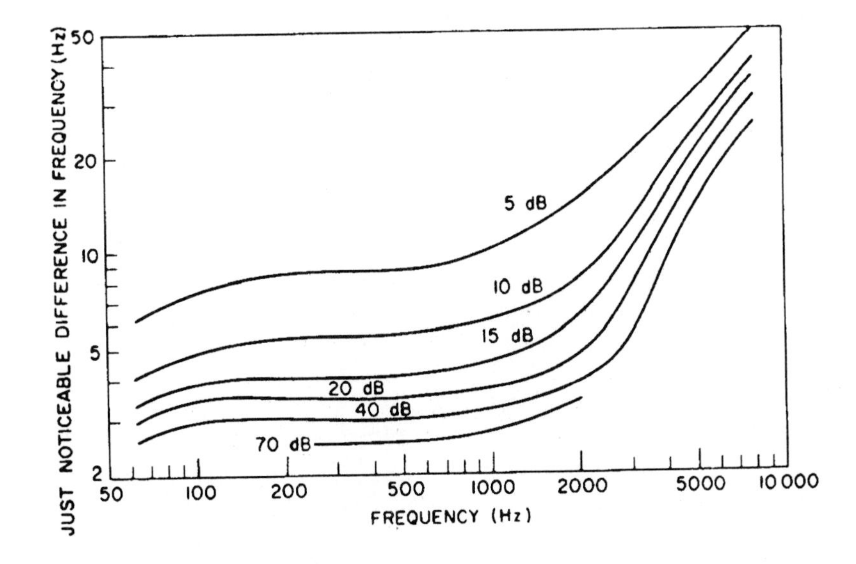

As can be seen from this table, small changes in pitch are easier to hear if they last longer than 0.1 second and are presented 30dB or more above threshold.

Table SNSNG-D10: Dependence on Duration of the JND in Frequency[2]

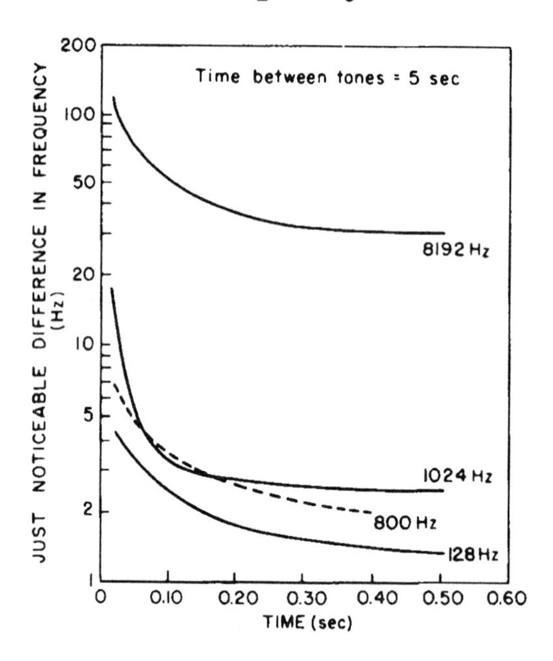

Table SNSNG-D11: JND in Sound-Pressure Level (SPL) for Pure Tones of Various Frequencies and for Wide-Band Noise[2]

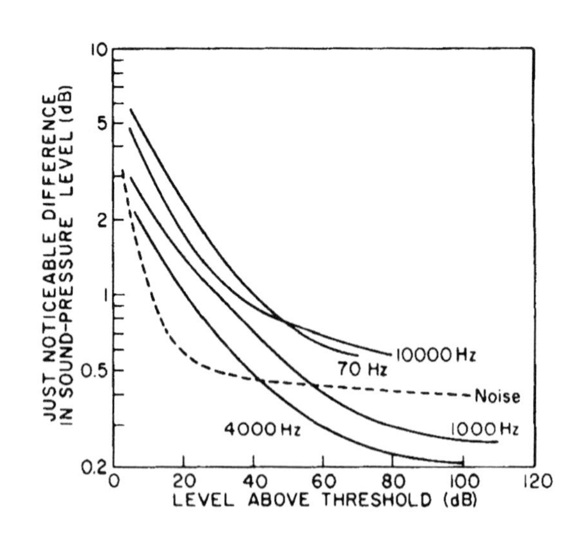

Table SNSNG-D12: Generally Accepted Values of Critical Bandwidth as a Function of Frequency[2]

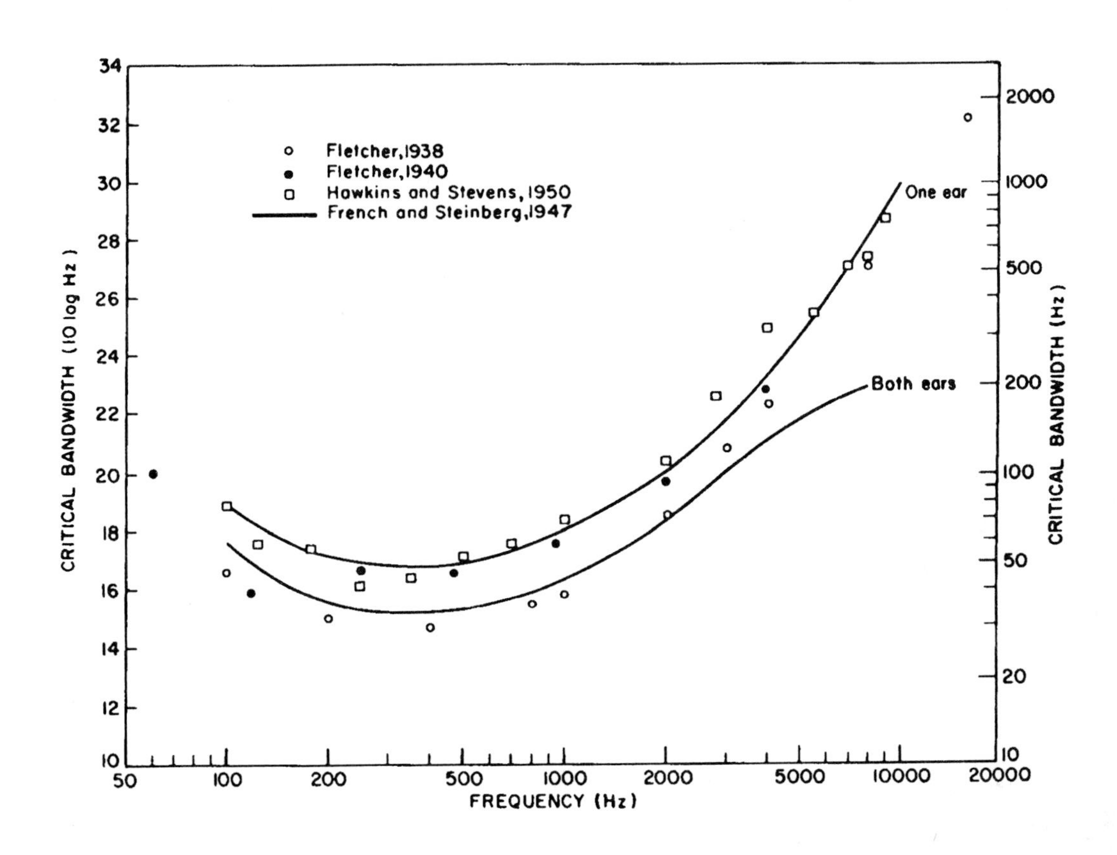

You can predict the masked threshold, by wide-band noise, of a pure tone if you know the spectrum level of the noise at the frequency of the masked tone. Measure the spectrum level of the wide-band masking noise at the frequency of the tone. Next, correct the level to the critical band level by adding the $\log_{10}$ of the critical bandwidth, which can be read from the left ordinal axis of this graph. The resulting value is the masked threshold if the resulting value is presented at 20dB above the absolute threshold of the pure tone.

Table SNSNG-D13: Masking Produced by Pure Tones[2]

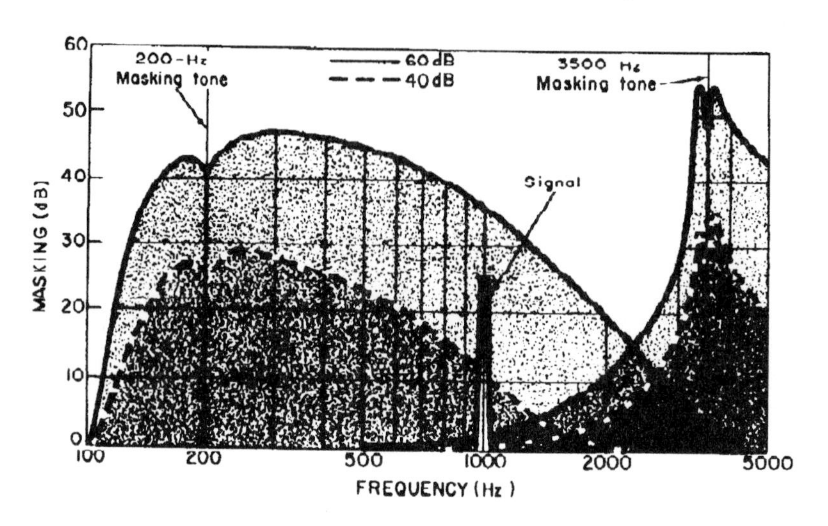

The greatest masking of a pure tone, or noise with a strong tonal part, is greatest near the given tonal frequency, but also extends to signals on both sides of the masking tone. However, if the tones are high in intensity, the masked threshold is raised for signals that contain components that are an integral multiple of the masking tone, more so than those that do not have such a harmonic relationship to the masking tone.

Table SNSNG-D14: Masking as a Function of Frequency for Masking by Pure Tones of Various Frequencies and Levels[2]

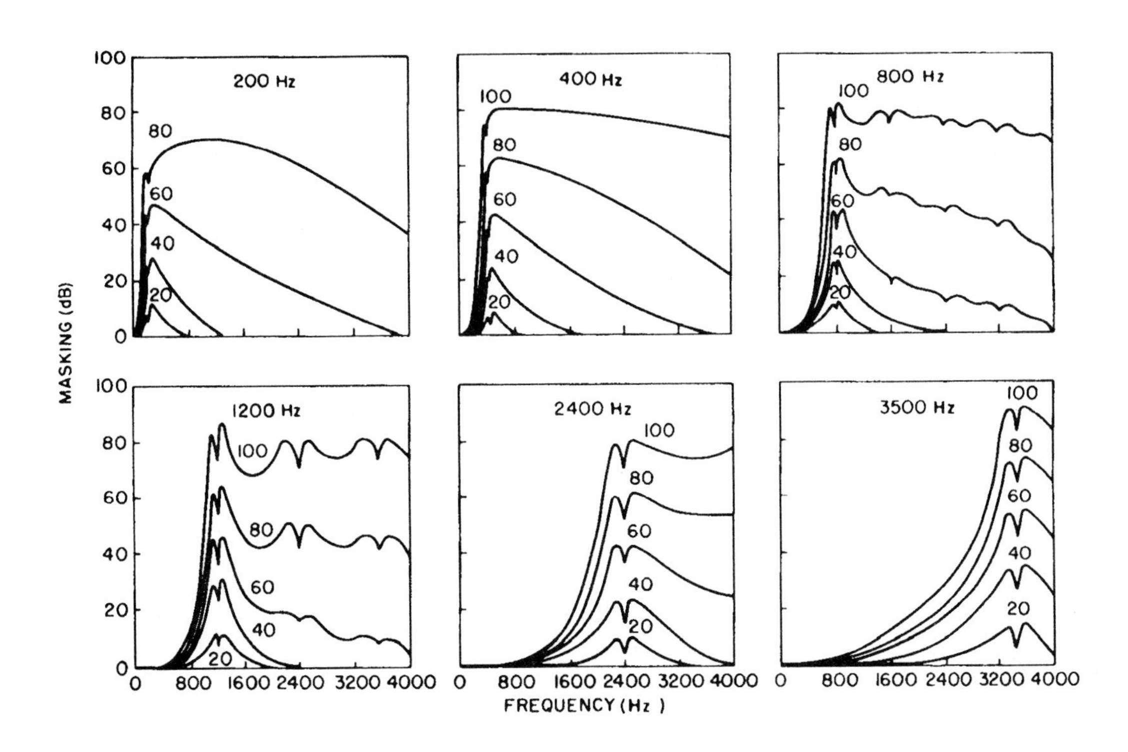

The number at the top of each graph is the frequency of the masking tone. The number on each curve is the level above threshold of each masking tone.

Table SNSNG-D15: Curves of Masking vs. Frequency for Masking by a Narrow Band of Noise[2]

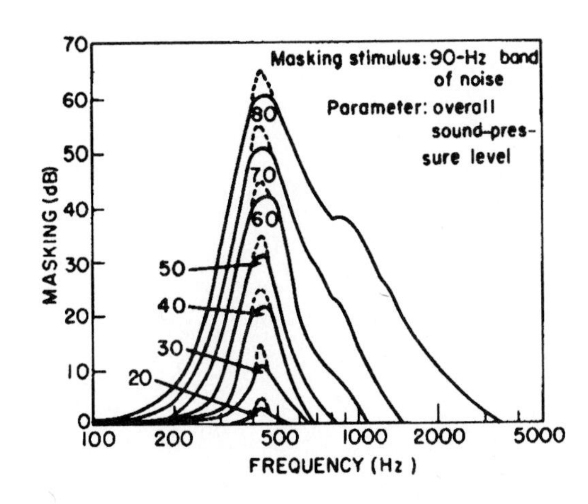

Table SNSNG-D16: Masked Thresholds of Pure Tones When Masked by a Wide Band Noise of Uniform Spectrum[2]

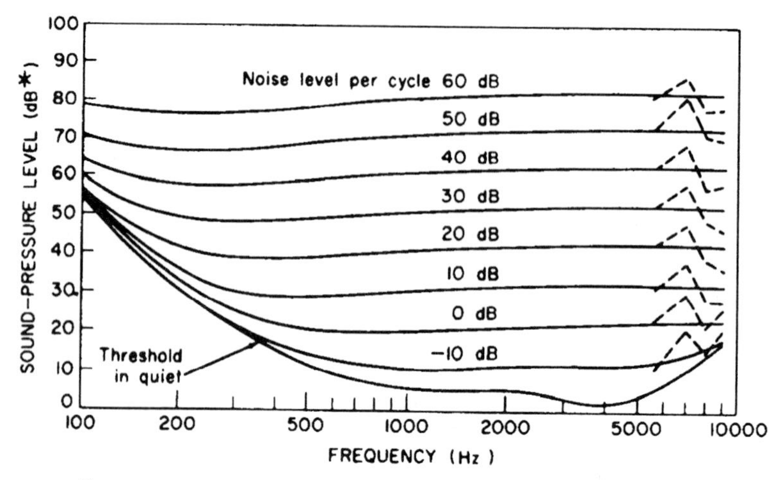

VESTIBULAR SENSE

Table SNSNG-D17: Thresholds for Sensing Rotation[2]

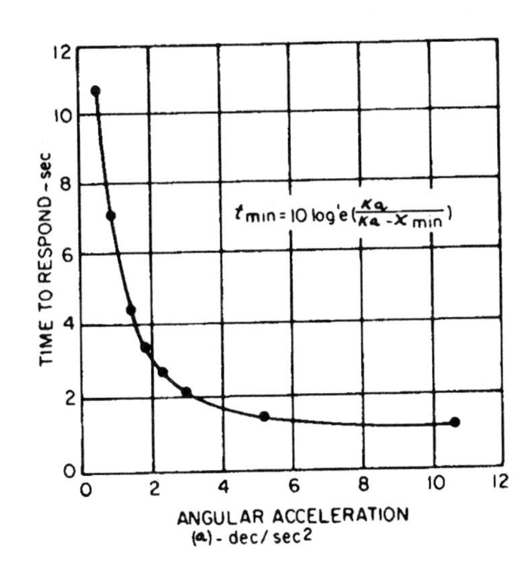

Table SNSNG-D18: Perception of Vertical by the Vestibular Sense[2]

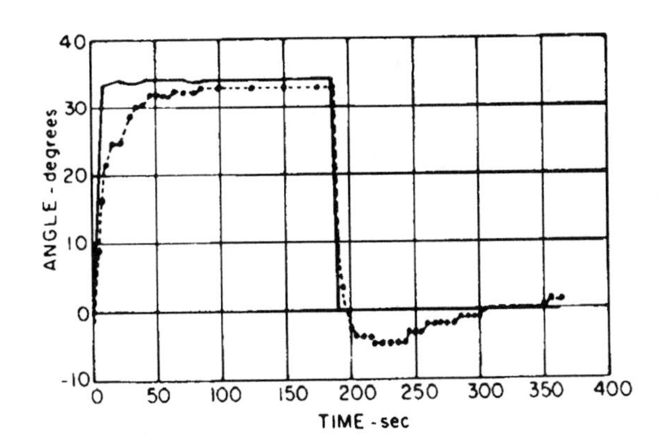

REFERENCES

1. Pulat, B.M. 1992. *Fundamentals of Industrial Ergonomics.* Englewood Cliffs, NJ: Prentice-Hall. Reprinted with Permission.

2. Van Cott, H.P., and Kinkade, R.G., 1972. *Human Engineering Guide to Equipment Design. (Revised)* Washington, DC.: US Government Printing Office. Library of Congress Number: 72-600054.

CHAPTER 5: DISPLAY TABLES

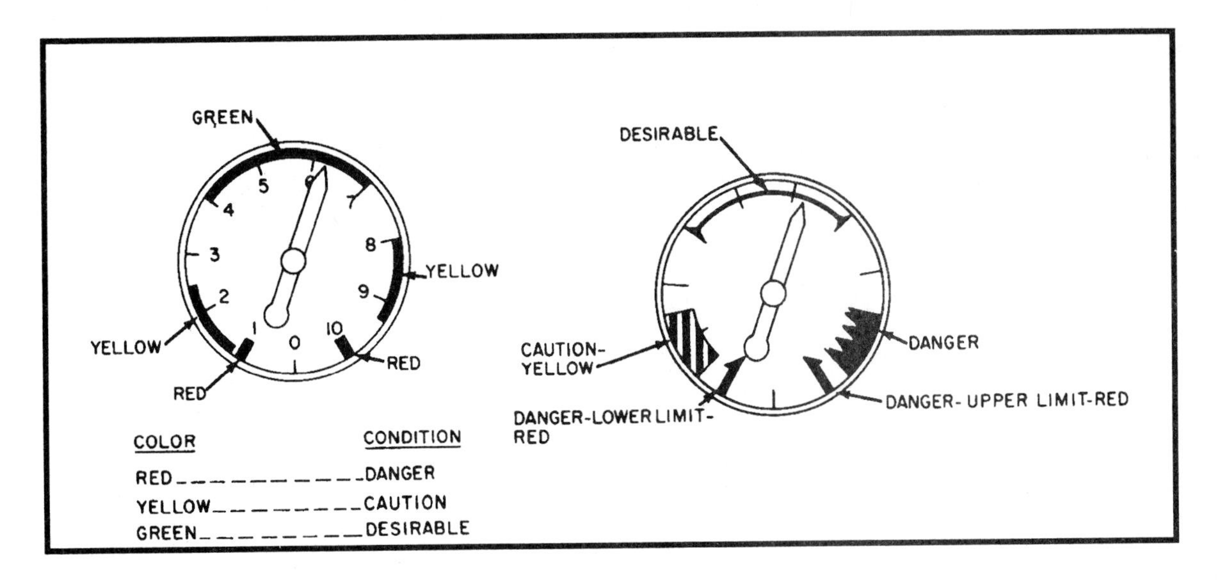

Section A: Display Guidelines

This section presents guidelines to aid in deciding when to use auditory versus visual displays. The information in this section is useful for the ergonomist or human factors engineer who must design display systems.

COMPARISON OF AUDITORY AND VISUAL DISPLAYS

Table DSPLY-A1: Use of Visual Versus Auditory Displays[1]

Use Visual Presentation if
• The person's job allows him or her to remain in one position
• The message does not call for immediate action
• The message is complex
• The message is long
• The message will be referred to later
• The auditory system of the person is overburdened
• The message deals with location in space
• The receiving location is too noisy

Use Auditory Presentation if
• The person's job requires him or her to move about continually
• The message calls for immediate action
• The message is simple
• The message is short
• The message will not be referred to later
• The visual system of the person is overburdened
• The message deals with events in time
• The receiving location is too bright or if preservation of dark adaptation is necessary

The upper part summarizes conditions in the workplace, or conditions related to the information to be communicated, that make visual presentation the preferred method. The lower part provides a comparable list for auditory presentation of information. Visual presentation is preferred for complex messages in noisy environments where response time is not critical. Auditory presentation is preferred for simple messages in areas where people move around frequently and where response time must be rapid.

Table DSPLY-A2: Task Conditions Affecting Signal Detectability During Extended Monitoring[1]

To increase the probability of detecting a signal

- Use simultaneous presentation of signals (audio and visual)
- Provide two operators for monitoring; allow them to communicate freely
- Provide 10 minutes of rest or alternate activity for every 30 minutes of monitoring
- Introduce artificial signals that must be responded to. These signals should be the same as real signals. Provide feedback to the operator on detection of the artificial signals

Factors that decrease the probability of signal detection

- Too many or too few signals to be detected and responded to
- Introduction of a secondary display-monitoring task
- Introduction of artificial signals for which a response is not required
- Instructions to the operator to report only signals of which there is no doubt

Conditions that make it easier to detect a signal, such as a defect in a product, are listed. Factors that make detection more difficult are also given. To improve inspection performance, one should implement the appropriate suggestions in the first part of the table and avoid the situations in the last part.

Table DSPLY-A3: Impact of Sensory System Parameters on Equipment Design[2]

Parameter	Implications of parameter for equipment design	Equipment affected
Detection sensitivity (lower threshold).	Defines minimal intensity and frequency of signals that can be detected by a sense organ.	Alarms, voice, and visual displays.
Detection sensitivity (upper limit).	Defines limit on intensity and frequency beyond which sensitivity is lost and/or damage may occur to sense organ	Alarms, ambient illumination, protective equipment (e.g., goggles, ear protectors), noise suppression.
Differential sensitivity (difference threshold).	Defines intensity or frequency by which: (a) signal A must be increased or decreased for the change to be detected, (b) signals A and B must differ to be detected.	Scope resolution, scale, and pointer design.
Sensitivity range (upper limit minus lower threshold).	Defines maximum "bandwidth" of a physical energy that can be used for signal presentation & display purposes.	Voice communications equipment (headsets, speakers); visual displays (e.g., sonar, radar, photogrammetry, etc.).
Information transmission capacity.	Determines maximum number and type of codes possible within a stimulus dimension. Determines maximum rate of information presentation. Determines maximum rate of operator decision-making.	Map, display board, and scope symbology; coded warning signals; information update rates; desirability of control dynamics to aid operator response; amount of information presented.
Speed	Determines maximum rate of information presentation, operator response speed, and system response.	Determines information presentation & update rate.
Reliability	Affects overall design, utility, and cost of system.	All man-machine interfaces.
Variability	Information presentation parameter values must be selected on basis of performance of "typical" operators.	All man-machine interfaces.

REFERENCES

1. Reprinted with permission from *Ergonomic Design for People at Work*, © Eastman Kodak Company, 1983, published by Van Nostrand Reinhold. Courtesy of Eastman Kodak Company.

2. Van Cott, H.P., and Kinkade, R.G., 1972. *Human Engineering Guide to Equipment Design. (Revised)* Washington, DC.: US Government Printing Office. Library of Congress Number: 72-600054.

CHAPTER 5: DISPLAY TABLES

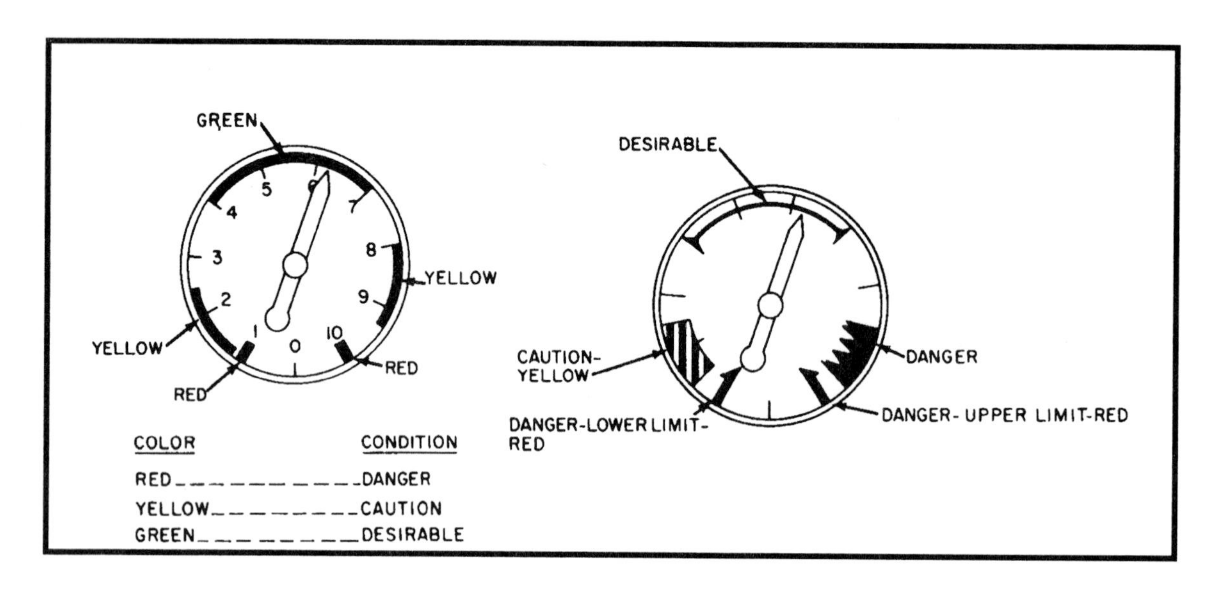

Section B: Visual Displays

This section presents guidelines to aid in deciding when to use visual displays and how to best design them. The information in this section is useful for the ergonomist or human factors engineer who must design visual display systems.

VISUAL DISPLAY CODING

Table DSPLY-B1: Comparison of Different Coding Methods[1]

Code	Number of code steps*		Evaluation	Comment
	Maximum	Recommended		
Color:				
Lights	10	3	Good	Location time short. Little space required. Good for qualitative coding. Larger alphabets can be achieved by combining saturation and brightness with the color code. Ambient illumination not critical factor.
Surfaces	50	9	Good	Same as above except ambient illumination must be controlled. Has broad application.
Shapes:				
Numerals & letters	Unlimited			Location time longer than for color or pictorial shapes. Requires good resolution. Useful for quantitative and qualitative coding. Certain symbols easily confused.
Geometric	15	5	Fair	Memory required to decode. Requires good resolution.
Pictorial	30	10	Good	Allows direct association for decoding. Requires good resolution. Good for qualitative coding only.
Magnitude:				
Area	6	3	Fair	Requires large symbol space. Location time good.
Length	6	3	Fair	Requires large symbol space. Good for limited applications.
Brightness	4	2	Poor	Interferes with other signals. Ambient illumination must be controlled.
Visual number	6	4	Fair	Requires large symbol space. Limited application.
Frequency	4	2	Poor	Distracting. Has merit when attention is demanded.
Stereo-depth	4	2	Poor	Limits population of users. Highly limited application difficult to instrument.
Inclination	24	12	Good	Good for limited application. Recommended for quantitative code only.
Compound codes	Unlimited		Good	Provides for large alphabets for complex information. Allows compounding of qualitative and quantitative codes.

* The maximum number assumes a high training and use level of the code. Also a 5% error in decoding must be expected. The recommended number assumes operational conditions and a need for high accuracy.

Table DSPLY-B2: Number of Different Characters Necessary to Develop a Set of Codes[2]

Use this table when code length is between one and four characters and the total number of codes required by the system is between two and one-hundred

Total Number of Codes *Required by the System*	Desired Code Length *(Number of Characters Per Code)*			
	1	*2*	*3*	*4*
2	2	2	2	2
3	3	2	2	2
4	4	2	2	2
5	5	3	2	2
6	6	3	2	2
7	7	3	2	2
8	8	3	2	2
9	9	3	3	2
10	10	4	3	2
11	11	4	3	2
12	12	4	3	2
13	13	4	3	2
14	14	4	3	2
15	15	4	3	2
16	16	4	3	2
17	17	5	3	3
18	18	5	3	3
19	19	5	3	3
20	20	5	3	3
21	21	5	3	3
22	22	5	3	3
23	23	5	3	3
24	24	5	3	3
25	25	5	3	3
26	26	6	3	3
27	27	6	3	3
28	28	6	4	3
29	29	6	4	3
30	30	6	4	3
40		7	4	3
50		8	4	3
60		8	4	3
70		9	5	3
80		9	5	3
90		10	5	4
100		10	5	4

Table DSPLY-B2: Number of Different Characters Necessary to Develop a Set of Codes[2] (Continued)

Use this table when code length is between two and ten characters and the total number of codes required by the system is between one hundred and 900,000

Total Number of Codes *Required by the System*	Desired Code Length *(Number of Characters per Code)*								
	2	*3*	*4*	*5*	*6*	*7*	*8*	*9*	*10*
100	10	5	4	3	3	2	2	2	2
200	15	6	4	3	3	3	2	2	2
300	18	7	5	4	3	3	3	2	2
400	20	8	5	4	3	3	3	2	2
500	23	9	5	4	3	3	3	3	2
600	25	9	5	4	3	3	3	3	2
700	27	9	6	4	3	3	3	3	2
800	29	10	6	4	4	3	3	3	2
900	30	10	6	4	4	3	3	3	2
1,000		10	6	4	4	3	3	3	2
2,000		13	7	5	4	3	3	3	2
3,000		15	8	5	4	4	3	3	3
4,000		16	8	6	4	4	3	3	3
5,000		18	9	6	5	4	3	3	3
6,000		19	9	6	5	4	3	3	3
7,000		20	10	6	5	4	4	3	3
8,000		20	10	7	5	4	4	3	3
9,000		21	10	7	5	4	4	3	3
10,000		22	10	7	5	4	4	3	3
20,000		28	12	8	6	5	4	4	3
30,000			14	8	6	5	4	4	3
40,000			15	9	6	5	4	4	3
50,000			15	9	7	5	4	4	3
60,000			16	10	7	5	4	4	4
70,000			17	10	7	5	5	4	4
80,000			17	10	7	6	5	4	4
90,000			18	10	7	6	5	4	4
100,000			18	10	7	6	5	4	4
200,000			22	12	8	6	5	4	4
300,000			24	13	9	7	5	5	4
400,000			26	14	9	7	6	5	4
500,000			27	14	9	7	6	5	4
600,000			28	15	10	8	6	5	4
700,000			29	15	10	8	6	5	4
800,000			30	16	10	8	6	5	4
900,000				16	10	8	6	5	4

Table DSPLY-B2: Number of Different Characters Necessary to Develop a Set of Codes[2] (Continued)

Use this table when code length is between one and four characters and the total number of codes required by the system is between two and one-hundred

Total Number of Codes *Required by the System*	Desired Code Length *(Number of Characters Per Code)*			
	1	*2*	*3*	*4*
2	2	2	2	2
3	3	2	2	2
4	4	2	2	2
5	5	3	2	2
6	6	3	2	2
7	7	3	2	2
8	8	3	2	2
9	9	3	3	2
10	10	4	3	2
11	11	4	3	2
12	12	4	3	2
13	13	4	3	2
14	14	4	3	2
15	15	4	3	2
16	16	4	3	2
17	17	5	3	3
18	18	5	3	3
19	19	5	3	3
20	20	5	3	3
21	21	5	3	3
22	22	5	3	3
23	23	5	3	3
24	24	5	3	3
25	25	5	3	3
26	26	6	3	3
27	27	6	3	3
28	28	6	4	3
29	29	6	4	3
30	30	6	4	3
40		7	4	3
50		8	4	3
60		8	4	3
70		9	5	3
80		9	5	3
90		10	5	4
100		10	5	4

VISUAL DISPLAY RECOMMENDATIONS

Table DSPLY-B3: Types of Information and Recommended Displays[2]

Information Type	Preferred Display	Comments	Examples in Industry
Quantitative Reading	Digital readout or counter	Minimum reading time Minimum error potential	Numbers of units produced on a production machine
Qualitative Reading	Moving pointer or graph	Position easy to detect, trends apparent	Temperature changes in a work area
Check Reading	Moving Pointer	Deviation from normal easily detected	Pressure gauges on a utilities console
Adjustment	Moving pointer or digital readout	Direct relation between pointer movement and motion of control, accuracy	Calibration charts on test equipment
Status Reading	Lights	Color-coded, indication of status (e.g., "on")	Consoles in production lines
Operating Instructions	Annunciator Lights	Engraved with action required, blinking for warnings	Manufacturing lines in major production systems

Table DSPLY-B4: Comparison of Basic Symbolic Indicator Types[1]

For—	Counter is—	Moving pointer is—	Moving scale is—
Quantitative reading.	Good (requires minimum reading time with minimum reading error).	Fair	Fair.
Qualitative and check reading.	Poor (position changes not easily detected).	Good (location of pointer and change in position is easily detected).	Poor (difficult to judge direction and magnitude of pointer deviation).
Setting	Good (most accurate method of monitoring numerical settings, but relation between pointer motion and motion of setting knob is less direct).	Good (has simple and direct relation between pointer motion and motion of setting knob, and pointer-position change aids monitoring).	Fair (has somewhat ambiguous relation between pointer motion and motion of setting knob).
Tracking	Poor (not readily monitored, and has ambiguous relationship to manual-control motion).	Good (pointer position is readily monitored and controlled, provides simple relationship to manual-control motion, and provides some information about rate).	Fair (not readily monitored and has somewhat ambiguous relationship to manual-control motion).
Orientation	Poor	Good (generally moving pointer should represent vehicle, or moving component of system).	Good (generally moving scale should represent outside world, or other stable frame of reference).
General	Fair (most economical in use of space and illuminated area, scale length limited only by number of counter drums, but is difficult to illuminate properly).	Good (but requires greatest exposed and illuminated area on panel, and scale length is limited).	Fair (offers saving in panel space because only small section of scale need be exposed and illuminated, and long scale is possible).

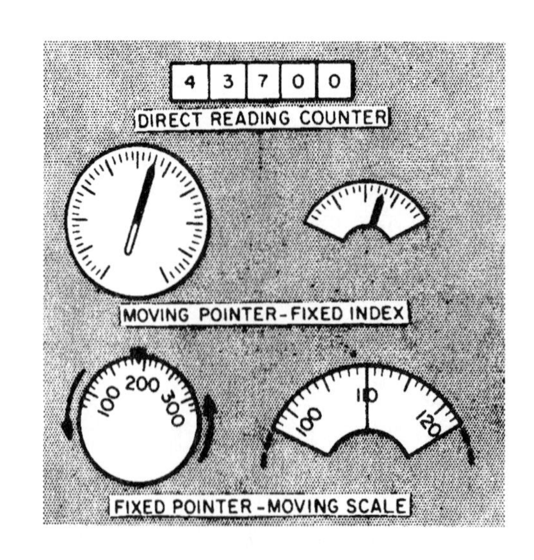

5 Display Tables

SCALE AND LEGEND DESIGN

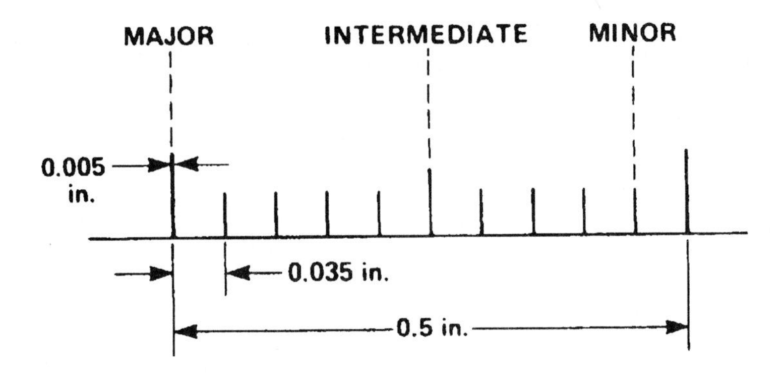

Recommended minimum scale dimensions for high illumination (28 in. viewing distance).

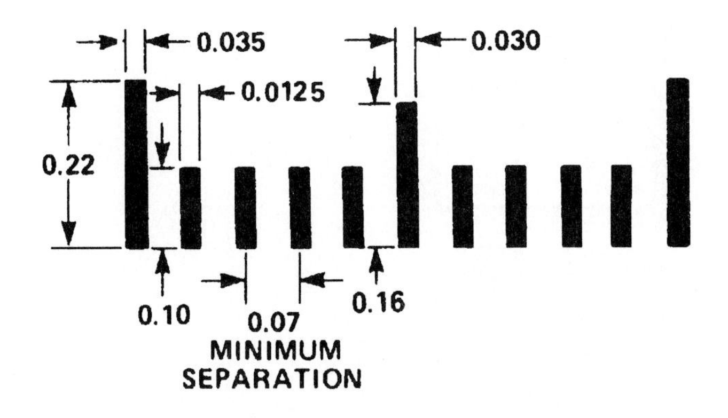

Recommended minimum scale dimensions for low illumination (28 in. viewing distance).

Figure DSPLY-B1: Recommended Scale Dimensions for High and Low Illumination [2]

Table DSPLY-B5: Recommended Scale Ranges and Intervals Values[1]

GRADUATION INTERVAL VALUE	RECOMMENDED SCALES	NUMBERED INTERVAL VALUE	GRADUATION MARKS		
			MAJOR	INTERMEDIATE	MINOR
0.1, 1, 10	2, 20, 200 — 3, 30, 300 — 4, 40, 400	1, 10, 100	X	X	X
	0 — 5, 50, 500 — 10, 100, 1000	5, 50, 500	X		X
	4, 40, 400 — 6, 60, 600 — 8, 80, 800 — 10, 100, 1000	2, 20, 200	X	X	
0.2, 2, 20	1, 10, 100 — 2, 20, 200 — 3, 30, 300	1, 10, 100	X		X
	4, 40, 400 — 6, 60, 600 — 8, 80, 800 — 10, 100, 1000	2, 20, 200	X	X	X
0.5, 5, 50	1, 10, 100 — 2, 20, 200 — 3, 30, 300 — 4, 40, 400 — 5, 50, 500	1, 10, 100	X	X	
	0 — 2, 20, 200 — 4, 40, 400	2, 20, 200	X	X	X
	0 — 5, 50, 500 — 10, 100, 1000	5, 50, 500	X	X	X

Table DSPLY-B6: Scale Numbering Recommendations [2]

Nature of markings	Height (in.)*	
	Low luminance†	High luminance‡
Critical markings, position variable (numerals on counters and settable or moving scales)	0.20–0.30	0.12–0.20
Critical markings, position fixed (numerals on fixed scales, control and switch markings, emergency instructions)	0.15–0.30	0.10–0.20
Noncritical markings (identification labels, routine instructions, any markings required only for familiarization)	0.05–0.20	0.05–0.20

* For 28-in. viewing distance. For other viewing distances, increase or decrease values proportionately.
† Between 0.03 and 1.0 ft.-L.
‡ Above 1.0 ft.-L.

Table DSPLY-B7: Examples of Good, Fair, and Poor Progressions for Scale Numbers [1]

Good					Fair					Poor				
0.1	0.2	0.3	0.4	0.5	0.2	0.4	0.6	0.8	1.0	0.25	0.5	0.75	1.0	
1	2	3	4	5	2	4	6	8	10	2.5	5	7.5	10	
10	20	30	40	50	20	40	60	80	100	25	50	75	100	
100	200	300	400	500	200	400	600	800	1000	250	500	750	1000	
0.5	1.0	1.5	2.0	2.5						0.4	0.8	1.2	1.6	1.8
5	10	15	20	25						4	8	12	16	18
50	100	150	200	250						40	80	120	160	180

5 Display Tables

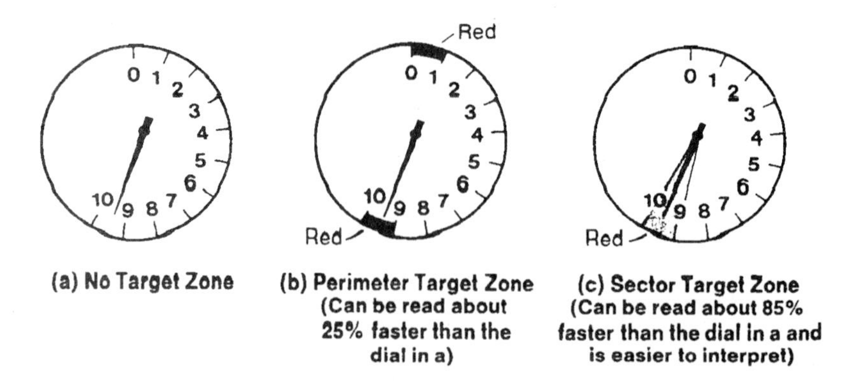

Figure DSPLY-B2: Target Zone Markings on Instruments[3]

Three dials are shown, two with markings to indicate abnormal functioning or conditions to which an operator has to respond. In part a the abnormal function zone is not marked, so an operator has to be trained to recognize when a potential problem exists. In part b the two zones of concern are marked by a red rectangle at the outer edge of the dial. The pointer can be seen against the light-colored dial; its tip points to the red zone when readings indicate abnormal function. In part c the entire dial is colored red within the zone of abnormal function, making it very obvious when the pointer falls in this sector. Response time is faster for the dial in part c than for those in part a and part b because interpretation of the meaning of the pointer when it is in the red sector is immediate and demands action.

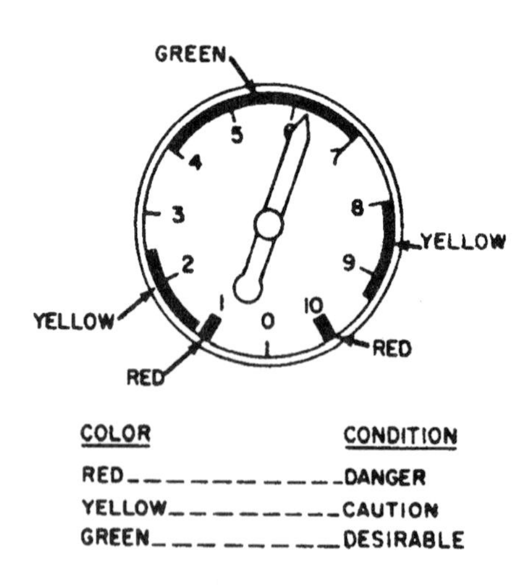

Figure DSPLY-B3: Color Coding of Instrument Range Markers[1]

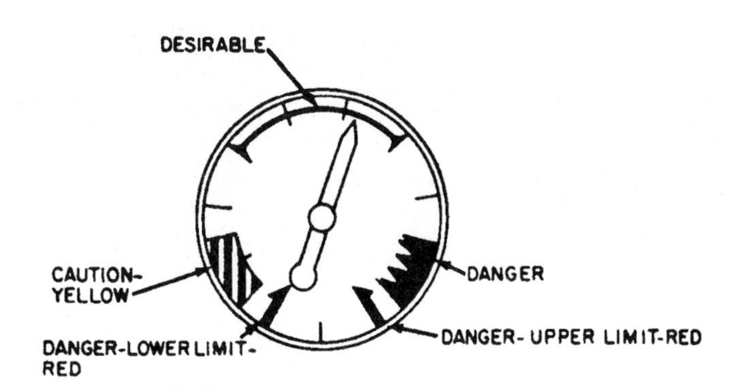

Figure DSPLY-B4: Shape Coding of Instrument Range Markers[1]

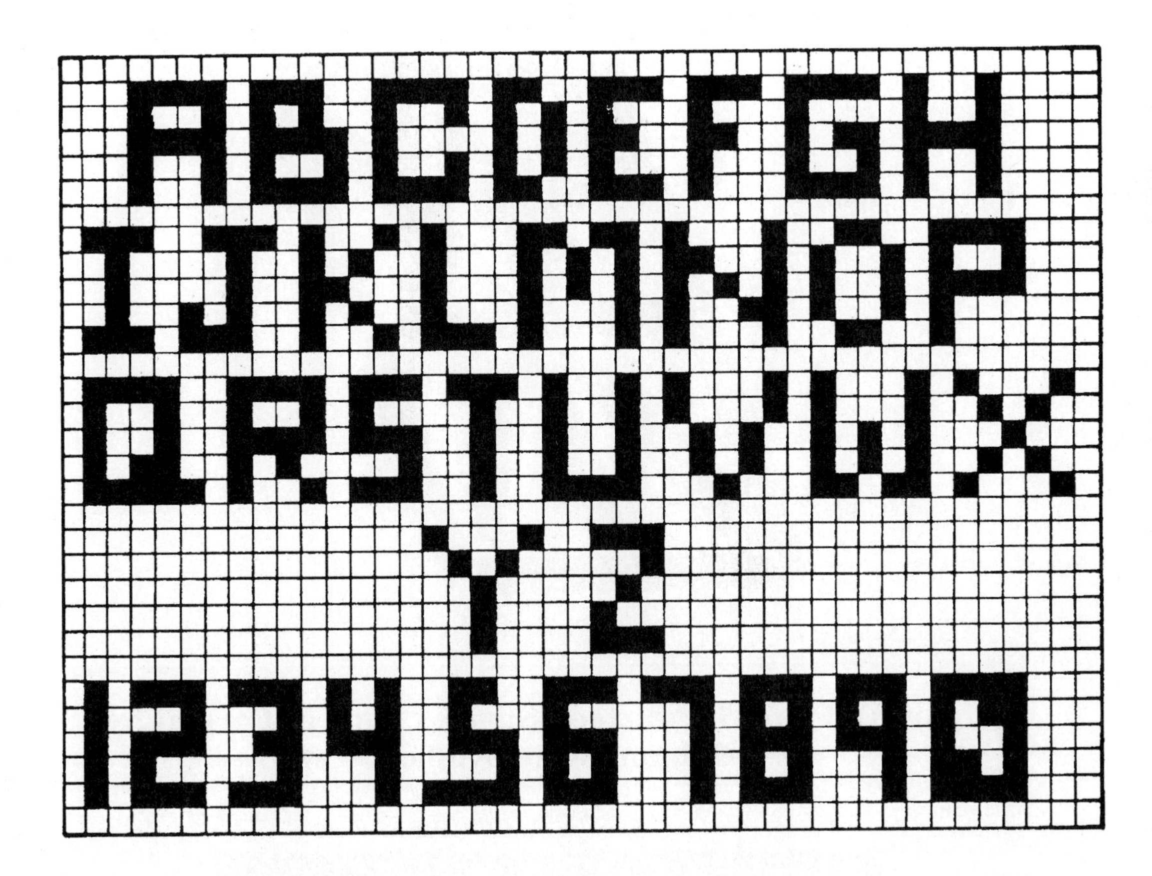

Figure DSPLY-B5: Recommended Letters and Numerals Composed of Cells (or PIXELS) Limited to a Resolution of Five Cells in Height[1]

Figure DSPLY-B6: Seven-Stroke Segment Display Format[1]

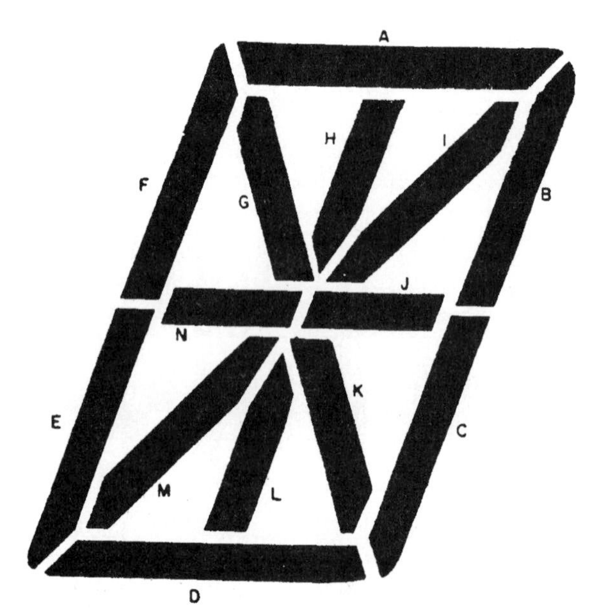

Figure DSPLY-B7: Fourteen-Stroke Segment Format for Alphanumeric Symbols[1]

Figure DSPLY-B8: Use of Certain Segments for Difficult Letters

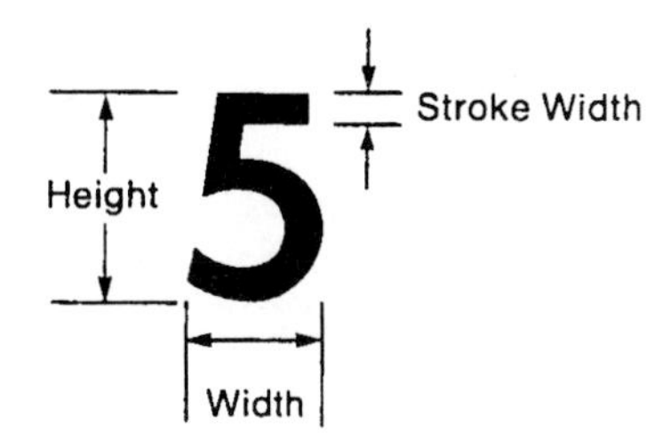

Figure DSPLY-B9: Definition of Font Characteristics[3]

Height is measured from the top to the bottom of the character and width across the widest part. Stroke width is the thickness of the line used to generate the letter or number.

When a printed label or message must be read quickly and easily, it is important to choose a plain and simple design of type font. There are some slightly more complex designs that can be easily read because they are familiar from wide use. USE OF ALL UPPER CASE LETTERS REDUCES LEGIBILITY. **LESS FAMILIAR DESIGNS MAY RESULT IN ERRORS ESPECIALLY IF THEY ARE READ IN HASTE.** FONTS DESIGNED PRIMARILY FOR AESTHETIC REASONS ARE VERY POOR CHOICES. OBVIOUSLY EXTREMES LIKE OLD ENGLISH SHOULD NEVER BE USED. **AVOID COMPLEX FONTS** **Keep It Simple.**

Figure DSPLY-B10: Examples of Common Types of Font[3]

Unadorned fonts in both uppercase and lowercase are easier and faster to read than more complex fonts.

Table DSPLY-B10: Letter or Number Height Versus Viewing Distance for Labels[3]

Viewing Distance	Critical Labels	Routine Labels
0.7 m (28 in.)	2 to 5 mm (0.1–0.2 in.)	1 to 4 mm (0.04–0.2 in)
0.9 m (3 ft.)	3 to 7 mm (0.1–0.3 in.)	2 to 5 mm (0.1–0.2 in.)
1.8 m (6 ft.)	7 to 13 mm (0.3–0.5 in.)	3 to 10 mm (0.1–0.4 in.)
6.1 m (20 ft.)	22 to 43 mm (0.9–1.7 in.)	11 to 33 mm (0.4–1.3 in.)

The distance from the operator to the display when it is read (Column 1) will determine how high the letters or numbers should be for legibility (Columns 2 and 3). Critical labels refer to key control or component identifiers and to position markers on such controls (Column 2). Routine labels refer to overall instrument identifiers or any markings required only for initial familiarization (Column 3).

Table DSPLY-B11: Legibility of Color Combinations in White Light[3]

Legibility	Color Combination
Very Good	Black characters on a white background Black on yellow
Good	Yellow on black White on black Dark blue on white Green on white
Fair	Red on white Red on yellow
Poor	Green on red Red on green Orange on black Orange on white
Very poor	Black on blue Yellow on white

The ease with which written information can be distinguished from its background is indicated in the first column for different color combinations (Column 2).

Table DSPLY-B12: Recommended Lighting for Indicators, Panels, or Charts[1]

Condition of use	Recommendations		
	Lighting technique	Luminance of markings (ft.-l)	Brightness adjustment
Indicator reading, dark adaptation necessary.	Red flood, integral or both, with operator choice.	0.02–0.1	Continuous throughout range.
Indicator reading, dark adaptation not necessary but desirable.	Red or low-color-temperature white flood, integral, or both, with operator choice.	0.02–1.0	Continuous throughout range.
Indicator reading, dark adaptation not necessary.	White flood	1–20	Fixed or continuous.
Reading of legends on control consoles, dark adaptation necessary.	Red integral lighting red flood, or both, with operator choice.	0.02–0.1	Continuous throughout range.
Reading of legends on control consoles, dark adaptation not necessary.	White flood	1–20	Fixed or continuous.
Possible exposure to bright flashes.	White flood	10–20	Fixed.
Very high altitude, daylight restricted by cockpit design.	White flood	10–20	Fixed.
Chart reading, dark adaptation necessary.	Red or white flood with operator choice.	0.1–1.0 (on white portions of chart).	Continuous throughout range.
Chart reading, dark adaptation not necessary.	White flood	5–20	Fixed or continuous.

VDT and Projected Displays

Table DSPLY-B13: Recommended VDT Characteristics[4]

Hardware Characteristics

1. Resolution is best with 9 to 10 scan lines per millimeter (228 to 254 lines per inch). This assures good image quality.
2. Choose a character color from the middle of the chromaticity diagram. Orange and green characters are recommended.
3. Luminance of the screen should be greater than 25 mL (millilambert).
4. Character regeneration rate of 60 cycles per second or more is desired.
5. The visual angle subtended at the eye of each character must be between 15 and 20 minutes of arc. At a viewing distance of 71 cm (28 in.), the minimum character height is 4.6 mm (0.18 in.). The visual angle in terms of minutes of arc can be calculatted as follows:

$$\text{visual angle (min)} = \frac{(57.3)(60)L}{D}$$

   Where L is the size of object measured perpendicular to the NLS (normal line of sight) and D is the distance from the eye to the object.
6. Surfaces adjacent to the scope should have a matte finish.
7. Scopes may be hooded for minimum indirect glare.
8. A minimum contrast of 88% with the background is recommended.
9. Use uppercase letters for headings and titles. Lowercase characters may be used for other text.
10. Scan lines should not be visible to the operator.
11. Optimum dot size seems to be between 0.8 and 1.2 mm (0.0315 to 0.047 in.). Dots should be large enough and spaced so that they fuse together.
12. A dot matrix size of 7 × 9 or larger is recommended.
13. The width/height ratio of characters should be between 0.7 and 0.8.
14. Characters should be equally focused and well defined at every point on the screen.
15. The cursor should be easily visible and should not affect the legibility of characters.
16. The cursor should blink at a rate of about 2 to 3 Hz.
17. Cursor control should be easily accomplished by either the right or left hand.
18. If an alarm is to be used to alert the operator, both auditory and visual alarms should be used. The auditory signal must be more than 200 Hz in frequency and less than 1000 Hz. A volume control is recommended. However, it should not shut off the alarm entirely.

Other

1. Orient VDTs to minimize indirect glare.
2. The CRT surface and the NLS should not make an angle greater than 30 degrees.

Table DSPLY-B14: Height of Numerals and Letters in Projected Displays[1]

	Minimum acceptable	Preferred minimum
Visual angle—minutes	10	15
Letter ht./max. observer distance	1/344	1/230
Letter ht./width of source image*	1/43	1/30

* Assumes maximum observer distance of 8× image width.

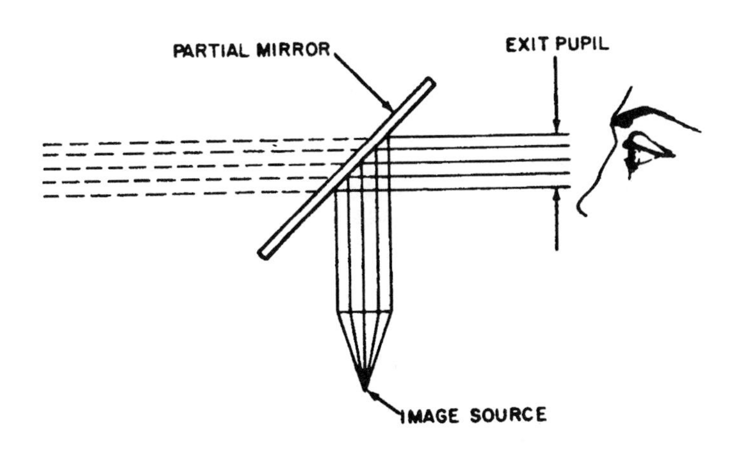

Figure DSPLY-B11: Basic Optics of a Collimated Display[1]

Table DSPLY-B15: Critical Angles for Arrangement of Observer (O), Screen Image, and Projector[1]

A- ANGLE OFF PROJECTION AXIS
B- MOST CRITICAL ANGLE FOR MAT AND LENTICULAR SCREENS - FOR O_1
C- MOST CRITICAL ANGLE FOR BEADED SCREENS - FOR O_1
D- MOST CRITICAL ANGLE FOR REAR PROJECTION SCREENS - FOR O_2

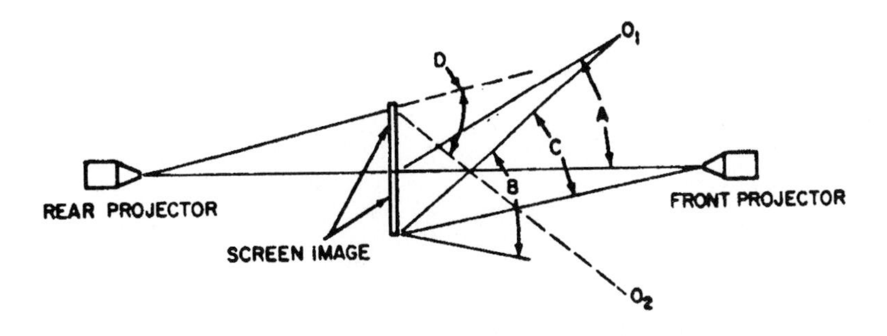

Table DSPLY-B16: General Recommendations for Group Viewing for Slides and Motion Pictures[1]

Factors	Optimum	Preferred limits	Acceptable limits
Ratio of $\frac{\text{viewing distance}}{\text{image width}}$	4	3–6	2–8
Angle off center line—degrees	0	20	30
* Image luminance—ft.-Lamberts (no film in projector).	10	8–14	5–20
Luminance variation across screen—ratio of maximum to minimum luminance.	1	1.5	3.0
Luminance variation as a function of seat position—ratio of maximum to minimum luminance.	1	2	4.0
Ratio of $\frac{\text{ambient light}}{\text{highest part of image}}$	0	0.002–0.01	0.2†

* For still projections higher values may be used.
† For line drawings, tables, not involving gray scale or color.

REFERENCES

1. Van Cott, H.P., and Kinkade, R.G., 1972. *Human Engineering Guide to Equipment Design. (Revised)* Washington, DC.: US Government Printing Office. Library of Congress Number: 72-600054.

2. Bailey, R.W. 1989. *Human Performance Engineering. (2nd Ed)* Englewood Cliffs, NJ: Prentice-Hall. Reprinted with Permission.

3. Reprinted with permission from *Ergonomic Design for People at Work*, © Eastman Kodak Company, 1983, published by Van Nostrand Reinhold. Courtesy of Eastman Kodak Company.

4. Pulat, B.M. 1992. *Fundamentals of Industrial Ergonomics.* Englewood Cliffs, NJ: Prentice-Hall. Reprinted with Permission.

CHAPTER 5: DISPLAY TABLES

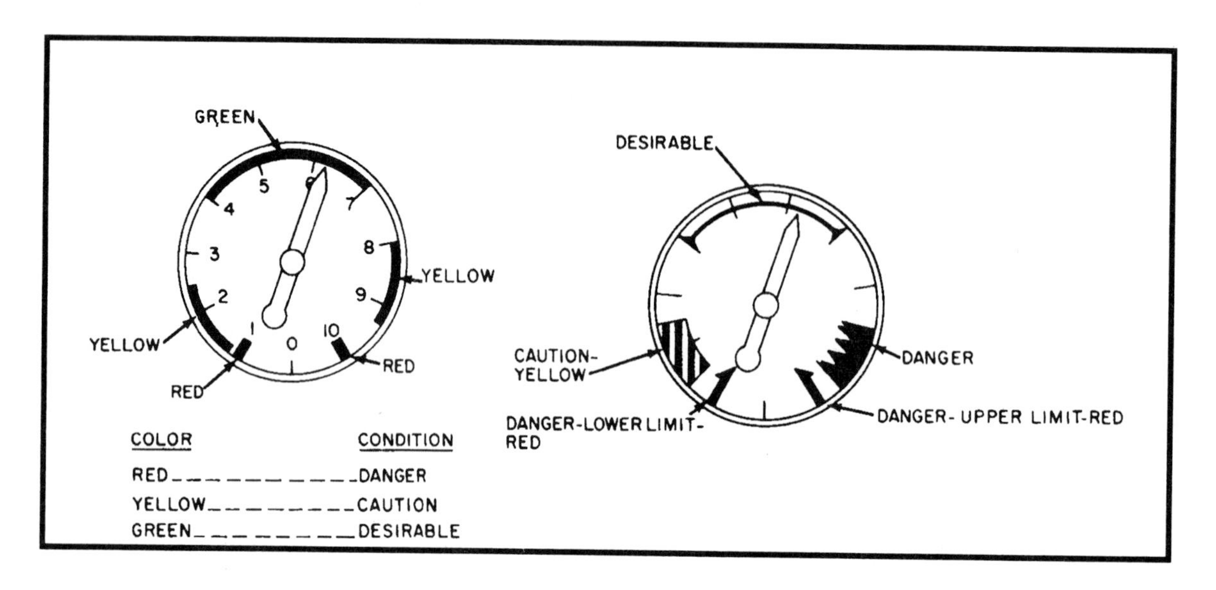

Section C: Auditory Displays

This section presents guidelines to aid in deciding when to use auditory displays and how to best design them. The information in this section is useful for the ergonomist or human factors engineer who must design auditory display systems.

AUDITORY DISPLAYS

Table DSPLY-C1: Types of Alarms, Their Characteristics, and Special Features[1]

Alarm	Intensity	Frequency	Attention-getting ability	Noise-penetration ability	Special features
Diaphone (foghorn).	Very high	Very low	Good	Poor in low-frequency noise. Good in high-frequency noise.	
Horn	High	Low to high.	Good	Good	Can be designed to beam sound directionally. Can be rotated to get wide coverage.
Whistle	High	Low to high.	Good if intermittent.	Good if frequency is properly chosen.	Can be made directional by reflectors.
Siren	High	Low to high.	Very good if pitch rises and falls.	Very good with rising and falling frequency.	Can be coupled to horn for directional transmission.
Bell	Medium	Medium to high.	Good	Good in low-frequency noise.	Can be provided with manual shutoff to insure alarm until action is taken.
Buzzer	Low to medium.	Low to medium.	Good	Fair if spectrum is suited to background noise.	Can be provided with manual stutoff to insure alarm until action is taken.
Chimes and gong.	Low to medium.	Low to medium.	Fair	Fair if spectrum is suited to background noise.	
Oscillator	Low to high.	Medium to high.	Good if intermittent.	Good if frequency is properly chosen.	Can be presented over intercom system.

5 Display Tables

Table DSPLY-C2: Summary of Design Recommendations for Auditory Displays I [2]

1. Increase signal-to-noise ratio for detectability. Signal levels of 8 to 12 dB above the masked threshold will provide full detectability.
2. For maximum detectability, the signal duration should be at least 300 ms. If the signal has to be shorter than this, the intensity should be increased.
3. Rapid response to a signal can be achieved by levels 15 to 18 dB above the masked threshold.
4. To minimize operator annoyance, the signal level must be less than 30 dB above the masked threshold.
5. Interference is at its maximum when the signal tone is near the frequency of the interfering tone.
6. Interference spreads to additional frequencies as the intensity of interfering tone increases.
7. When it is below the signal frequency, the effect of the interfering tone is maximum, as opposed to above.
8. For absolute discriminations, keep the signal frequency at 1000 to 4000 Hz.
9. For auditory coding, use five intensity levels or less.
10. Four to seven levels should not be exceeded for frequency coding.
11. Four prominent frequency components are recommended for a signal to minimize masking effects and maximize the number of different distinct signal codes that can be generated.
12. At least four of the first 10 harmonics should be prominent. The prominent frequency components for signals should be in the range 1000 to 4000 Hz.
13. To acknowledge warning, provide the signal with a manual shutoff.

Table DSPLY-C3: Summary of Design Recommendations for Auditory Displays II [1]

Conditions	Design recommendations
1. If distance to listener is great—	1. Use high intensities and avoid high frequencies.
2. If sound must bend around obstacles and pass through partitions—	2. Use low frequencies.
3. If background noise is present—	3. Select alarm frequency in region where noise masking is minimal.
4. To demand attention—	4. Modulate signal to give intermittent "beeps" or modulate frequency to make pitch rise and fall at rate of about 1–3 cps.
5. To acknowledge warning—	5. Provide signal with manual shutoff so that it sounds continuously until action is taken.

REFERENCES

1. Van Cott, H.P., and Kinkade, R.G., 1972. *Human Engineering Guide to Equipment Design. (Revised)* Washington, DC.: US Government Printing Office. Library of Congress Number: 72-600054.

2. Pulat, B.M. 1992. *Fundamentals of Industrial Ergonomics.* Englewood Cliffs, NJ: Prentice-Hall. Reprinted with Permission.

CHAPTER 5: DISPLAY TABLES

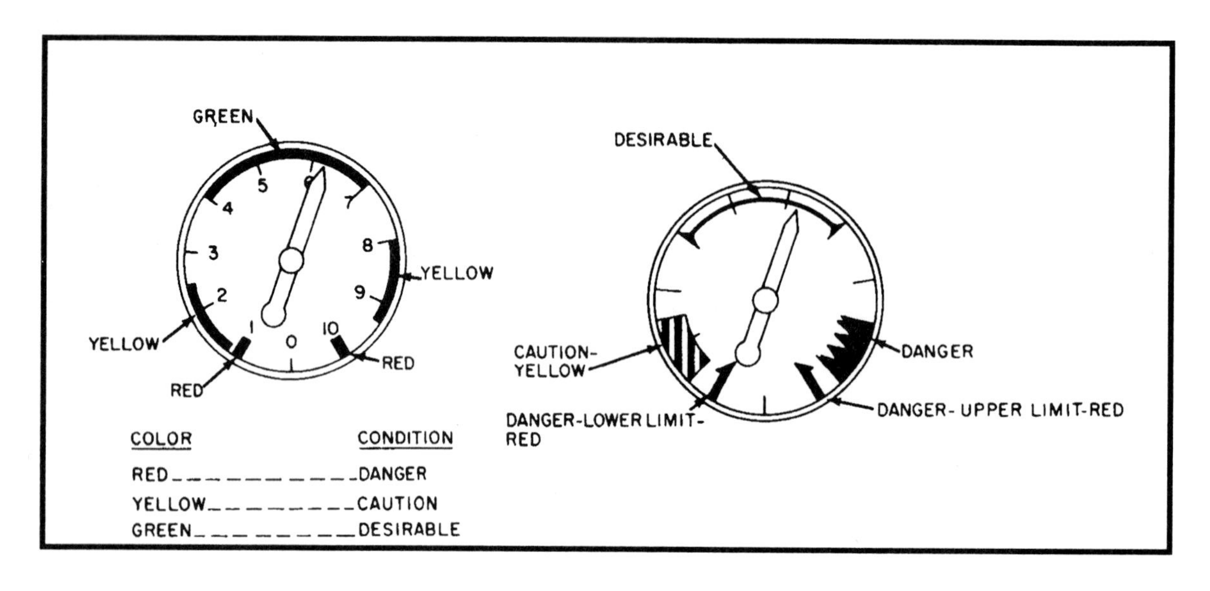

Section D: Speech Displays

This section presents guidelines to aid in deciding when how to use speech. The information in this section is useful for the ergonomist or human factors engineer who must design speech systems.

SPEECH DISPLAYS

Table DSPLY-D1: Preferred Speech Interference Levels (PSIL) as a Function of Distance and Ease of Communication[4]

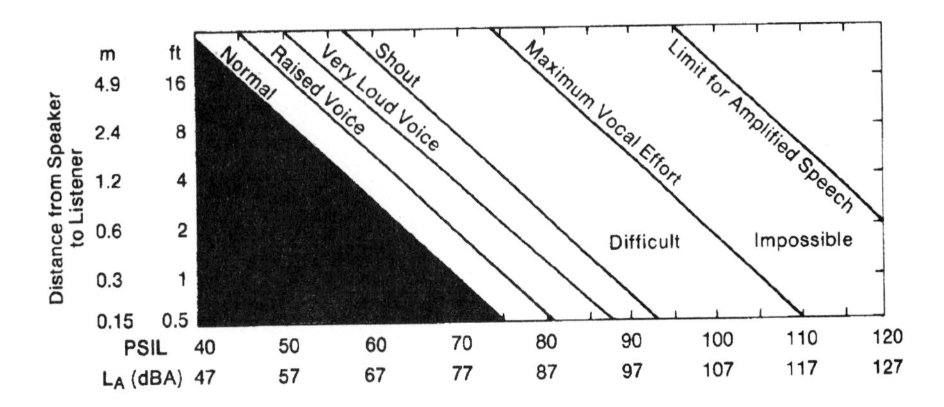

The ambient sound pressure level (L_A, in decibels, or PSIL, the preferred speech interference level) is given on the horizontal axis. The distance between the speaker and listener (on the vertical axis) will determine to how high a level the speaker must raise his or her voice in different noise conditions. The curve relating ambient noise to distance for communication for normal speaking voice levels is shown on the left. The second, third, and fourth curves illustrate the increasing voice levels needed to be heard in the ambient noise. The fifth curve (second from the right side) indicates the maximum vocal level for conversations in a noisy environment. The curve on the right shows the upper limits of ambient noise and distance for understanding amplified speech.

Table DSPLY-D2: Preferred Speech Interference Levels (PSIL) for Communicating by Telephone [4]

PSIL	Telephone Use
60 dB or less	Satisfactory
60 to 75 dB	Difficult
Above 75 dB	Impossible

Table DSPLY-D3: Relation Between Word Length (# of Letters) and the Signal-to-Noise Ratio Required for 50% Intelligibility [1]

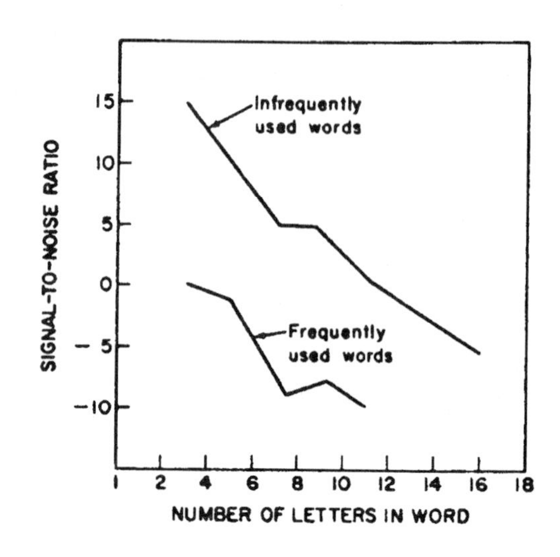

Infrequently used words require higher signal-to-noise ratios than frequently used words, when word length is held constant.

Table DSPLY-D4: Twenty Frequency Bands of Equal Contribution to Speech Intelligibilty (Male Voices) [1]

Band No.	Limits (Hz)	Mid-frequency (Hz)	Band No.	Limits (Hz)	Mid-frequency (Hz)
1	200–330	270	11	1660–1830	1740
2	330–430	380	12	1830–2020	1920
3	430–560	490	13	2020–2240	2130
4	560–700	630	14	2240–2500	2370
5	700–840	770	15	2500–2820	2660
6	840–1000	920	16	2820–3200	3000
7	1000–1150	1070	17	3200–3650	3400
8	1150–1310	1230	18	3650–4250	3950
9	1310–1480	1400	19	4250–5050	4650
10	1480–1660	1570	20	5050–6100	5600

Table DSPLY-D5: Preferred Noise Criterion (PNC) Curves[4]

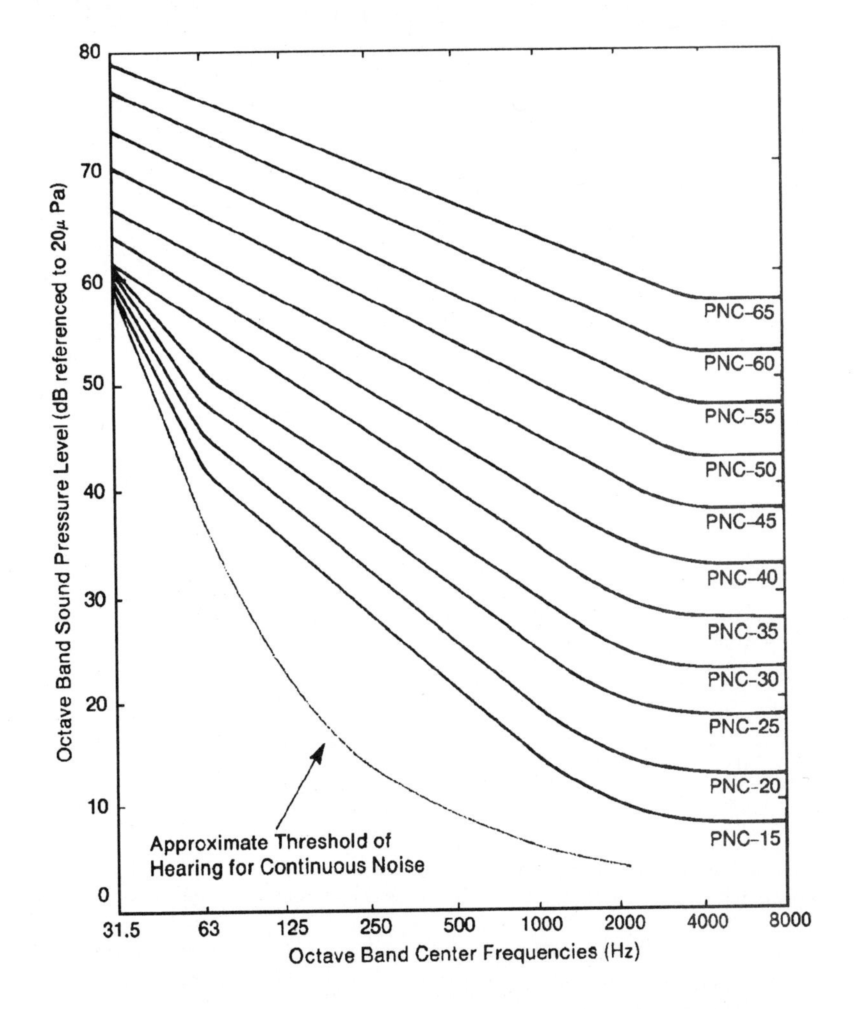

The relationships between sound-level intensity (octave band sound pressure, in decibels, dB, referenced to 20 micropascals, μPa, on the vertical axis) and frequency (represented by the nine center frequencies on the horizontal axis) are displayed for different conditions of hearing (PNC curves). The curves range from the lower threshold of hearing for continuous noise to a PNC–65 curve, where there is significant interference with communication. The higher the frequency of the noise, the lower its intensity must be to bring it to the appropriate PNC curve for hearing or communication.

Table DSPLY-D6: Recommended PNC Curves and Sound Pressure Levels for Several Categories of Activities [4]

Acoustical Requirements	PNC Curve*	Approximate† L_A (dBA)
Listening to faint musical sounds or distant microphone pickup used	10 to 20	21 to 30
Excellent listening conditions	Not to Exceed 20	Not to Exceed 30
Close microphone pickup only	Not to Exceed 25	Not to Exceed 34
Good listening conditions	Not to Exceed 35	Not to Exceed 42
Sleeping, resting, and relaxing	25 to 40	34 to 47
Conversing or listening to radio and TV	30 to 40	38 to 47
Moderately good listening conditions	35 to 45	42 to 52
Fair listening conditions	40 to 50	47 to 56
Moderately fair listening conditions	45 to 55	52 to 61
Just acceptable speech and telephone communication	50 to 60	56 to 66
Speech not required but no risk of hearing damage	60 to 75	66 to 80

* PNC curves are used in many installations for establishing noise spectra.
† These levels (L_A) are to be used only for approximate estimates, since the overall sound pressure level does not give an indication of the spectrum.

The PNC, or preferred noise criterion, curves (column 2) and approximate sound pressure levels (L_A in dBA, column 3) for several hearing conditions (column 1) are given. Voice sound frequencies are used to determine the approximate sound pressure levels. At higher PNC curves it becomes more difficult to hear speech or music. All of these curves represent noise exposures lower than those that may cause hearing damage.

Table DSPLY-D7: Correction for Level of Speech Presented by a Loudspeaker in a Reverberant or Semireverberant Room [1]

OA SPL of speech (dB)	Amount to be subtracted from speech level (dB)
85	0
90	2
95	4
100	7
105	11
110	15
115	19
120	23
125	27
130	30

Table DSPLY-D8: Relation of Sound-Pressure Level to Frequency[1]

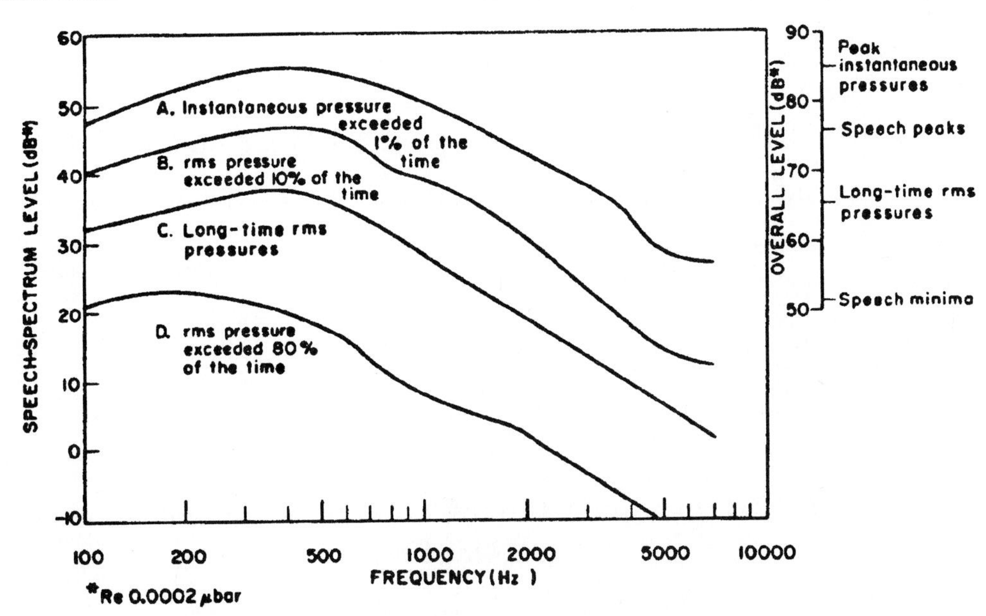

RMS is the root mean square pressure. If you take the time interval over which squared sound pressures are averages and make it directly correspond to the interval during which particular vowels and consonants are spoken, you can relate these measurements to the individual sounds spoken. See Table DSPLY-D9.

Table DSPLY-D9: Typical R.M.S. Pressure Levels of Fundamental Speech Sounds[1]

Key word	Sound*	Pressure level (dB)	Key word	Sound*	Pressure level (dB)
talk	o'	28.2	chat	ch	16.2
top	a	27.8	me	m	15.5
ton	o	27.0	jot	i	13.6
tap	a'	26.9	azure	zh	13.0
tone	o	26.7	zip	z	12.0
took	u	26.6	sit	s	12.0
tape	'a	25.7	tap	t	11.7
ten	e	25.4	get	g	11.7
tool	u	25.0	kit	k	11.1
tip	i	24.1	vat	v	10.8
team	e	23.4	that	th	10.4
err	r	23.2	bat	b	8.0
let	l	20.0	dot	d	8.0
shot	sh	19.0	pat	p	7.7
ring	ng	18.6	for	f	7.0
me	m	17.2	thin	th	0

* Spoken by an average talker at a normal level of effort

5 Display Tables

Table DSPLY-D10: SPLs of Speech 1 M From the Talker[1]

Measure of sound pressure	Whisper (dB)	Normal level (dB) Minimum	Average	Maximum	Shout (dB)
Peak instantaneous pressures	70	79	89	99	110
Speech peaks	58	67	79	87	98
Long-time rms pressures	46	55	65	75	86
Speech minima	30	39	49	59	70

Table DSPLY-D11: Distribution of Talker Levels for Persons Using the Telephone[1]

Percent of talkers	Level range (dB*)
7	Below 54
9	54–57
14	57–60
18	60–63
22	63–66
17	66–69
9	69–72
4	72–75
0	Above 75

* Above sound pressure of 0.0002 μbar at a point 1 m from the talker's lips

Table DSPLY-D11: Speech-Intensity Levels Around the Head of a Speaker[1]

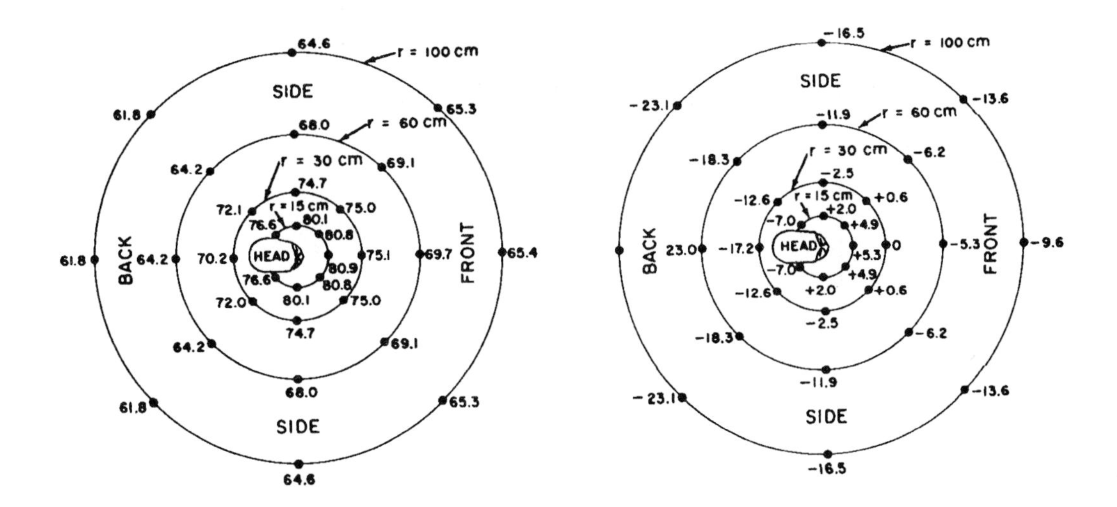

Numbers in left-hand diagram are for whole speech and are in decibels relative to 0.0002μbar; numbers in right-hand diagram are for band of speech of 2800 to 4000 Hz and are in decibels relative to level 30 cm. in front of lips.

Table DSPLY-D12: Effective Speech Level in Decibels as a Function of the Actual Speech Level Used by a Talker [1]

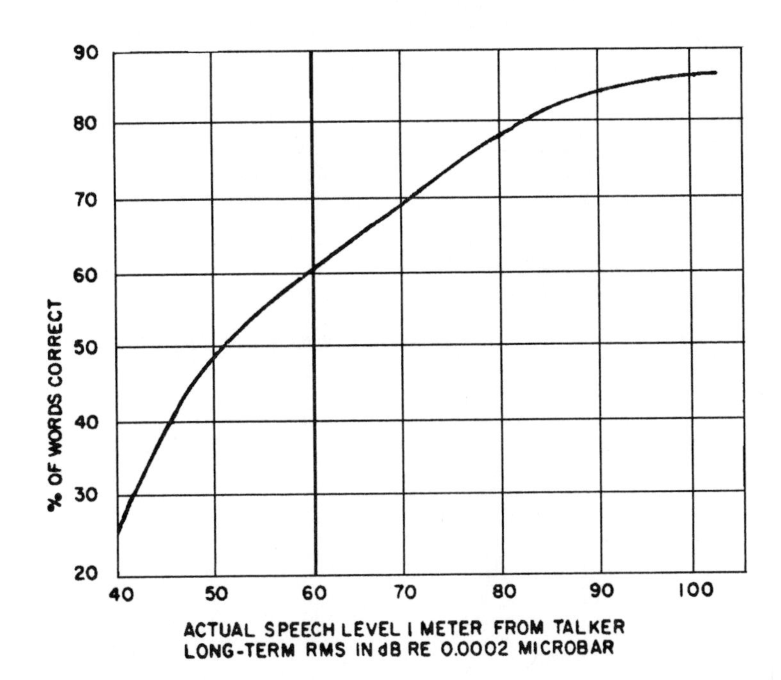

Table DSPLY-D13: Speech Level as a Function of Ambient Noise Level Surrounding a Talker [1]

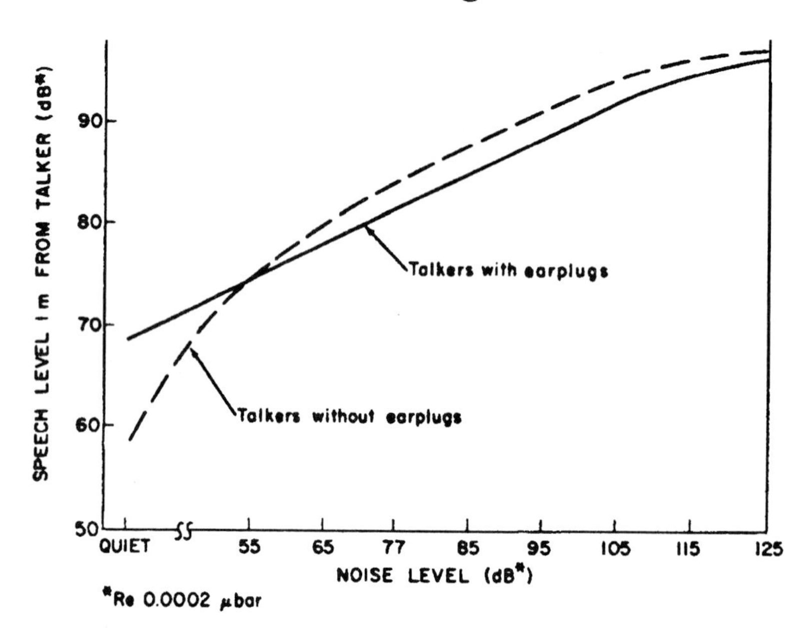

Table DSPLY-D14: Intelligibility of Words When Perceived With and Without Visual Cues From Observing the Talker [1]

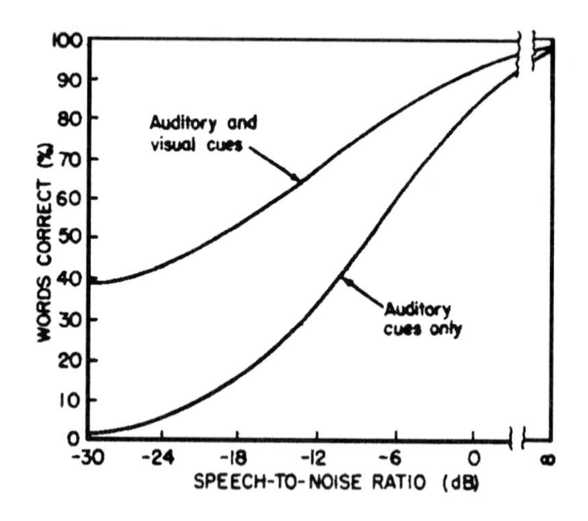

Table DSPLY-D15: Effect of Earplugs on Intelligibility of Speech Components in Quiet Environments[1]

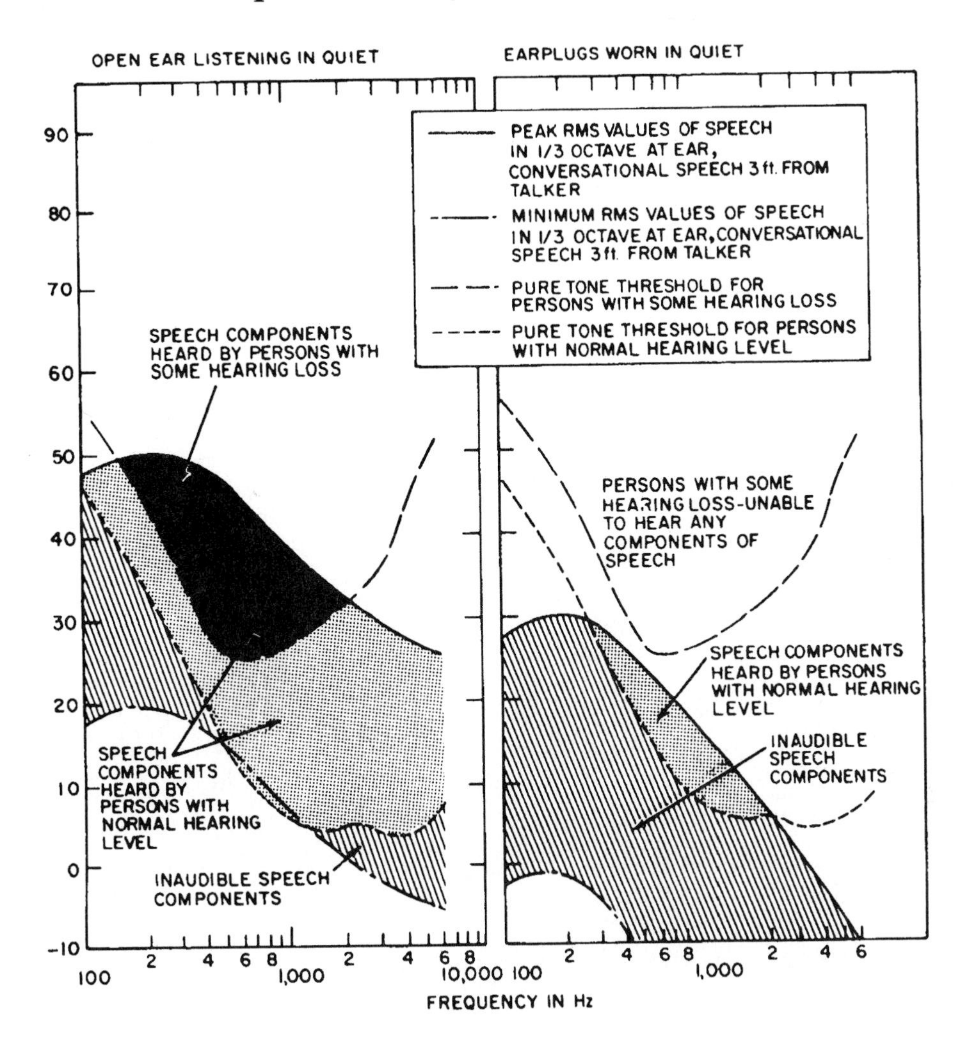

Table DSPLY-D16: Effect of Earplugs on Intelligibility of Speech Components in Moderate Noise [1]

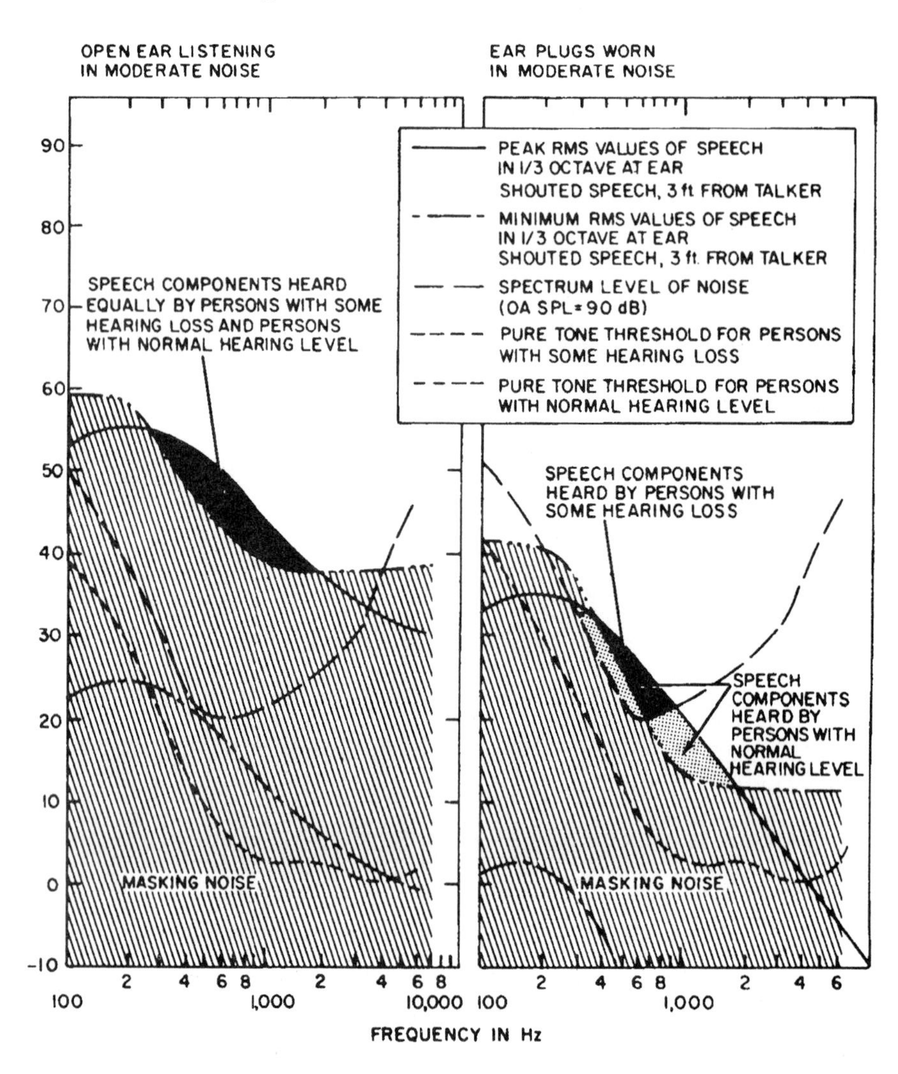

Table DSPLY-D17: Effect of Earplugs on Intelligibility of Speech Components in Intense Noise [1]

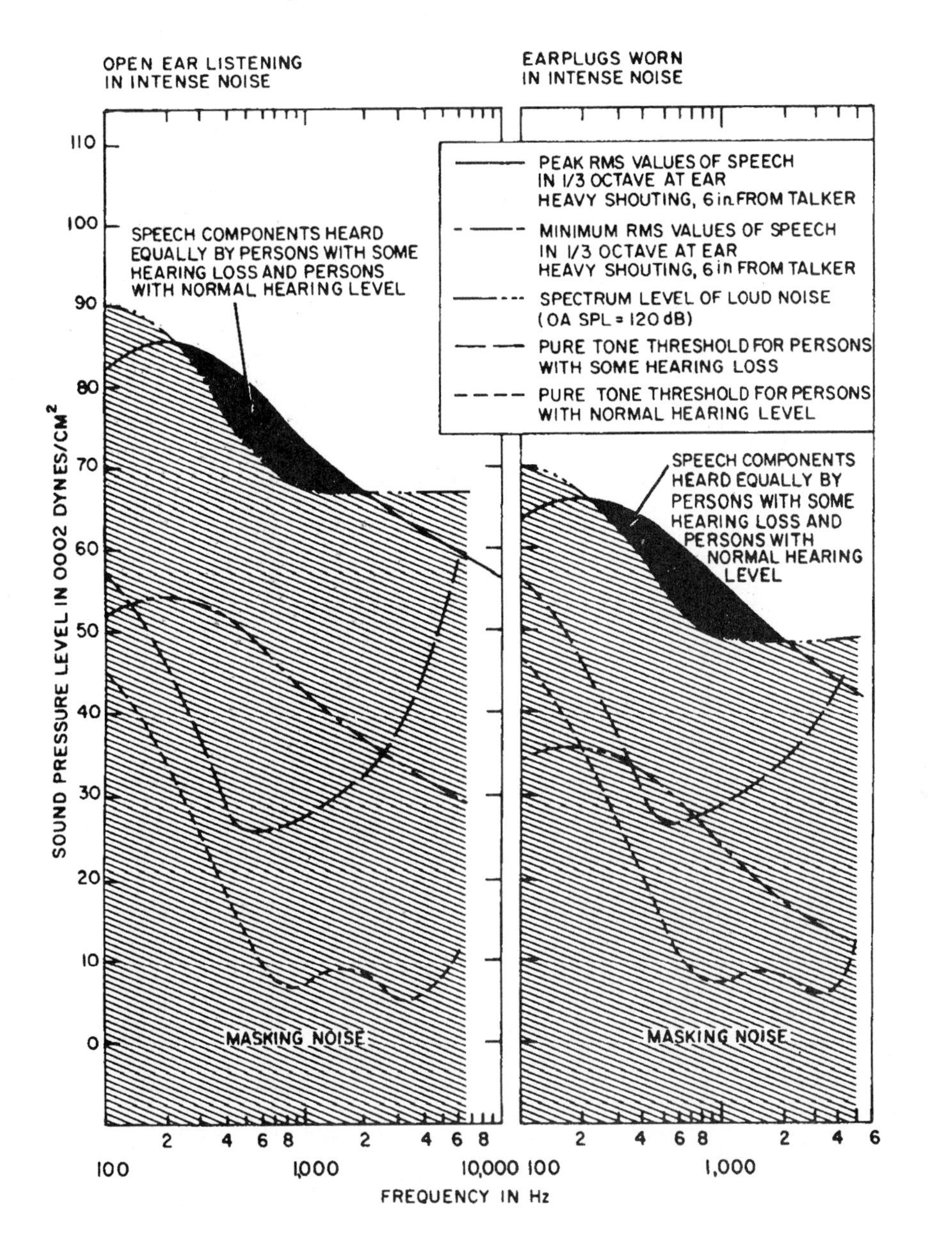

5 Display Tables

REFERENCES

1. Van Cott, H.P., and Kinkade, R.G., 1972. *Human Engineering Guide to Equipment Design. (Revised)* Washington, DC.: US Government Printing Office. Library of Congress Number: 72-600054.

2. Pulat, B.M. 1992. *Fundamentals of Industrial Ergonomics.* Englewood Cliffs, NJ: Prentice-Hall. Reprinted with Permission.

3. Bailey, R.W. 1989. *Human Performance Engineering. (2nd Ed)* Englewood Cliffs, NJ: Prentice-Hall. Reprinted with Permission.

4. Reprinted with permission from *Ergonomic Design for People at Work,* © Eastman Kodak Company, 1983, published by Van Nostrand Reinhold. Courtesy of Eastman Kodak Company.

CHAPTER 6: WORK PHYSIOLOGY TABLES

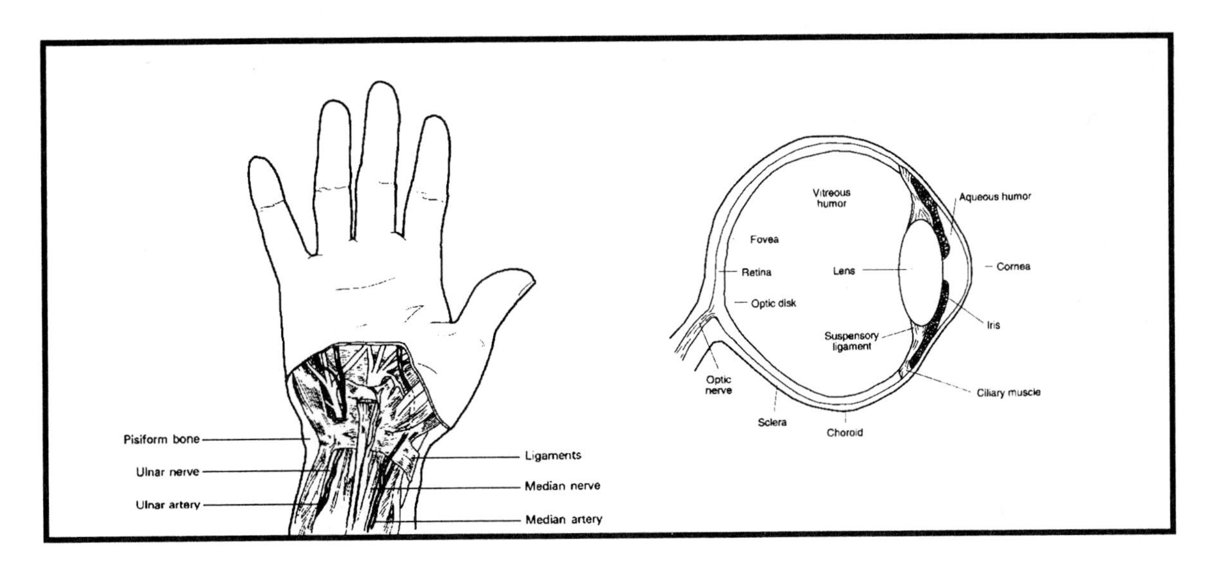

Section A: Anatomy and Physical Capacity

This section presents information on anatomy and physiology as it relates to performing physical work. The information in this section is useful for the ergonomist or human factors engineer who must design workplaces where the job requires physical effort and/or repeated activities.

ANATOMY OF THE HAND

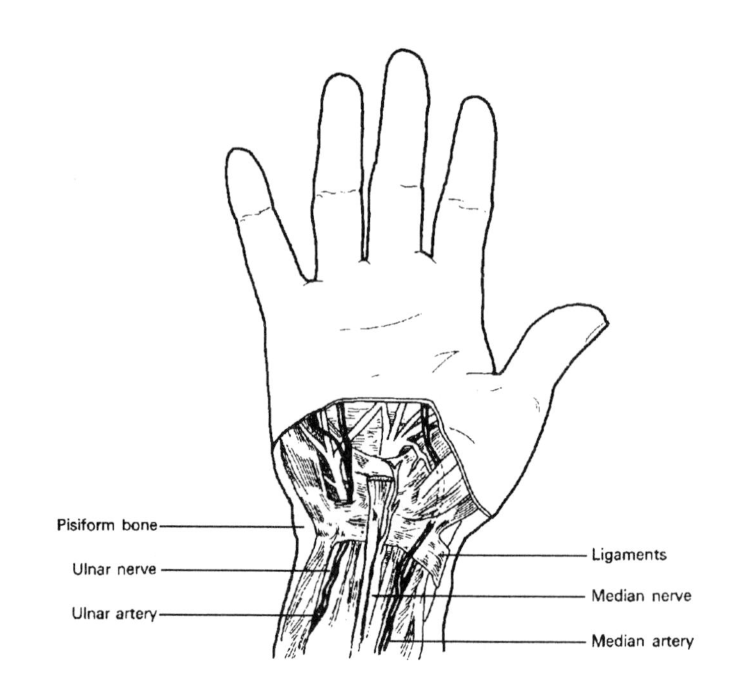

Figure PHYSLGY-A1: Anatomy of the Hand[1]

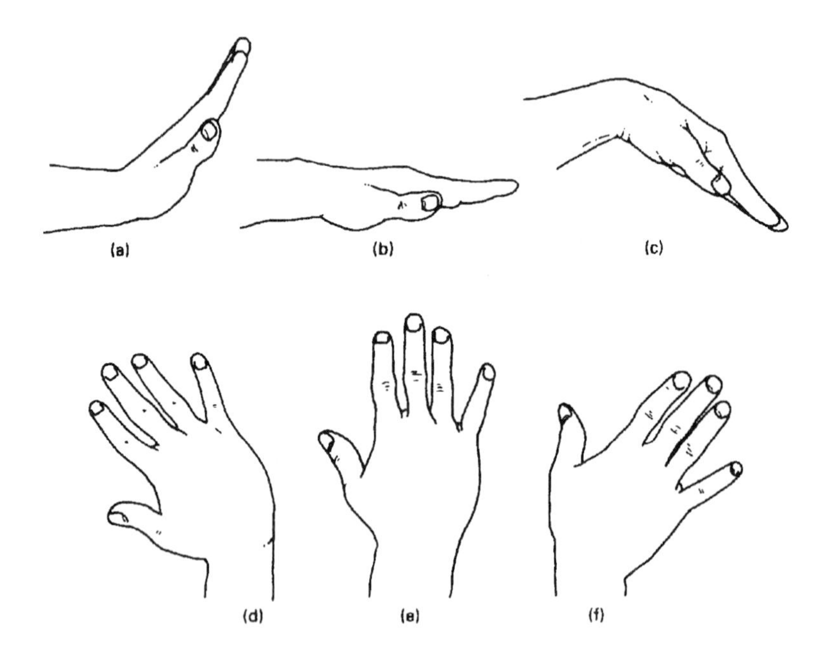

PHYSLGY-A2: Figure Various Postures of the Hand: (a) Dorsiflxior, (b) Neutral, (c) Palmer Flexion, (d) Radial Deviation, (e) Neutral, (f) Ulnar Deviation[1]

PHYSICAL CAPACITY

Table PHYSLGY-A1: Human Activity Continuum and the Primary Measure of Physical Strain[1]

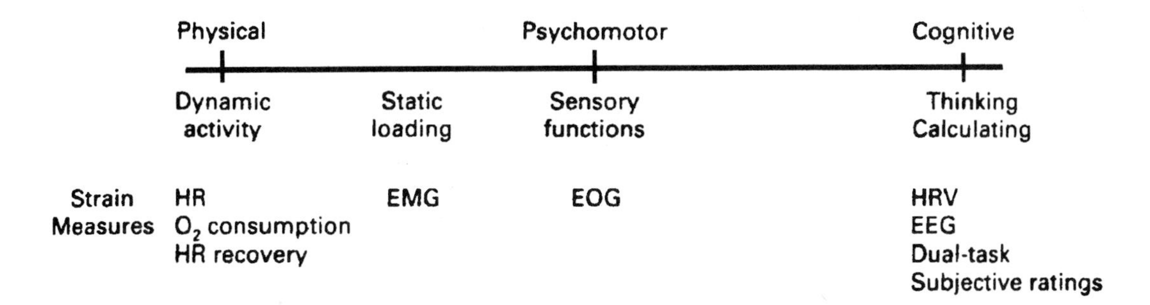

Table PHYSLGY-A2: Rest-Time Requirements[1]

$R_T = 0$	for $K < S$
$R_T = \dfrac{\left(\dfrac{K}{S} - 1\right) \times 100 + \dfrac{T(K - S)}{K - BM}}{2}$	for $S \leq K < 2S$
$R_T = \dfrac{T(K - S)}{K - BM} \times 1.11$	for $K \geq 2S$
$BM_F = 1.4$	$BM_M = 1.7$

Where R_T = Allowed Rest Time (min), K = Energy Cost of the work (kcal/min), S= Accepted Standard (4 kcal/min for females and 5 kcal/min for males), T = Total Expected Time on task, and BM = Basal Metabolism (kcal/min)

Table PHYSLGY-A3: Age Multiplier for Rest Periods[1]

Age	Multiplier
20–30	1.0
40	1.04
50	1.1
60	1.2
65	1.25

Rest period allowances determined through the equations presented in the above table should be multiplied by the appropriate age multiplier.

Table PHYSLGY-A4: Work Classification by Energy Cost[1]

Work grade	Energy expenditure (kcal/min)	O_2 consumption (liters/min)
Severe	12.5 <	2.5 <
Very heavy	10.0–12.5	2–2.5
Heavy	7.5–10.0	1.5–2
Moderate	5.0–7.5	1.0–1.5
Light	2.5–5.0	0.5–1.0

Table PHYSLGY-A5: Energy Requirements for Industrial Jobs[1]

Activity	Average energy requirement (kcal/min)
Screwdriving (horizontal)	0.5
Light engineering work	2.0
Light assembly (seated)	2.2
Printing	2.3
Laboratory work	2.5
Bench soldering	2.7
Walking on level (3 km/h)	2.8
Cleaning windows	3.1
Shoe repair and manufacturing	3.8
Walking on level (4.5 km/h)	4.0
Bricklaying	4.0
Tractor plowing	4.2
Agricultural work	4.8
Milking by hand	4.9
Heavy washing (housework)	5.0
Weeding	5.1
Walking on level (6 km/h)	5.2
Pushing wheelbarrow	6.0
Binding	6.2
Handsawing wood	6.8
Chopping wood	8.0
Mowing	8.3
Shoveling (7-kg weight)	8.5
in front of heat	10.2
Tending the heat furnace	11.5

Table PHYSLGY-A6: Metabolic Demands of Industrial Tasks [1]

Light	Moderate	Heavy	Very Heavy	Extremely Heavy
70–140 W (60–120 kcal/hr)	>140–280 W (>120–240 kcal/hr)	>280–350 W (>240–300 kcal/hr)	>350–420 W (>300–360 kcal/hr)	>420 W (>360 kcal/hr)
small-parts assembly typing keypunching inspecting operating a milling machine drafting armature winding hand typesetting operating a drill press desk work small-parts finishing	industrial sewing bench work filing machine tending small-size packing operating a lathe operating medium-size presses machining small-sized sheet metal work electronics testing plastic moulding operating a punch press operating a crane laying stones and bricks sorting scrap	making cement industrial cleaning joining floorboards plastering power truck operation handling light cases to and from a pallet road paving painting handling operations metal casting cutting or stacking lumber large-size packing	shoveling (>7 kg) ditch digging hewing and loading coal handling moderately heavy (>7 kg) cases to and from a pallet tree planting handling heavy (>11 kg) units frequently (>4 per minute)	stoking a furnace ladder or stair climbing coal car unloading lifting 20-kg cases 10 times per minute

[1] The values given include basal metabolism.

Several industrial jobs and tasks are listed according to their average energy demands, ranging from light (column 1) to extremely heavy (column 5) work. The range of energy expenditures is shown at the top of each column, in watts and kilocalories per hour. Tasks in the light- and moderate-effort categories (columns 1 and 2) are more likely to be done for a full 8-hour shift. Heavy to extremely heavy tasks (columns 3–5) are usually alternated with more sedentary paperwork or standby activities.

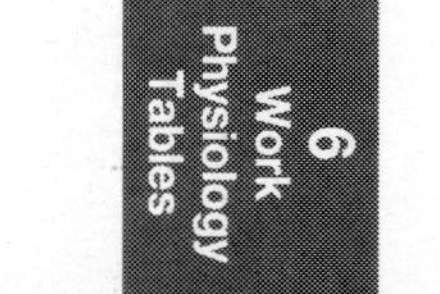

Table PHYSLGY-A8: National Safety Council's Data on Injuries by Part of Body[1]

Part of body	Percent of injuries
Eyes and head	11
Trunk	29
Fingers, arms, hands	31
Legs, feet, toes	20
General	9

REFERENCES

1. Pulat, B.M. 1992. *Fundamentals of Industrial Ergonomics.* Englewood Cliffs, NJ: Prentice-Hall. Reprinted with Permission.

2. Reprinted with permission from *Ergonomic Design for People at Work*, © Eastman Kodak Company, 1983, published by Van Nostrand Reinhold. Courtesy of Eastman Kodak Company.

CHAPTER 6: WORK PHYSIOLOGY TABLES

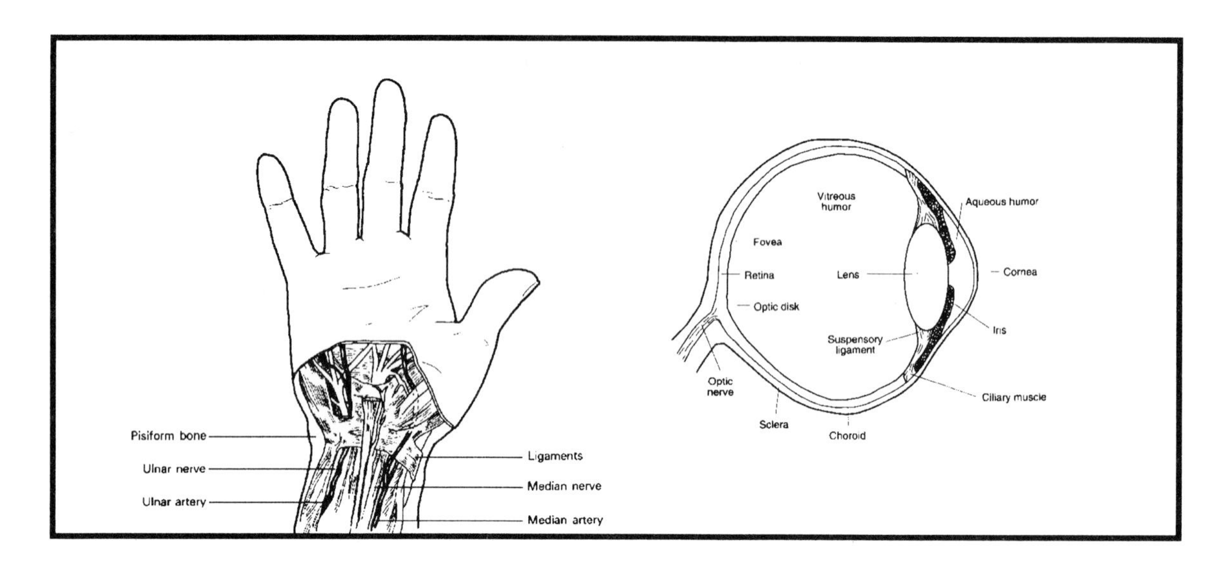

Section B: Manual Materials Handling

This section presents information concerning lifting and manual materials handling. The information in this section is useful for the ergonomist or human factors engineer who must jobs and workplaces that will involve lifting and other types of physical exertion.

JOB SEVERITY INDEX

Table PHYSLGY-B1: Prediction Models for Maximum Weight of Lift Plus Body Weight [1]

Lifting Range	*Constant Term*	*Sex Code Coefficient*	*Weight Code Coefficient*	*Arm Strength Coefficient*	*Age Coefficient*	*Shoulder Height Coefficient*	*Back Strength Coefficient*	*Abdominal Depth Coefficient*	*Dynamic Endurance Coefficient*
#1	−32.683	−12.832	10.980	0.065	−0.250	0.555	0.025	2.226	0.796
#2	−65.857	−7.321	5.402	0.084	−0.270	0.651	0.035	2.931	1.181
#3	−18.690	−4.734	7.326	0.095	−0.404	0.344	0.031	2.817	0.646
#4	−24.982	−8.398	5.299	0.120	−0.274	0.348	0.048	2.849	0.641
#5	−35.866	−8.567	7.823	0.135	−0.226	0.042	0.008	2.334	0.960
#6	−16.956	−8.870	9.217	0.043	−0.268	0.401	0.041	2.143	0.494

NOTE: Range #1 = floor to knuckle, #2 = floor to shoulder, #3 = floor to reach, #4 = knuckle to shoulder, #5 = knuckle to reach, #6 = shoulder to reach. Sex code = 1 for females, 0 for males. Age is in years, weights and strengths are in kg, body measurements are in centimeters. Weight code = 1, if body weight is above 61.14 kg for females and 76.99 kg for males; 0, otherwise. Dynamic endurance is an elapsed time in minutes.

6 Work Physiology Tables

Table PHYSLGY-B2: Lifting Range Determination[1]
(KL = Knuckle Level)

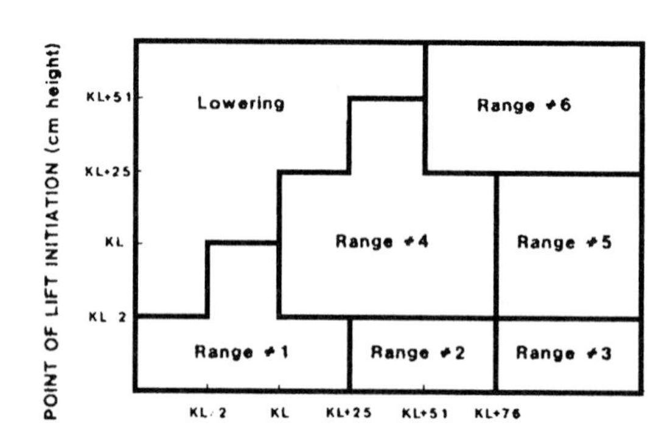

Table PHYSLGY-B3: Lifting Frequency Multipliers[1]
(A- Lifting Ranges 4 and 5; B- Lifting Ranges 1,2,3,and 6)

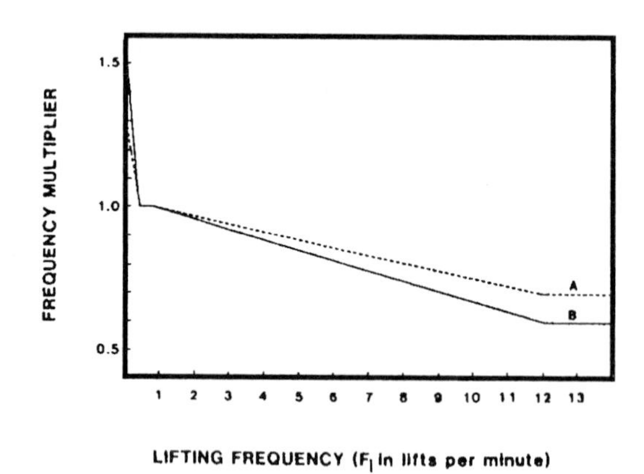

Table PHYSLGY-B4: Center of Gravity Multipliers[1]

(A- Lifting Range 6; B- Lifting Ranges 4 and 5; C- Lifting Ranges 1,2,3)

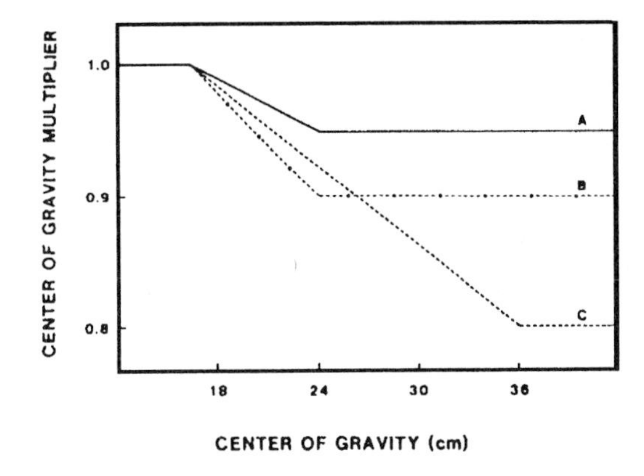

Table PHYSLGY-B5: Injury Rates Observed[1]

(Back Injury Caused by Lifting)

Group	JSI Range	Number of Workers	Number of Exposure Hours	Injury Rate*	Disabling Injury Rate**	Severity Rate***
1	0.00 to 0.29	43	105 010	3.8	3.8	3.0
2	0.30 to 0.57	43	106 473	1.9	0.0	0.0
3	0.57 to 0.81	39	103 665	7.7	0.0	0.0
4	0.81 to 1.07	36	107 132	3.7	3.7	11.0
5	1.09 to 1.42	47	105 546	0.0	0.0	0.0
6	1.42 to 1.76	55	106 901	15.0	11.2	17.3
7	1.77 to 2.09	50	106 213	17.0	13.2	16.7
8	2.10 to 2.61	45	105 650	18.9	9.5	18.8
9	2.62 to 3.65	47	106 134	18.8	15.1	9.8
10	3.66 and above	48	105 157	17.1	11.4	120.8

* Injury rate = Number of injuries/100 FTE; (100 FTE = 200 000 exposure hours = 100 worker years)
** Disabling injury rate = Number of lost-time injuries/100 FTE
*** Severity rate = Number of days lost/number of lost-time injuries

Table PHYSLGY-B6: Cumulative Injury Rate Versus JSI[2]

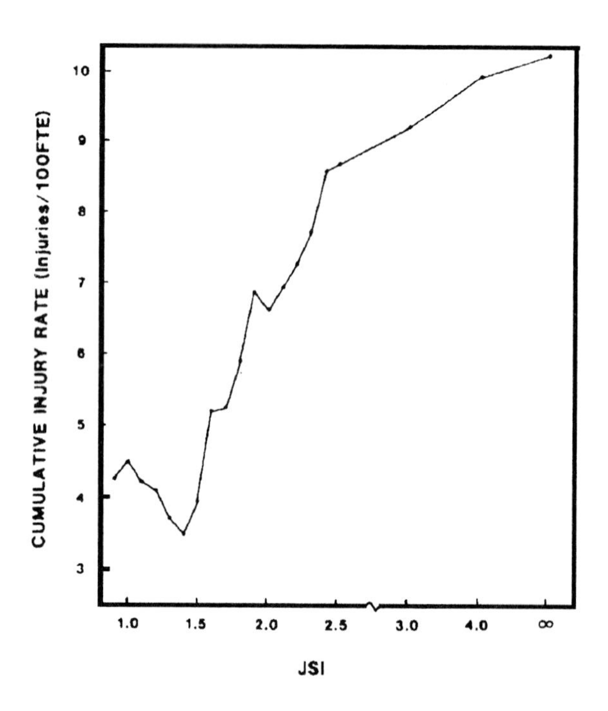

Table PHYSLGY-B7: Cumulative Disabling Injury Rate Versus JSI[1]

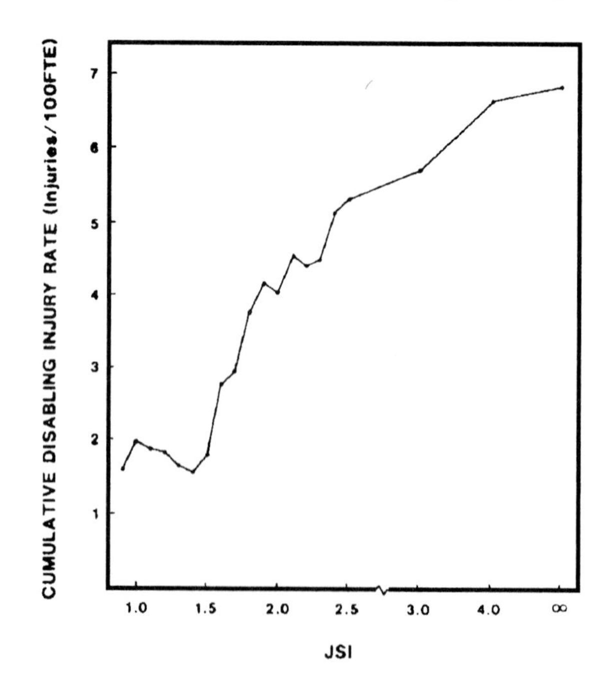

Table PHYSLGY-B8: Cumulative Severity Rate Versus JSI[1]

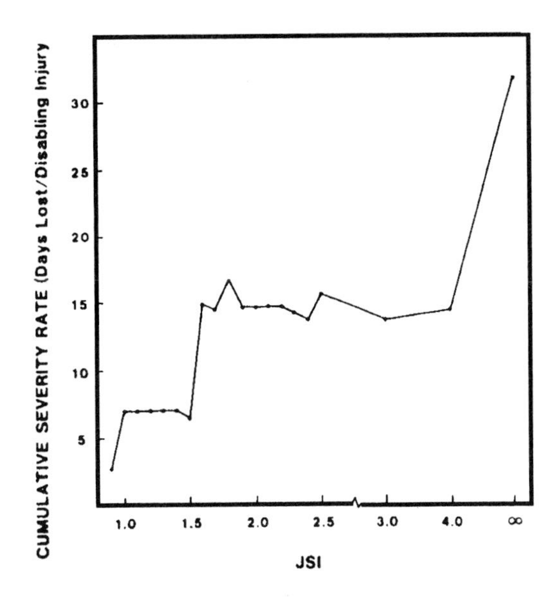

Table PHYSLGY-B9: Injury and Expense Rates Observed in Various Job-Stress Categories [1]

(Back Injuries Caused by Lifting)

Percentage of Population Overstressed	*Number of Hours*	*Injury Rate**	*Disabling Rate**	*Severity Rate*	*Expense Rate**
$\% \leq 5$	305 333	5.24	2.62	39.75**	9 208***
$5 < \% \leq 75$	510 485	10.97	7.05	15.11	35 092
$\% > 75$	342 063	15.70	11.57	51.07	36 337

* Per 100 FTE
** Excluding one very serious injury, this rate is 1.75.
*** Expense data for second study only.

LIFTING AND MANUAL MATERIALS HANDLING

Table PHYSLGY-B10: Container Weight Lifted as a Function of Container Width[2]

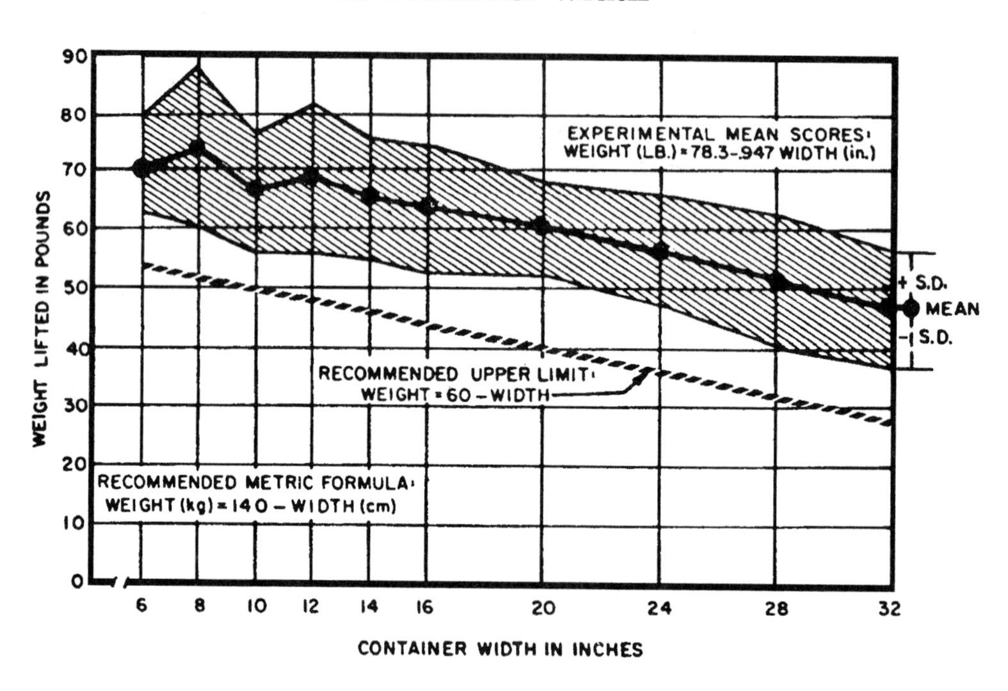

Table PHYSLGY-B11: Maximal Weights Lifted in Boxes of Various Widths Lifted With the Right Hand[2]

(N = 30)*

Container width (in.)	Weight lifted (lb)† (means)	S.D.
6	70	8.6
8	73	14.4
10	66	9.7
12	69	12.7
14	65	9.6
16	64	11.1
20	60	7.0
24	56	9.2
28	50	11.1
32	48	9.8

* Two groups of 15 male subjects each, cautioned not to overstrain.

† All boxes 10 inches high, with a handle centered on top. All boxes lifted from floor to table top, 30 in. high.

Table PHYSLGY-B12: Maximal Weights Lifted to Various Heights by American Male Air Force Personnel[2]

(N = 19)

Height lifted (ft)*	Percentiles (lb)			S.D.
	5th	50th	95th	
1	142	231	301	47
2	139	193	259	40
3	77	119	172	31
4	55	81	112	19
5	36	58	83	16

* Subjects lifted a maximally weighted ammunition case (25½ × 10¾ × 6 in.) from the floor and placed it on platforms of various heights.

Table PHYSLGY-B13: "Reasonable" Weights Lifted Without Strain to Various Heights by American Males[2]

(N = 75)

Stature ranges of the subjects	Height lifted* (ft)	Average weight (lb)	S.D. (lb)
63.3–66.6	1½	124	20.8
	3½	73	11.6
	5¼	53	8.6
68.9–69.6	1½	138	22.6
	3½	92	12.5
	5¼	65	10.8
71.6–74.9	1½	146	30.9
	3½	96	14.2
	5¼	67	8.5

Subjects lifted a box (12 × 12 × 6 in.), equipped with two handles, from the floor and placed it on platforms of different heights. The box was weighted so that the subjects felt this was a reasonable weight which they could lift repeatedly without strain.

Table PHYSLGY-B14: NIOSH Guide for Occasional Lifts in the Sagital Plane[3]

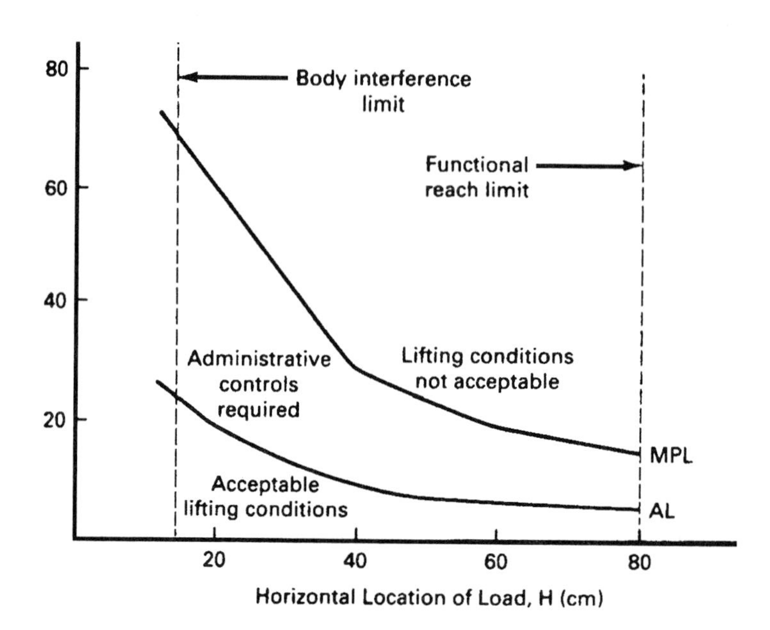

Table PHYSLGY-B14: Lift Range Adjustments[2]

Point of lift initiation[a]	Point of lift termination[a]	Range assignment[b]
0 in. to KL/2	0 in. to KL + 10 in.	1. FK
	KL + 10 in. to KL + 30 in.	2. FS
	KL + 30 in. and above	3. FR
KL/2 to KL	KL/2 to KL	1. FK
	KL to KL + 30 in.	4. KS
	KL + 30 in. and above	5. KR
KL to KL + 10 in.	KL to KL + 30 in.	4. KS
	KL + 30 in. and above	5. KR
KL + 10 in. to KL + 20 in.	KL + 10 in. to KL + 20 in.	4. KS
	KL + 20 in. and above	6. SR
KL + 20 in. and above	KL + 20 in. and above	6. SR

[a]KL, knuckle level.

[b]Range FK is the floor-to-knuckle range; FS is the floor-to-shoulder range; FR is the floor-to-reach range; KS is the knuckle-to-shoulder range; KR is the knuckle-to-reach range; SR is the shoulder-to-reach range.

Table PHYSLGY-B15: Initial Capacity Calculation[2]

Range of lift	Frequency of lift* (lifts/min) 0.1 < FY < 1.0	1.0 ≤ FY ≤ 12.0
	Male	
1. FK	57.2 × (FY) ** (−0.184697)	57.2 − 2.0 × (FY − 1)
2. FS	51.2 × (FY) ** (−0.184697)	51.2 − 2.0 × (FY − 1)
3. FR	49.1 × (FY) ** (−0.184697)	49.1 − 2.0 × (FY − 1)
4. KS	52.8 × (FY) ** (−0.138650)	52.8 − 2.0 × (FY − 1)
5. KR	50.0 × (FY) ** (−0.138650)	50.0 − 2.0 × (FY − 1)
6. SR	48.4 × (FY) ** (−0.138650)	48.4 − 2.0 × (FY − 1)
	Female	
1. FK	37.4 × (FY) ** (−0.187818)	37.4 − 1.1 × (FY − 1)
2. FS	31.1 × (FY) ** (−0.187818)	31.1 − 1.1 × (FY − 1)
3. FR	28.1 × (FY) ** (−0.187818)	28.1 − 1.1 × (FY − 1)
4. KS	30.8 × (FY) ** (−0.156150)	30.8 − 1.1 × (FY − 1)
5. KR	27.3 × (FY) ** (−0.156150)	27.3 − 1.1 × (FY − 1)
6. SR	26.4 × (FY) ** (−0.156150)	26.4 − 1.1 × (FY − 1)

*FY, frequency of lift (lifts/min); **, exponentation (e.g., FY to the power of −0.184697). The initial lifting capacity is 57.2 lb for males and 37.4 lb for females based on the mean capacity for lift based on published data from M. M. Ayoub and S. N. Snook for the various ranges of lift for the 50th percentage and 1.0 lift/min.

Table PHYSLGY-B16: Box Size Adjustment[2]

Range of lift	Box size* (in. in the sagittal plane) 12 in. ≤ BX ≤ 18 in.	BX > 18 in.
	Male	
1. FK	CAP + 1.65 × (18 − BX)	CAP + 0.8 × (18 − BX)
2. FS	CAP + 1.65 × (18 − BX)	CAP + 0.8 × (18 − BX)
3. FR	CAP + 1.65 × (18 − BX)	CAP + 0.8 × (18 − BX)
4. KS	CAP + 1.10 × (18 − BX)	CAP + 0.8 × (18 − BX)
5. KR	CAP + 1.10 × (18 − BX)	CAP + 0.8 × (18 − BX)
6. SR	CAP + 1.10 × (18 − BX)	CAP + 0.8 × (18 − BX)
	Female	
1. FK	CAP + 1.10 × (18 − BX)	CAP + 0.4 × (18 − BX)
2. FS	CAP + 1.10 × (18 − BX)	CAP + 0.4 × (18 − BX)
3. FR	CAP + 1.10 × (18 − BX)	CAP + 0.4 × (18 − BX)
4. KS	CAP + 0.55 × (18 − BX)	CAP + 0.2 × (18 − BX)
5. KR	CAP + 0.55 × (18 − BX)	CAP + 0.2 × (18 − BX)
6. SR	CAP + 0.55 × (18 − BX)	CAP + 0.2 × (18 − BX)

*CAP, capacity of the lift
BX, box size (in.)

Table PHYSLGY-B17: Final Adjustments Due to Lift Frequency and Population Percentiles[2]

Range of lift	Frequency[a]	
	0.1 ≤ FY < 1.0	1.0 ≤ FY ≤ 12.0
	Male	
1. FK	CAP + Z × 16.86 × (FY) ** (−0.174197)	CAP + Z × (16.86 − 0.5964 × (FY − 1))
2. FS	CAP + Z × 15.09 × (FY) ** (−0.174197)	CAP + Z × (15.09 − 0.5338 × (FY − 1))
3. FR	CAP + Z × 14.47 × (FY) ** (−0.174197)	CAP + Z × (14.47 − 0.5119 × (FY − 1))
4. KS	CAP + Z × 14.67 × (FY) ** (−0.156762)	CAP + Z × (14.67 − 0.5534 × (FY − 1))
5. KR	CAP + Z × 13.89 × (FY) ** (−0.156762)	CAP + Z × (13.89 − 0.5240 × (FY − 1))
6. SR	CAP + Z × 13.45 × (FY) ** (−0.156762)	CAP + Z × (13.45 − 0.5074 × (FY − 1))
	Female	
1. FK	CAP + Z × 6.87 × (FY) ** (−0.251605)	CAP + Z × (6.87 − 0.1564 × (FY − 1))
2. FS	CAP + Z × 5.71 × (FY) ** (−0.251605)	CAP + Z × (5.71 − 0.1300 × (FY − 1))
3. FR	CAP + Z × 5.16 × (FY) ** (−0.251605)	CAP + Z × (5.16 − 0.1175 × (FY − 1))
4. KS	CAP + Z × 5.66 × (FY) ** (−0.258700)	CAP + Z × (5.66 − 0.1289 × (FY − 1))
5. KR	CAP + Z × 5.01 × (FY) ** (−0.258700)	CAP + Z × (5.01 − 0.1141 × (FY − 1))
6. SR	CAP + Z × 4.85 × (FY) ** (−0.258700)	CAP + Z × (4.85 − 0.1104 × (FY − 1))

[a]CAP, capacity of lift
Z, Z score of population percentage
FY, frequency of lift (lifts/min); **, exponent.

REFERENCES

1. Liles, D.H., Deivanayagam, S., Ayoub, M.M., and Mahajan, P. 1984 A job severity index for the evaluation and control of lifting injury. *Human Factors.* 26(6), 1984. Reprinted with permission from *Human Factors,* Vol. 26, No. 6, 1984. Copyright 1984 by the Human Factors Society, Inc. All rights reserved.

2. Van Cott, H.P., and Kinkade, R.G., 1972. *Human Engineering Guide to Equipment Design. (Revised)* Washington, DC.: US Government Printing Office. Library of Congress Number: 72-600054.

3. Pulat, B.M. 1992. *Fundamentals of Industrial Ergonomics.* Englewood Cliffs, NJ: Prentice-Hall. Reprinted with Permission.

CHAPTER 7: INFORMATION PROCESSING TABLES

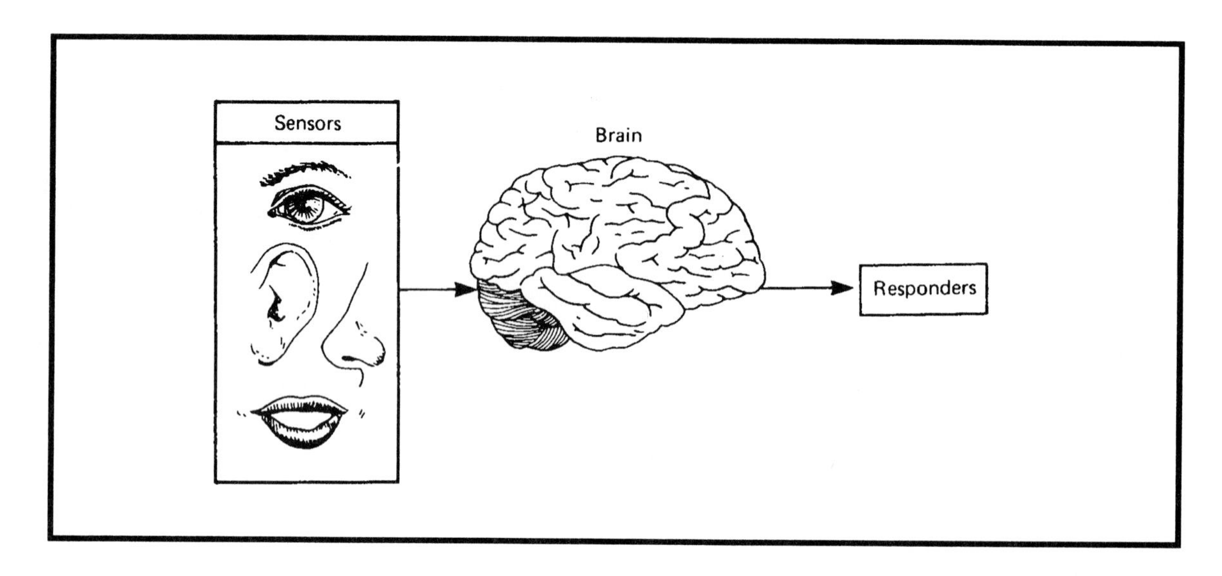

Section A: Reaction Time and Vigilance

This section presents information on human information processing as it relates to work. The information in this section is useful for the ergonomist or human factors engineer who must design workplaces where the job requires the worker to react to a stimulus or detect an infrequent stimulus.

Table INFO-A1: Humans Versus Machines[1]

Humans Excel In
• Sensitivity to a wide variety of stimuli
• Ability to react to unexpected, low-probability events
• Ability to exercise judgment where events cannot be completely defined
• Perception of patterns and making generalizations about them

Machines Excel In
• Performing routine, repetitive, or very precise operations
• Exerting great force, smoothly and with precision
• Operating in environments that are hostile to humans or beyond human tolerance
• Being insensitive to extraneous factors
• Repeating operations very rapidly, continuously, and precisely the same way over a long period
• Doing many different things at one time

The capabilities of humans are compared to those of machines in order to identify whether an industrial task is best handled by a person or by automated equipment. Humans are best at tasks that require judgment and integration of information, whereas machines are best at routine tasks that have to be done precisely, rapidly, and continuously in nonoptimal environments.

Table INFO-A2: Typical Reaction Times[2]

Delays	*Typical Times (msec)*
Sensory receptor	1-38
Neural transmission to brain	2-100
Cognitive-processing delays (brain)	70-300
Neural transmission to muscle	10-20
Muscle latency and activation time	30-70
Total	113-528

Table INFO-A3: Range of Simple and Complex Reaction Times for the Senses Under Various Conditions[3]

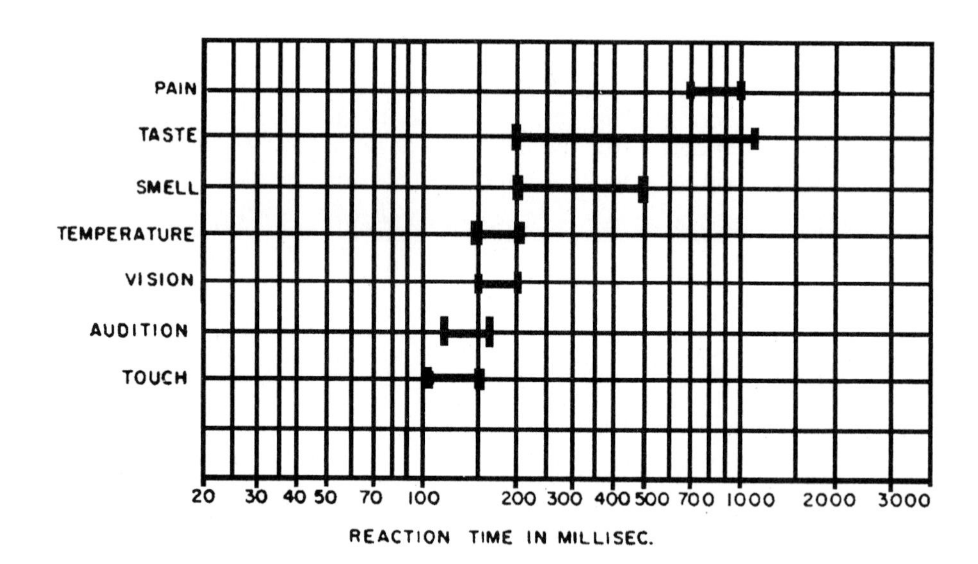

Table INFO-A4: Reaction Time to Magnitude of Stimulus Change[3]

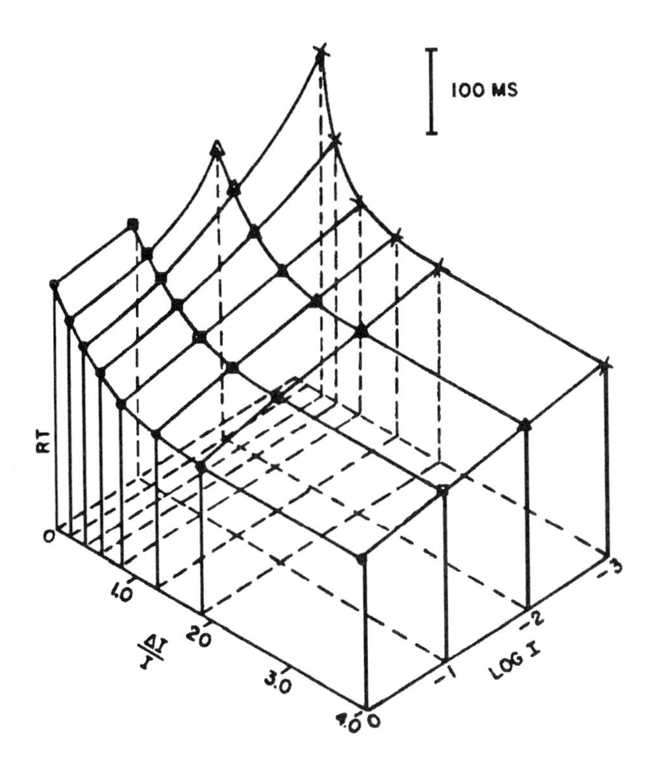

Table INFO-A5: Reaction Time as a Function of the Number of Equiprobable Signals[3]

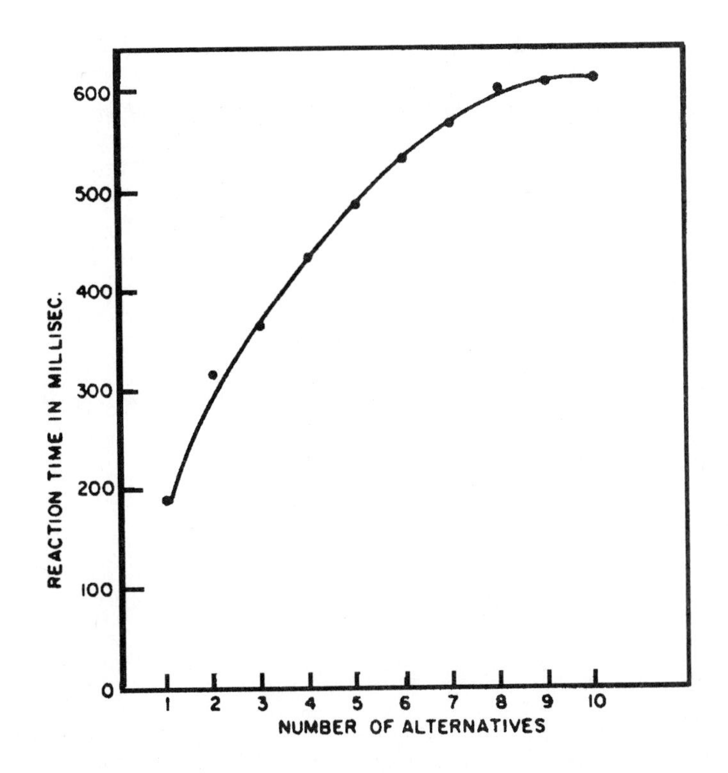

VIGILANCE

Table INFO-A6: Task Conditions Affecting Signal Detectability [3]

Improved probability of detection	
Simultaneous presentation of signals to dual channels	
Men monitoring display in pairs; members of pairs permitted to speak with one another; 10 minutes rest each 30 minutes of work; random schedule inspection by supervisor.	
Introduction of artificial signals during vigilance period to which a response is required.	
Introduction of knowledge of results of artificial signals	
Artificial signals identical to real signals	
Decreased probability of correct detections	
Introduction of artificial signals for which a response is not required.	
Excessive or impoverished task load on operator	
Introduction of a secondary display monitoring task	
Operator reports only signals of which he is sure.	
Change in probability of detection with time	
A short pretest period followed by infrequently appearing signals during vigilance.	High initial probability of detection, falling off rapidly.
Few pretest signals before vigilance period.	Reduces decrement in probability of detection with time.
Prolonged continuous vigilance	Decreases probability of correct signal detection.

Table INFO-A7: Human Performance on a Vigilance Task[4]

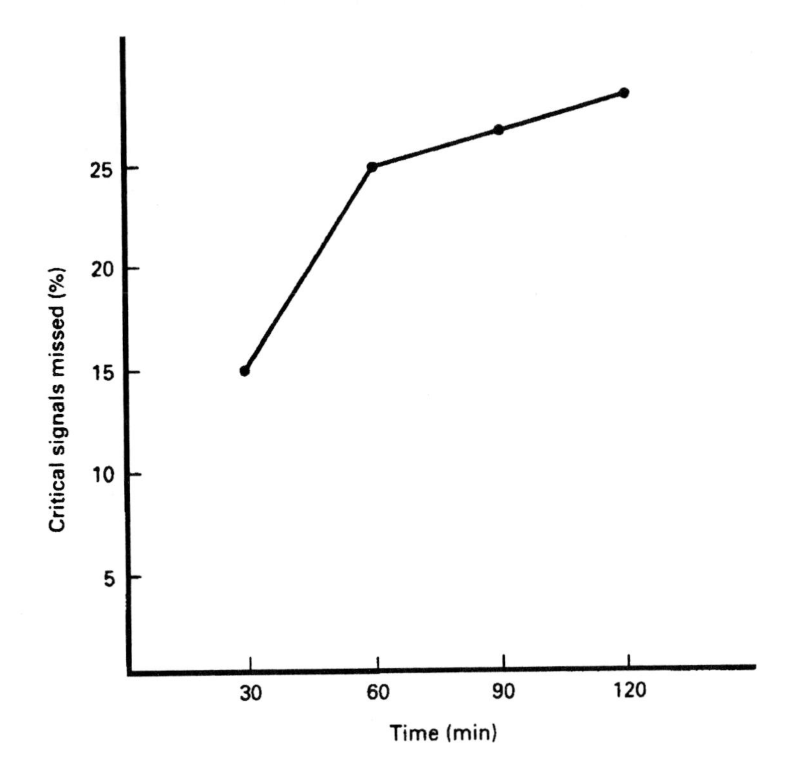

REFERENCES

1. Reprinted with permission from *Ergonomic Design for People at Work*, © Eastman Kodak Company, 1983, published by Van Nostrand Reinhold. Courtesy of Eastman Kodak Company.

2. Bailey, R.W. 1989. *Human Performance Engineering. (2nd Ed)* Englewood Cliffs, NJ: Prentice-Hall. Reprinted with Permission.

3. Van Cott, H.P., and Kinkade, R.G., 1972. *Human Engineering Guide to Equipment Design. (Revised)* Washington, DC.: US Government Printing Office. Library of Congress Number: 72-600054.

4. Pulat, B.M. 1992. *Fundamentals of Industrial Ergonomics.* Englewood Cliffs, NJ: Prentice-Hall. Reprinted with Permission.

CHAPTER 7: INFORMATION PROCESSING TABLES

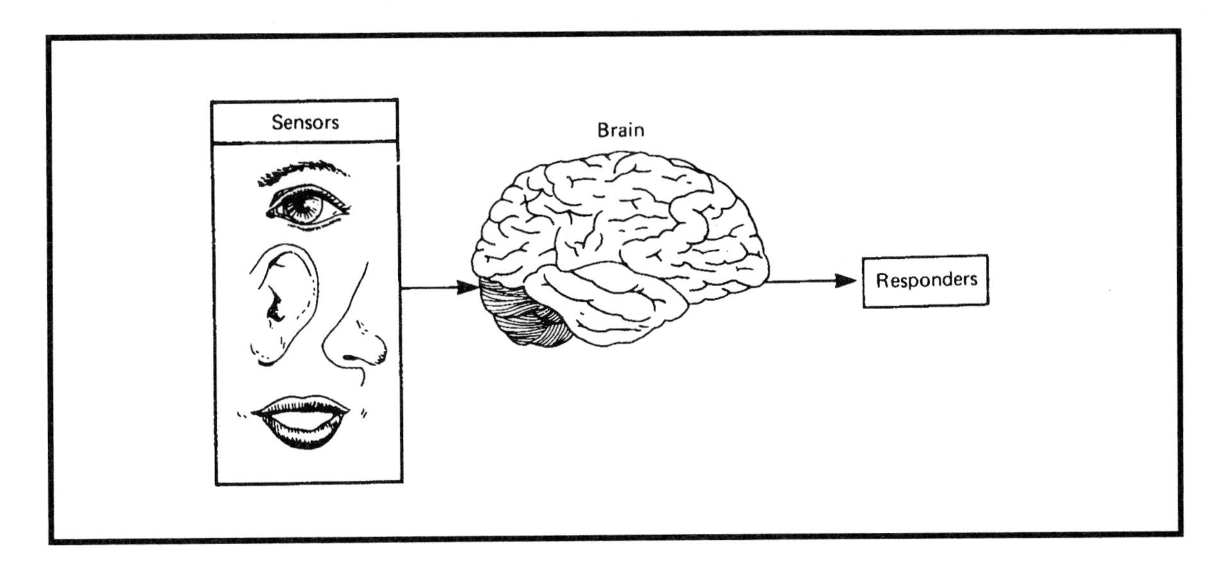

Section B: Manual Input and Human Error

This section presents information concerning human errors made during data entry. The section focuses on data entry since this is often an area that has a great potential for eliciting human error.

DATA ENTRY ERRORS

Table INFO-B1: Representative Manual Entry Rates[1]

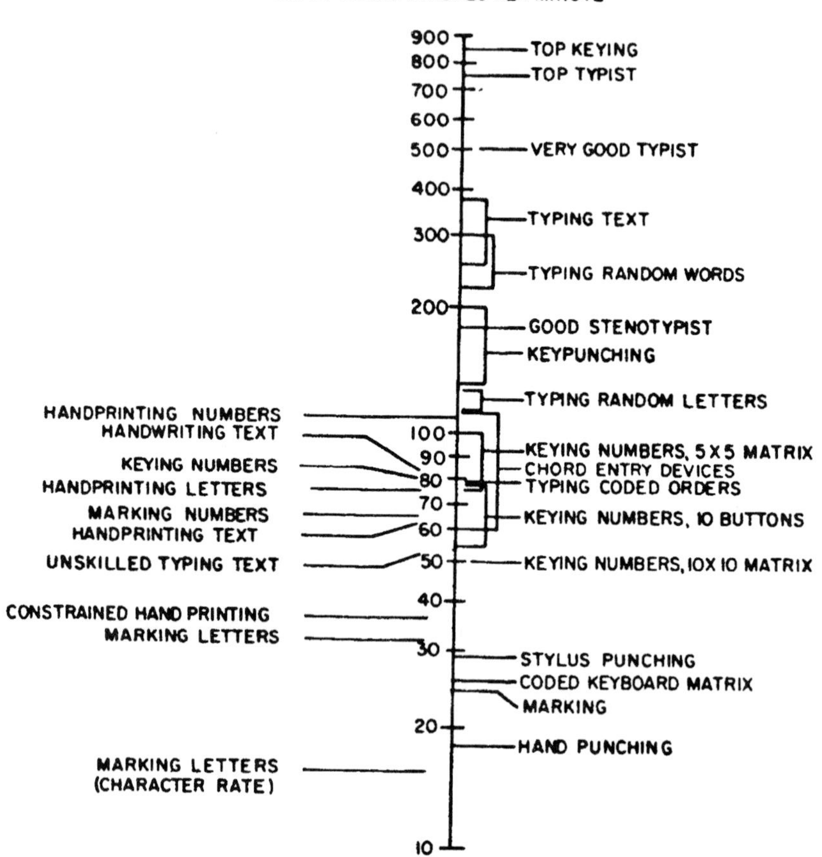

Table INFO-B2: Listing of Visual Confusions to Help Identify Perceptual Stage Errors[2]

Response	*Stimulus*	*Response*	*Stimulus*
B	R	P	D
C	F	S	J
C	G	T	I
D	O	T	J
D	P	U	J
F	E	U	V
F	P	V	U
G	C	V	Y
G	Q	W	N
K	Z	X	Y
L	Z	Y	V
M	H	Y	X
N	H	1	I
O	C	2	Z
O	D	5	J
O	G	5	S
O	Q	6	G

Table INFO-B3: Listing of Auditory Confusions (Sound-Alike Characters) to Help Identify Perceptual Stage Errors[2]

A-J	C-E	E-V	H-8	N-A	T-2	4-5
A-K	C-P	F-S	I-R	O-A	V-3	6-8
A-L	C-T	F-X	I-4	P-Q	Z-7	
A-N	D-B	G-P	I-5	P-T	O-4	
A-O	D-E	G-Q	I-9	P-V	O-8	
B-C	D-T	G-T	J-2	Q-T	1-7	
B-D	D-V	G-U	K-N	Q-U	1-8	
B-E	E-G	G-V	L-O	Q-E	1-9	
B-G	E-P	H-S	L-R	R-4	3-8	
B-P	E-T	H-X	M-7	S-X	4-1	

Table INFO-B4: Motor Confusions for a QWERTY Keyboard[2]

Response	*Stimulus*	*Response*	*Stimulus*	*Response*	*Stimulus*
A	S	L	K	X	C
B	V	M	N	Y	T
B	N	N	B	Y	U
C	X	N	M	Z	X
C	V	O	I	1	2
D	S	O	P	2	1
D	F	P	O	2	3
E	W	Q	W	3	2
E	R	R	E	3	4
F	D	R	T	4	5
F	G	S	A	4	3
G	F	S	D	5	6
G	H	T	R	5	4
H	G	T	Y	6	7
H	J	U	Y	6	5
I	U	U	I	7	8
I	O	V	C	7	6
J	H	V	B	8	9
J	K	W	Q	8	7
K	J	W	E	9	0
K	L	X	Z	9	8
				0	9

Table INFO-B5: Examples of Industrial Data Entry Error [3]

Task	Error	Opportunity for Error
Labeling	Mislabel	Illegible product ID* Inadequate label storage or retrieval system
Packaging	Mix Product	Simultaneous handling of similar product Switched control cards Product change: improper clearing of line Shift change: communications failure
Order Picking	Mix Order	Similar items adjacent to each other Similar product IDs*
Sorting	Mix Kinds	Incorrectly identified Poor handwriting Distraction during the task
Transcribing, Keying	Substitute, Transpose, Omit	Poor handwriting Memory overload Incompatibility of format Look-alike characters
Following Instructions	Misinterpret	Poor comprehensibility Poor legibility Poor readability
Filling Out Forms	Enter Information at Wrong Place, Enter Wrong Information	Too much information on form Poor layout of form Lack of instructions
Looking Up Tables	Mistrack Across Columns	Excessive spacing between columns Lack of tracking aids (spaces or lines)
Monitoring Control Panels	Misread Dial, Misjudge Trend	Parallax problem (difficult to line up) Look-alike dials with different scales Information overload: inadequate sampling Inadequate knowledge of system

* ID = identification number or code.

The first two columns specify the task and the error that is most likely to occur. The third column (Opportunity for Error) gives characteristics of the information display or work layout that contribute to errors in common industrial tasks.

Table INFO-B6: Error Opportunities in Coding[3]

Error Type	Preventive Feature in Code Design	Reason
Omission or addition of characters to code	Use uniform length and composition	Omission or addition will result in code immediately recognized as nonexistent
Substitution between numbers and letters	Use consistent location for numbers and letters	Substitution will result in code immediately recognized as nonexistent
Transposition of letters	Use a familiar acronym or pronounceable word (instead of random letters) that is visually and audibly distinct	It will be remembered as one element rather than individual elements, as random letters are remembered
Transposition of numbers	Introduce a rule for the relationship between adjacent numbers in the string	Transposition will yield a code with a pair of digits out of order
Illegibility	Use consistent number and letter locations Control handwriting by providing individual box for each character	Poor handwriting more easily deciphered

Common types of errors in written information are shown in column 1. Techniques to reduce the probability of making each error are given in column 2; an explanation of how they prevent errors is given in column 3.

REFERENCES

1. Van Cott, H.P., and Kinkade, R.G., 1972. *Human Engineering Guide to Equipment Design. (Revised)* Washington, DC.: US Government Printing Office. Library of Congress Number: 72-600054.

2. Bailey, R.W. 1989. *Human Performance Engineering. (2nd Ed)* Englewood Cliffs, NJ: Prentice-Hall. Reprinted with Permission.

3. Reprinted with permission from *Ergonomic Design for People at Work*, © Eastman Kodak Company, 1983, published by Van Nostrand Reinhold. Courtesy of Eastman Kodak Company.

CHAPTER 8: METHODS AND EQUATIONS

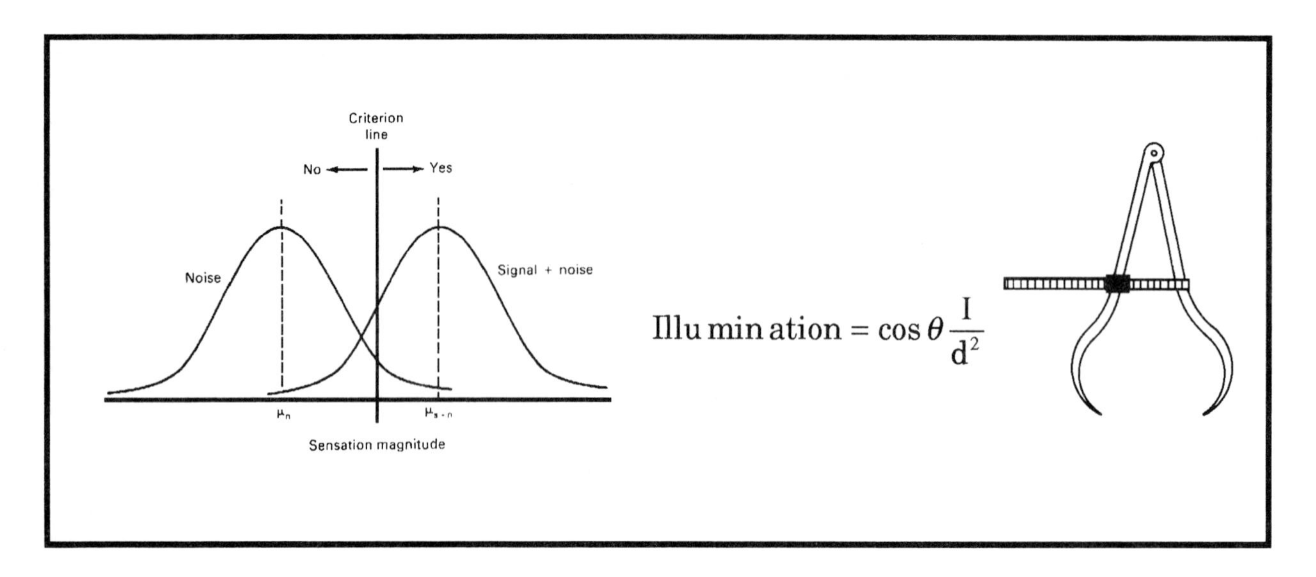

Section A: Statistics and Psychophysics

This section presents statistical tables and a summary of psychophysical methods -- as well as tables of digrams, trigrams, and letter usage frequencies. The information in this section is useful for the ergonomicist or human factors engineer who is involved with research.

PSYCHOPHYSICS

Table MTDS-A1: Methods of Psychophysics[1]

Method	Procedure	Statistical index	Problems to which most applicable
Adjustment (average error)	Observer adjusts signal intensity until it is subjectively equal to, or in some relation to, a criterion.	Average of settings (average error of settings measures operator precision).	Absolute thresholds; equality; equal ratio or interval.
Minimal change (limits)	Signal is varied up and down in magnitude. Observer reports when it meets criterion.	Average value of signal at transition point of observer's judgment.	All thresholds; equality.
Constant stimuli	Comparison stimuli are paired at random with a fixed standard stimulus. Observer reports whether each comparison stimulus is greater or less than standard.	JND equals stimulus distance between 50 and 75 percent points on psychometric function.	All thresholds; equality; equal ratio or interval.
Paired comparison	Stimuli (need not be physical energies) are presented in pairs, exhausting all combinations of stimuli taken two at a time. Observer reports which of each pair is greater in respect to given attribute or criterion.	Proportion of judgments calling one stimulus greater than another. Proportions may be translated into scale values via the assumption of a normal distribution of judgments.	Order on a scale (e.g., of comfort, desirability, etc.). Equal interval.
Rating scale (absolute judgment).	Each of a set of stimuli is given an "absolute" rating or index in terms of some selected attribute. Rating may be numerical or descriptive.	Average or median rating assigned by observers.	Order on a scale. Also used to determine useable codes (e.g., color coding).

Memory

Table MTDS-A2: Frequency Listing of the 26 English Letters Based on a Sample of 20,000 English Words[2]

Letter	*Percent of Total*
E	13.3
T	9.8
A	8.1
H	7.7
O	6.6
S	6.1
N	6.0
R	5.9
I	5.1
L	4.5
D	4.3
U	3.1
W	2.9
M	2.5
C	2.4
G	2.2
Y	2.1
F	1.8
B	1.6
P	1.5
K	1.1
V	1.0
J	.2
X	.1
Q	.1
Z	.1
	100.0

Table MTDS-A3: Listing of 200 Digrams Based on 20,000 English Words[2]

Digram	Frequency	Digram	Frequency	Digram	Frequency	Digram	Frequency
TH	3774	CA	368	IE	189	EW	106
HE	3155	NO	349	FR	188	EF	103
AN	1576	LO	344	EM	187	WN	103
ER	1314	YO	339	TR	187	FT	102
ND	1213	KE	337	EC	181	AP	100
HA	1164	OO	336	CK	178	NA	100
RE	1139	EL	332	AM	177	BL	98
OU	1115	LA	332	SU	175	GR	98
IN	1110	TO	331	EV	172	NC	98
HI	824	SH	328	PL	169	PI	97
OR	812	IL	324	SS	168	GI	96
AR	802	AI	322	HT	165	DS	95
EN	799	AY	319	IV	165	GA	94
AT	785	RS	318	MI	165	HR	91
NG	771	ET	316	CT	154	EP	90
ED	767	RI	309	FE	154	RU	89
ST	754	AC	308	TT	154	BR	88
AS	683	IC	304	YE	152	IO	88
VE	683	US	299	PO	149	OI	88
EA	670	CO	298	DI	148	AU	87
AL	664	GE	289	NS	148	EX	87
ES	630	LD	289	UG	148	UM	87
SE	626	MO	289	AK	146	FF	86
ON	598	RA	289	FA	145	IK	86
WA	595	GH	288	RY	145	MU	85
LE	591	CE	285	AB	144	TW	84
TE	583	WE	285	PA	142	DR	83
IT	558	PE	280	AG	138	KN	82
LL	546	UN	278	OP	138	LU	81
ME	530	LY	276	DO	137	YS	81
NE	512	IR	272	BA	136	NL	80
RO	504	WO	264	OV	136	OF	80
UT	492	ID	260	GO	135	BI	78
HO	487	TA	259	NI	135	MP	77
IS	484	BU	256	RD	133	HU	75
WH	472	IM	255	TU	132	TL	75
EE	470	TI	252	EI	127	LT	74
FO	429	UL	247	KI	127	CR	71
OM	417	BO	240	OK	123	RL	71
BE	415	AV	233	LS	121	UE	71
OT	415	IG	233	TY	121	FL	69
CH	412	OL	218	OD	119	RR	69
UR	402	SI	214	NY	115	PU	68
OW	398	SO	213	UC	114	AF	67
MA	394	TS	209	PR	110	CI	67
LI	390	FI	205	VI	109	OB	67
AD	382	SA	196	CL	108	QU	66
NT	378	OS	195	SP	107	OA	65
DE	375	RT	190	DA	106	RM	65
WI	374	EY	189	RN	106	UI	65

Table MTDS-A4: Listing of 200 Trigrams Based on 20,000 English Words[2]

Trigram	Frequency	Trigram	Frequency	Trigram	Frequency	Trigram	Frequency
THE	2565	WER	119	ACE	80	EAT	65
AND	959	ATE	118	AID	80	ERY	65
ING	526	HOU	114	IND	80	HOW	65
HAT	479	OVE	114	URE	80	NIN	65
THA	438	NOW	112	COU	79	OSE	65
HER	414	WHO	112	LEA	79	RES	65
HIS	354	OUN	111	TUR	79	STE	65
FOR	353	COM	110	IDE	78	TIM	65
YOU	326	EVE	110	TIN	78	ION	64
WAS	304	HIN	109	ART	77	OOD	64
ALL	270	OUG	109	EAS	77	PLE	64
THI	259	USE	109	EST	77	RIE	64
ERE	255	ERS	108	VEN	77	WIL	64
ITH	238	AKE	107	HEM	76	NGE	63
ARE	228	MOR	107	LON	76	THR	63
WIT	227	WAY	106	ANT	75	TTE	63
OUT	225	INT	103	END	75	CHA	62
VER	221	STA	101	LED	75	HES	62
OUR	209	ABO	99	MEN	75	SHO	62
ONE	205	HIC	99	YEA	75	TEN	62
EAR	197	UND	99	HEA	74	DAY	61
AVE	194	AIN	98	PLA	74	ILE	61
NOT	191	ICH	98	ACK	73	MOS	61
OME	191	OWN	97	ARD	73	NEW	61
TER	179	OLD	96	GET	73	ONL	61
BUT	178	UST	96	ROU	73	ACT	60
HAD	173	ONG	95	SAI	73	BEC	60
GHT	163	WOR	94	ARS	72	LAS	60
IGH	161	BOU	93	LES	71	ANG	59
ORE	153	AME	91	PER	71	ICE	59
HAV	147	AST	91	UCH	71	ROV	59
ILL	146	CAN	91	AYS	70	ECT	58
OUL	145	HAS	89	EIR	70	EET	58
IVE	143	OST	89	RED	70	FIN	58
MAN	143	WOU	89	CAR	69	STO	58
SHE	143	ANY	88	HEI	69	AIR	57
ULD	143	KIN	88	SOM	69	EVE	57
OTH	140	WHA	88	THO	69	ITT	57
ENT	139	REE	87	NTO	68	MAR	57
FRO	138	BEE	86	TOO	68	ACH	56
HEN	133	IKE	86	AGE	67	EED	56
HEY	131	TED	86	CAM	67	RSE	56
WHE	131	ELL	85	NCE	67	TLE	56
ROM	130	LOO	84	ORT	67	IRS	55
EEN	128	OOK	84	CAL	66	ITE	55
HAN	128	SEE	84	DER	66	OIN	55
REA	128	EAD	83	FTE	66	PRO	55
UGH	128	LIK	83	IME	66	WAN	55
HIM	126	ITS	81	LLE	66	ADE	54
WHI	125	KED	81	NLY	66	DOW	54

STATISTICAL TABLES

Table MTDS-A5: Z Scores for Population Percentages[3]

Population (%)	Z Score
95	−1.645
90	−1.282
80	−0.841
70	−0.527
60	−0.255
50	0.0
40	0.255
30	0.527
20	0.841
10	1.282
5	1.645

Table MTDS-A6: Normal ("Bell-Shaped") Distribution [1]

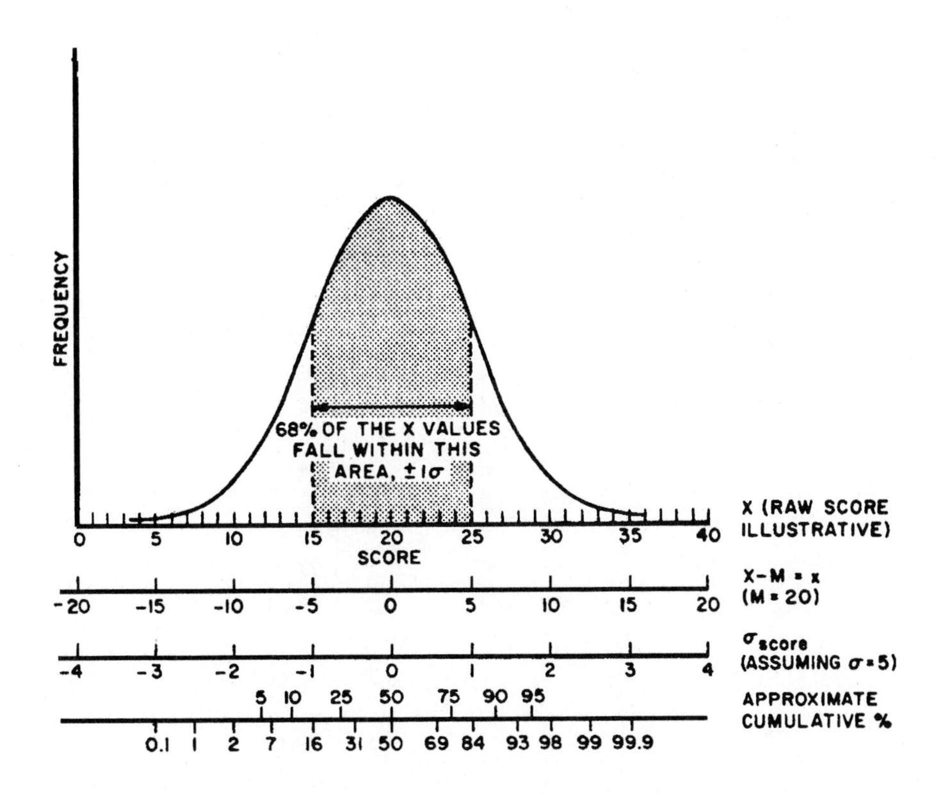

Many sets of measurements of human characteristics approximate the normal distribution. Here X is a measure of a characteristic and Y is the relative frequency of different values of X.

Table MTDS-A7: Values of the t-Distribution[2]

	t value
1	12.71
2	4.30
3	3.18
4	2.78
5	2.57
6	2.45
7	2.37
8	2.31
9	2.26
10	2.23
11	2.20
12	2.18
13	2.16
14	2.15
15	2.13
16	2.12
17	2.11
18	2.10
19	2.09
20	2.09
21	2.08
22	2.07
23	2.07
24	2.06
25	2.06
30	2.04
40	2.02
60	2.00
120	1.98
over 120	1.96

Values in the table represent the 0.05 probability level. In order to conclude the existance of a reliable difference between two groups, the obtained (calculated) t value must be greater than the tabled t value.

Table MTDS-A8: Values of the F Distribution[2]

	1	2	3	4	5	6	7	8	9	10
2	18.51	19.00	19.10	19.25	19.30	19.33	19.35	19.37	19.38	19.40
3	10.13	9.56	9.28	9.12	9.01	8.94	8.89	8.85	8.81	8.79
4	7.71	6.94	6.69	6.39	6.36	6.16	6.09	6.04	6.00	5.96
5	6.61	5.79	5.41	5.19	5.05	4.95	4.88	4.82	4.77	4.74
6	5.99	5.14	4.76	4.53	4.39	4.28	4.21	4.15	4.10	4.06
7	5.59	4.74	4.35	4.12	3.97	3.87	3.79	3.73	3.68	3.64
8	5.32	4.48	4.07	3.84	3.69	3.58	3.60	3.44	3.39	3.36
9	5.12	4.26	3.86	3.63	3.48	3.37	3.29	3.23	3.18	3.14
10	4.96	4.00	3.71	3.48	3.33	3.22	3.14	3.07	3.02	2.98
11	4.84	3.48	3.59	3.36	3.20	3.09	3.01	2.95	2.90	2.85
12	4.75	3.89	3.49	3.26	3.11	3.00	2.91	2.85	2.80	2.75
13	4.67	3.81	3.41	3.18	3.03	2.92	2.83	2.77	2.71	2.67
14	4.60	3.74	3.34	3.11	2.96	2.85	2.76	2.70	2.65	2.60
15	4.54	3.68	3.29	3.06	2.90	2.79	2.71	2.64	2.59	2.54
16	4.49	3.63	3.24	3.01	2.85	2.74	2.66	2.59	2.54	2.49
17	4.45	3.59	3.20	2.96	2.81	2.70	2.61	2.55	2.49	2.45
18	4.41	3.58	3.16	2.93	2.77	2.66	2.58	2.51	2.46	2.41
19	4.38	3.52	3.13	2.90	2.74	2.63	2.55	2.48	2.42	2.38
20	4.35	3.49	3.10	2.87	2.71	2.60	2.52	2.45	2.30	2.35
25	4.24	3.39	2.99	2.76	2.60	2.49	2.40	2.34	2.28	2.24
30	4.17	3.32	2.92	2.69	2.53	2.42	2.33	2.27	2.21	2.16
40	4.08	3.28	2.84	2.61	2.45	2.34	2.25	2.18	2.12	2.08
60	4.00	3.16	2.76	2.53	2.37	2.25	2.17	2.10	2.04	1.99
120	3.92	3.07	2.68	2.45	2.29	2.17	2.09	2.02	1.96	1.91

Values in the table represent the 0.05 probability level. In order to conclude the existance of a reliable difference between two groups, the obtained (calculated) F value must be greater than the tabled F value.

Table MTDS-A9: Values of the CHI-Square Distribution[2]

	Chi-square Values
1	3.84
2	5.99
3	7.82
4	9.49
5	11.07
6	12.59
7	14.07
8	15.51
9	16.92
10	18.31
11	19.68
12	21.03
13	22.36
14	23.68
15	25.00
16	26.30
17	27.59
18	28.87
19	30.14
20	31.41
21	32.67
22	33.92
23	35.17
24	36.42
25	37.65
26	38.89
27	40.11
28	41.34
29	42.56
30	43.77

Values in the table represent the 0.05 probability level. In order to conclude the existance of a reliable difference between two groups, the obtained (calculated) χ^2 value must be greater than the tabled χ^2 value.

REFERENCES

1. Van Cott, H.P., and Kinkade, R.G., 1972. *Human Engineering Guide to Equipment Design. (Revised)* Washington, D.C.: US Government Printing Office. Library of Congress Number: 72-600054.

2. Bailey, R.W. 1989. *Human Performance Engineering. (2nd Ed)* Englewood Cliffs, NJ: Prentice-Hall. Reprinted with Permission.

3. Pulat, B.M. 1992. *Fundamentals of Industrial Ergonomics.* Englewood Cliffs, NJ: Prentice-Hall. Reprinted with Permission.

CHAPTER 8: METHODS AND EQUATIONS

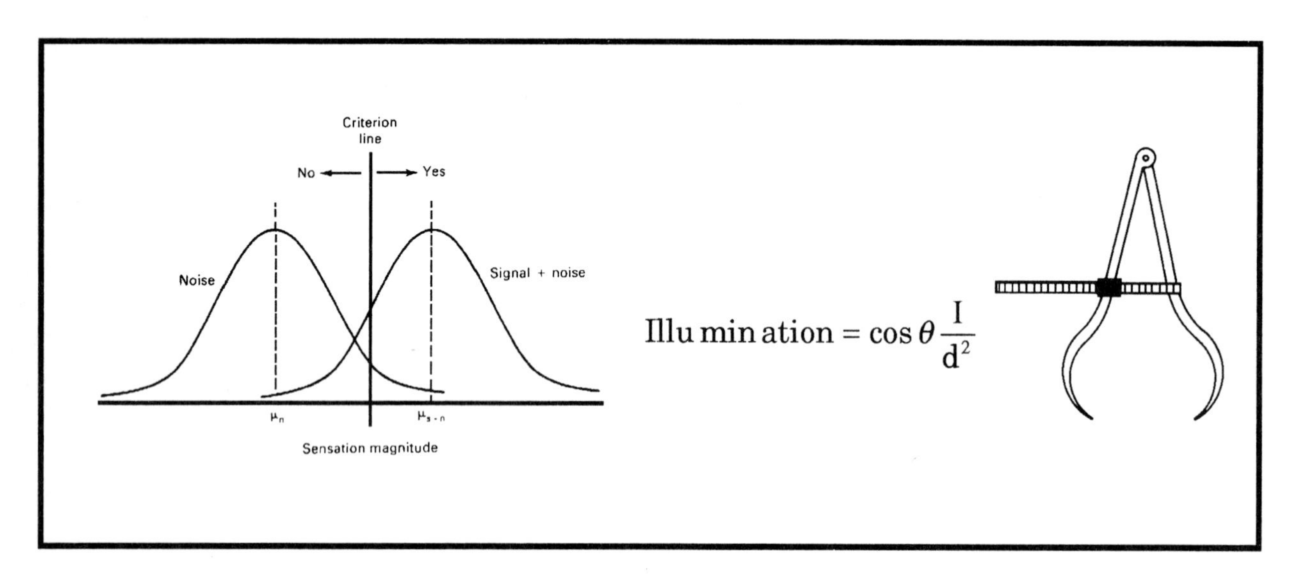

Section B: Useful Formulas and Equations

This section presents commonly used formulas and equations, such as the equations for computing visual angle, contrast, work-rest cycles, and JSI.

VISUAL PROCESSING AND DISPLAYS

EQ1: Visual Angle

$$\text{Visual Angle (min. of arc)} = \frac{(57.3)(60)L}{D}$$

Where:

L = the size of the object measured perpendicular to the line of sight

D = the distance from the eye to the object

EQ2: Visual Angle (Alternate)

$$\theta = \arctan\frac{\text{size of target}}{\text{distance of target}}$$

EQ3-EQ4: Illumination (fc)

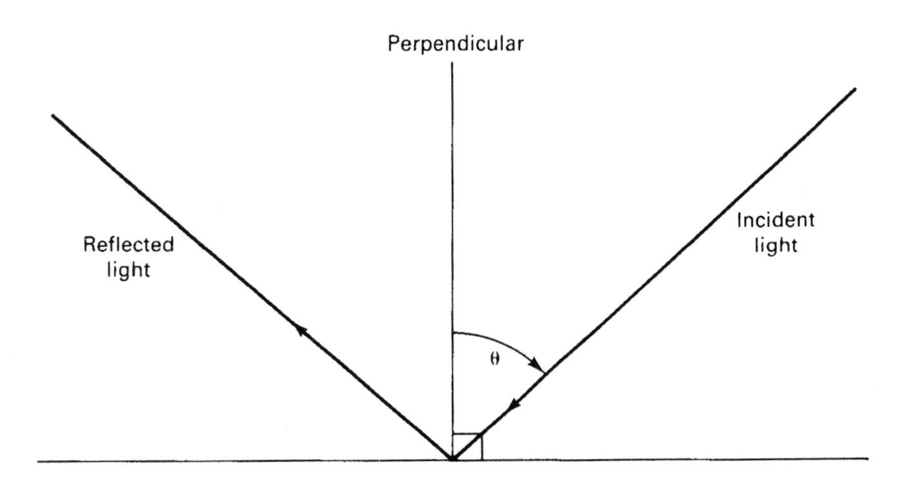

EQ3:

$$\text{Illumination} = \frac{I}{d^2}$$

(For Light Impinging on the Surface at a Right Angle)

EQ4:

$$\text{Illumination} = \cos\theta\frac{I}{d^2}$$

(For Light Impinging on the Surface at an Angle)

Where:

I = intensity of source measured in international candles

d = Distance between the illuminated surface and the light source.

θ = The angle between a line perpendicular to the surface on which illumination is being calculated and the direction of the light.

EQ5: Reflectance (%)

$$\text{Reflectance (\%)} = 100 \text{ X } \frac{\text{Luminance (ft.-L.)}}{\text{Illumination (ft.c.)}}$$

EQ6: Contrast (%)

$$\text{Contrast (\%)} = 100 \text{ X } \frac{L_b - L_t}{L_b}$$

Where:

L_b = Background Luminance

L_t = Target Luminance

EQ7: Color Contrast[5]

$$\Delta E = \left[(L^*)^2 + (u^*) + (v^*)^2 \right]^{\frac{1}{2}}$$

Where:

ΔE = Color Contrast

L* = Luminance

u* & v* = Color's coordinates in the UCS numerical color space.

EQ8: Luminance

$$\text{Lu min ance} = \frac{\text{Illu min ance} \times \text{Re flec tan ce}}{\pi}$$

EQ9: Minimum Required Display Size (D)

$$D = \frac{1.54LRM}{T}$$

Where:

D = Display size in the axis that R is being displayed (in.)

L = Display viewing distance (in.)

R = Range on ground being displayed (statute min.)

M = Minimum target visual angle for detection (min. of visual arc)

T = Greatest target dimension (ft.)

EQ10: Gamma Correction for Display Voltage[5]

$$V_k = \left(\frac{I_k}{\text{constant}}\right)^{\frac{1}{\gamma}}$$

Where:

V_k = Voltage necessary to obtain a particular intensity (k)

I = Constant $(V)^{\gamma}$

Gamma Correction is an important engineering tool used for displays that require precise color control. Gamma Correction transforms the relationship between display drive voltage and intensity into a linear one. Gamma Correction is used to determine the degree of contrast of the display image (its gamma) and a constant for each color gun. To perform gamma correction, make precise photometric measurements of the display. Next, store the values in a computer-generated look-up table. Finally, use these values when manipulating Color Spaces. Color Spaces such as CIE XYZ require that the display be gamma corrected.

EQ-11: Image Viewing Time (T)

$$T = \frac{0.049D}{SV}$$

Where:

T = Viewing time (sec.)

S = Scale factor $\frac{\text{distance on display}}{\text{distance on ground}}$

R = Maximum display dimension (in.)

V = Vehicle velocity (knots)

EQ-12: Recognition Distance for Colored Lights

$$D = 2000 \text{ X } I$$

Where:

D = Viewing distance (ft.)

I = Intensity in candles, for a similar unit with a clear lens

EQ13: Intensity Adjustment for Flashing Colored Lights

$$I_E = \frac{t \text{ X } I}{0.09 + t}$$

Where:

I_E = Effective intensity in candles

I = Intensity of steady light in candles

t = flash duration in seconds

EQ14: Display Lettering Height (in.)

$$\text{Character Height (in)} = 0.0022D + K_1 + K_2$$

$$\text{Character Height (cm)} = 0.000866D + K_1 + K_2$$

Where:

D = Viewing Distance

K_1 = Correction for illumination and viewing conditions

K_2 = Importance Weight

Markings	Above 1 fc/ favorable reading	Above 1fc/Unfavorable Reading Below 1fc/Unfavorable Reading	Below 1fc Unfavorable Reading
Nonimportant (K_2=0)	K_1 = 0.15 cm K_1 = 0.06 in	K_1 = 0.4 cm K_1 = 0.16 in	K_1 = 0.66 cm K_1 = 0.26 in
Important (K_2=0.19 cm) (K_2=0.075 in)			

AUDITORY PROCESSING AND DISPLAYS

EQ15: Sound Pressure Level (SPL.)

$$L = 20 \log \frac{p_1}{p_0} = 10 \log \left(\frac{p_1}{p_0}\right)^2$$

Where:

L = The sound pressure level (SPL) of sound pressure p_1

p_0 = Reference pressure

EQ16: Received Speech Level

$$\frac{P_1}{P_2} = \frac{d_1}{d_2}$$

Where:

P_1 = The sound pressure level at distance d_1

P_2 = The sound pressure level at distance d_2

WORKPLACE

EQ17: Control-Display Ratio

$$C/D = \frac{\frac{a}{360} \times 2\pi L}{\text{display movement}}$$

Where:

a = The angular movement of the control in degrees

L = The length of the lever arm

The optimum C/D ratio range for knobs is 0.2-0.8, while for levers the optimum C/D ratio range is 2.5-4.0. The ratio is affected by display size, tolerance, and time delay.

EQ18: Job Severity Index (JSI)[2]

$$JSI = \sum_{i=1}^{n} \left(\frac{hours_i}{hours_t} \times \frac{days_i}{days_t} \right) \sum_{i=1}^{m_i} \left(\frac{F_j}{F_i} \times \frac{WT_j}{CAP_i} \right)$$

Where:

n = The number of each task group

$hours_i$ = Exposure hours/day for Group i

$days_i$ = Exposure days/week for Group i

$hours_t$ = Total hours/day for job

$days_t$ = Total days/week for job

m_i = The number of tasks in Group i

WT_j = The maximum required weight of lift for Task j

CAP_j = The adjusted capacity of the person working at Task j

F_j = Lifting Frequency for Task j

F_i = Total lifting frequency for Group i

EQ19: Accident Frequency Rate[4]

$$\text{Accident Frequency} = \frac{\text{Number of Accidents} \times 1{,}000{,}000}{\text{Worker - Hours of Job Exposure}}$$

EQ20: Disabling Injury Frequency Rate[4]

$$\text{Disabling Injury Frequency} = \frac{\text{Number of Disabling Injuries} \times 1{,}000{,}000}{\text{Worker - Hours of Job Exposure}}$$

EQ21: Disabling Injury Severity Rate[4]

$$\text{Disabling Injury Severity} = \frac{\text{Total Days Charged} \times 1{,}000{,}000}{\text{Worker - Hours of Job Exposure}}$$

EQ22-EQ24: Work Rest Cycle Equations[4]

EQ22: $R_T = 0$ for $K < S$

EQ23: $$R_T = \frac{\left(\frac{K}{S} - 1\right) \times 100 + \frac{T(K-S)}{K-BM}}{2}$$ for $S \le K < 2S$

EQ24: $$R_T = \frac{T(K-S)}{K-BM} \times 1.11$$ for $K \ge 2S$

Where:

R_T = Allowed Rest Time in Minutes

T = Total Expected Duration of the Task in Minutes

K = Energy Cost of the Work in Kilocalorites (kcal/min)

8 Methods and Equations

S = Standard Cost (4 kcal/min for females, 5 kcal/min for males)

BM = Basal Metabolism (kcal/min)

The results from these calculations should be multiplied by the multipliers in **Table PHSYLGY-A3** (pg 321) to adjust the rest periods for the age of the worker.

EQ25-EQ26: NIOSH Lifting Equations for Predicting Hazard Level[4]

EQ25: $$AL = 40\left(\frac{15}{H}\right)(1-0.004|V-75|)\left(0.07+\frac{7.5}{D}\right)\left(1-\frac{F}{F_{max}}\right)$$

(Action Level)

EQ26: $MPL = 3AL$ **(Maximum Permissible Limit)**

Where:

AL = Action Limit. Measured in kilograms and refers to the load lifted. This is the lower limit

MPL = Maximum Permissible Limit. Measured in kilograms and refers to the load lifted This is the upper limit. MPL is 3 times the AL.

H = Horizontal Distance (cm) from the load's center of mass at the origin of the vertical lift to the midpoint between the ankles (lumbar spine). H will assume a minimum value of 15 cm (body interference) and a maximum of 81 cm (reach distance).

V = Vertical Distance (cm) from the load of center mass to the floor at the origin of the lift. No minimum value is listed. Maximum value is 177 cm (upward reach).

D = Vertical Travel Distance (cm) of the object as measured by the difference between the initial and destination locations of the load's center of mass. A minimum value of 25 cm and a maximum value of (203-V) are assumed.

F = Average Frequency of lifting (lifts/min). A minimum value of 0.2 is assumed. Maximum values are:

Duration of Lifting Period	V > 76cm (Standing)	V ≤ 76cm (Stooping)
1 Hour (Occasionally)	18	15
8 Hours (Continuously)	15	12

A worker lifting loads below AL would not be expected to suffer any ill effects. Administrative controls are necessary when tasks are between AL and MPL; these controls consist of careful worker selection and training. Injuries are likely to occur if MPL is exceeded.

ENVIRONMENT

EQ27: Body's Resistance to Shock (Ohm's Law)[3]

$$I = \frac{V}{Z}$$

Where:

I = Current in Amperes (A)

V = Applied Voltage in Volts (V)

Z = Impedance or Resistance in Ohms (Ω)

EQ28 Body's Heat Balance Equation[3]

$$\pm S = M \pm C \pm R - E$$

Where:

S = Heat Storage in, or Loss From, the Body

M = Metabolic Heat Gain

C = Convective Heat Gain/Loss

R = Radiative Heat Gain/Loss

E = Evaporative Heat Loss

NIOSH and WHO guidelines recommend a rise of 1° C(1.8° F) in core body temperature to be the upper limit for exposures of 1-2Hr. durations, representing a storage of 250 kilojoules (kJ) or 60 kilocalories (kcal).

EQ29: Wet-Bulb Globe Temperature[3]

$$WBGT = 0.7NWB + 0.3GT$$

(Use if difference between radiant heat and temperature are negligible and job is performed indoors)

$$WBGT = 0.7NWB + 0.2GT + 0.1DBT$$

(Outdoors)

Where:

WBGT = Wet-Bulb Globe Temperature, in degrees Celsius

NWB = Natural Wet-Bulb Temperature, in °C. Temperature of a wet wick exposed to natural air currents.

NWB = WB for air velocities greater than 2.5 meters/sec (8ft/sec), where WB is the psychometric wet-bulb temperature; NWB = 0.1DBT + 0.9WB for air velocity between 0.3 and 2.5 meters/sec.

GT = Globe Temperature in °C.

DBT = Dry-Bulb Temperature in °C

WBGT is used to evaluate or design jobs performed in a hot environment. WBGT is the accepted index as a composite measure of thermal environment. If air velocity is less than 1.5 meters/sec, WBGT should not exceed 30 °C for light work (metabolic rate less than 198 kcal/hr), 27.8°C for moderate work (metabolic rate between 198 and 301 kcal/hr), and 26°C for heavy work (metabolic rate greater than 301 kcal/hr). These values should be used with caution and are suggestions only.

EQ30: Botsball Index[3]

$$WBGT = 1.01BB + 2.6$$

Where:

BB = BB index, which is obtained from a special thermometer that combines air temperature, humidity, air speed, and variables related to radiation. For more information about this measure, see: Botsford, J.H. 1971. A Wet-Globe Thermometer for Environmental Heat Measurement. *American Industrial Hygiene Journal* , 32, pp. 1-10. Also see: Beshir, M.Y., Ramsey, J.D., Burford, C.L. 1982. Threshold Values for the Botsball: A Field Study for Occupational Heat. *Ergonomics,* 25(3), pp. 247-254.

EQ31: Insulation Requirements Prediction Equation[3]

$$Icl = 5.55 \frac{Tsk - Ta}{0.75M}$$

Where:

Icl = Clothing Insulation Factor (See Page 184)

Tsk = Skin Temperature

Ta = Ambient Dry-Bulb Temperature in degrees Celsius

M = Metabolic Workload in kilocalories (kcal) per Square Meter per Hour

REFERENCES

1. Van Cott, H.P., and Kinkade, R.G., 1972. *Human Engineering Guide to Equipment Design. (Revised)* Washington, DC.: US Government Printing Office. Library of Congress Number: 72-600054.

2. Liles, D.H., Deivanayagam, S., Ayoub, M.M., and Mahajan, P. 1984 A job severity index for the evaluation and control of lifting injury. Reprinted with permission from *Human Factors.* Vol. 26, No. 6, 1984. Copyright 1984 by the Human Factors Society, Inc. All rights reserved.

3. Eastman Kodak Company. 1983. *Ergonomic Design for People at Work*, Vol. 1., New York: Van Nostrand Reinhold. Reprinted by permission of Eastman Kodak Company, © 1983. All Rights Reserved.

4. Pulat, B.M. 1992. *Fundamentals of Industrial Ergonomics.* Englewood Cliffs, NJ: Prentice-Hall. Reprinted with Permission.

5. Thorrell, L.G., and Smith, W.J. 1990. *Using Computer Color Effectively: An Illustrated Reference,* Englewood Cliffs, NJ: Prentice-Hall. Reprinted with Permission.

INDEX

A

Anthropometry

C

D

G

I

J

L

M

Index

R

S

T

V

W